Green Chemistry
and
Applications

T0135994

Editors

Aidé Sáenz-Galindo
Adali Oliva Castañeda-Facio
School of Chemistry
Autonomous University of Coahuila
Saltillo, Coahuila, México

Raúl Rodríguez-Herrera
Food Research Department
School of Chemistry
Autonomous University of Coahuila
Saltillo, Coahuila, México

CRC Press
Taylor & Francis Group
Boca Raton London New York

CRC Press is an imprint of the
Taylor & Francis Group, an **informa** business

A SCIENCE PUBLISHERS BOOK

Cover credit: Cover illustrations reproduced by kind courtesy of the editors.

CRC Press
Taylor & Francis Group
6000 Broken Sound Parkway NW, Suite 300
Boca Raton, FL 33487-2742

First issued in paperback 2022

Version Date: 20200601

ISBN-13: 978-0-367-51000-8 (pbk)
ISBN-13: 978-0-367-26033-0 (hbk)

DOI: 10.1201/9780429291166

Visit the Taylor & Francis Web site at
http://www.taylorandfrancis.com

and the CRC Press Web site at
http://www.routledge.com

Library of Congress Cataloging-in-Publication Data

Names: Sáenz-Galindo, Aidé, 1979- editor. I Facio, Adali Oliva
    Castañeda, 1978- editor. I Rodríguez-Herrera, Raúl, 1960- editor.
Title: Green chemistry and applications / editors, Aidé Sáenz-Galindo,
    Adali Oliva Castañeda-Facio, School of Chemistry, Autonomous University
    of Coahuila, Saltillo, Coahuila, México, Raúl Rodríguez-Herrera,
    Food Research Department, School of Chemistry, Autonomous University of
    Coahuila Saltillo, Coahuila, México.
Description: Boca Raton, FL : CRC Press, 2020. I Includes bibliographical
    references and index.
Identifiers: LCCN 2020000047 I ISBN 9780367260330 (hardcover)
Subjects: LCSH: Green chemistry.
Classification: LCC TP155.2.E58 G725 2020 I DDC 660.028/6--dc23
LC record available at https://lccn.loc.gov/2020000047

Preface

The problem of preserving a sustainable environment is an issue that involves everyone, from the domestic to the industrial level. For this reason, it is very important to know what to do about this problem. Climate change, pollution, of air and different water resources, from lakes, lagoons, rivers, seas or oceans, global warming, the extinction of flora and fauna, are some of the problems that human beings are currently facing. Different strategies and actions have been implemented to mitigate these types of problems. Green chemistry is an excellent option to deal with this type of problems, it is currently considered a philosophy at local, national and international level, and is applicable in countless areas.

This book sets out in detail the 12 principles of green chemistry, applied in different areas in order to demonstrate the importance of using this methodology as a viable strategy for the care and conservation of the environment. Each of the 12 principles is covered in different book chapters. This book addresses areas of importance, such as preventing pollution, economy in use of synthetic processes, one chapter on renewable raw materials with compounds of great added value, such as plants, agro-industrial waste, among others, as well as the use of enzymes such as accelerators or moderators of reaction speeds, safe design from chemical processes at the laboratory to the industrial level, the use of unconventional energy sources ranging from ultrasound and microwave, as well as the process design that reduce the production of reaction by products, the use of catalysts with long useful life times for various synthetic processes, different strategies to perform continuous monitoring of the process to avoid accidents. In addition, each chapter of this book, captures and schematizes examples that involve the principles of green chemistry, ranging from simple cases which can be easily reproducible to more elaborate processes.

Aidé Sáenz-Galindo
Adali Oliva Castañeda-Facio
Raúl Rodríguez-Herrera

Acknowledgments

This work was financially supported by The Secretary of Agriculture, Fishing and Livestock, through the Project: FON.SEC. SAGARPA-CONACYT CV-2015-4-266936. In addition, this project had partial financial support by the Autonomous University of Coahuila, Mexico.

Contents

Chapter 1

Introduction

Adali Oliva Castañeda-Facio*

Facultad de Ciencias Químicas. Universidad Autónoma de Coahuila.
Blvd-Venustiano Carranza s/n Colonia República,
Saltillo, Coahuila, Mexico. CP. 25290.

In the last century, environmental quality has deteriorated because of deforestation, illegal hunting and fishing, pollution, overpopulation and excessive use of fossil fuel materials, among other causes. Due of these reasons, there is a need to find alternatives that will lead to conservation and environmental sustainability. Development of fossil fuel processing has generated a watershed in modern chemical industry. Inspite of the different benefits and applications of exploitation of fossil fuel, a great amount of pollution associated with the use or dump of products, suproducts and residues derived from these materials, has promoted the emergence of new philosophies such as green chemistry, which has great relevance in terms of design of products and processes in order to reduce generation of toxic substances in addition to increasing the efficiency of material resources and energy. Use of cleaner technologies which reduce consumption of materials and increase use of renewable resources are excellent alternatives to preserve the environment.

The technical and scientific progress that has been made to obtain chemical products and the enormous productivity of the chemical industry have allowed elaboration of compounds and materials that intervene in our lives and that are valued by all of us, however, over the years, procurement of these products has been done without taking into account the negative consequences in different aspects of human life, and the environment, because of production of non-biodegradable materials, release of pollutants and toxic substances, generating contaminants. Because of these reasons, the challenge to change the current trend is of a very broad scope, due to society having to be satisfied with different materials for transportation,

*For Correspondence: adali.castaneda@uadec.edu.mx

electrical, food, pharmaceutical, metallurgical, chemical and agricultural industry, among others, and that each of these industries contribute to the emission and generation of chemical contaminants. For example, Sulfur Oxides (SOx), Nitrogen Oxides (NOx), carbon monoxide (CO) emitted into the atmosphere, originate due to production of electrical energy from coal combustion. Another industry that generates contamination is one that focuses on plastics, where waste from or as plastic products can be easily seen in the environment. In addition, residues of pesticides used in agriculture can be found in river systems or even in groundwater generating a serious contamination problem. Some other types of contaminants are hydrocarbons, aerosols, halogen, arsenic, heavy metals, among others.

In the 50's, industry waste and residues were released directly to the environment, however, this situation changed, thanks to the warning given with the book "Silent Spring " written by Rachel Carson in 1962, where with an indepth investigation and a long term vision, the author managed to capture the consequences of an accelerated technological advance in the search to control unwanted species in agriculture, using synthetic pesticides, without taking into account their effect on human health, animals and the environment. It could be said that this study played a very important role for the start of the environmental movement, since after the publication of this book a commission was appointed to regulate the use of pesticides and later led to creating the Environmental Protection Agency (EPA), resulting in the declaration of laws on environmental protection worldwide. However, until 1987, the United Nations World Commission on Environment and Development influenced by the 1980 "World Conservation Strategy" of the International Union for Conservation of Nature (IUCN), published the report on principles of sustainable development as "development that meets the needs of the present without compromising the ability of future generations to meet their own needs." Since the founding of the Environmental Protection Agency in 1990, it seeks to avoid the generation of pollutants through legislation, making clear that, there must be a social, economic and environmental balance and that economic growth, rational use of natural resources and environment are related to each other. This is how new concepts for the creation of products and chemical processes advance, preventing pollution instead of seeking a remedy to it, giving rise to green chemistry, developed by Paul Anastas and John Warner in 1997, who proposed to eliminate pollution from its origin.

Green chemistry bases its principles on the development of technology and innovation, also on continuous improvement to match environmental goals such as the financial ones of modern industries. The approaches of green chemistry are based on a protocol consisting of 12 basic principles that can be applied in various fields such as the chemical industry, pharmaceutical, agriculture and medicine among others.

The principles of green chemistry are listed below:

1. *Prevention*: It is more useful to avoid or reduce waste production than to treat or clean after it has been created.
2. *Atom Economy*: Synthetic methods should maximize incorporation of each material used in the process.

3. *Less Hazardous Chemical Syntheses*: It consists of developing processes that generate minimal toxicity and environmental impact.

4. *Designing Safer Chemicals*: Chemical products should be designed with minimal toxicity.

5. *Safer Solvents and Auxiliaries*: The auxiliary substances of the chemical processes (solvents, buffers, separation additives, among others), must be innocuous and should be reduced to a minimum.

6. *Design for Energy Efficiency*: Energy requirements should be recognized for their environmental and economical impacts and should be minimized. Synthetic methods should be conducted at room temperature and pressure.

7. *Use of Renewable Feedstocks*: The starting materials used in a process, should proceed from renewable sources, insofar as it is economically and technically feasible.

8. *Reduce Derivatives*: The synthesis must be designed with the minimum use of group protectors to avoid extra steps and reduce waste.

9. *Catalysis*: Catalytic reagents (as selective as possible) are superior to stoichiometric reagents.

10. *Design for Degradation*: Chemical products should be designed so that at the end of their function they break down into innocuous degradation products and do not remain in the environment.

11. *Real-time Analysis for Pollution Prevention*: Analytical methodologies need to be further developed to allow for real-time, in-process monitoring and control prior to formation of hazardous substances.

12. *Inherently Safer Chemistry for Accident Prevention*: Design chemical processes, using methods and substances that reduce accidents (emissions, explosions, fires, among others), and minimize the damage when an accident occurs.

Each of the last principles will be discussed extensively in this book's different chapters, showing concrete examples. The chapters provide recent and interesting information by bringing together experts in multiple disciplines of green chemistry, describing a variety of topics in various fields of research also offering vanguard references. The *Prevention* Chapter focuses on preventing waste because it is more favorable to man and the environment and, ultimately, less costly than the treatment and destruction of waste once it has arisen. It is also more suitable than the treatment of waste and destruction.

The prevention of waste is more favorable and less expensive for the environment and human beings. In the *Atom Economy* chapter, several examples of reactions are shown to maximize the use of raw materials and minimize the production of waste. On the other hand, the use of palm oil and the different processes for obtaining biomass are explained extensively in the chapter on *Less Hazardous Chemical Syntheses*. In addition, this chapter mentions the applications of Pyroligneous acid which is one of the byproducts obtained from palm oil and that can be used for different applications on health and agriculture. Another essential point in the relationship between structure, activity and toxicity of chemical substances to make them safer is analyzed in the chapter *Designing Safer Chemicals*, where they also discuss some rules established through different mechanisms to design safer substances and chemicals.

Principle 5 deals with Safer Solvents and Auxiliaries, and is the key aspect that is presented in the chapter *Use of the green chemistry for the extraction of bioactive compounds from vegetable sources*, where the use of plant materials as solvents is discussed, as they contain biologically active substances. The chapter on *Design for Energy Efficiency* emphasizes the use of alternative energy such as ultrasound and microwaves for synthesis of new materials. The chapters *Use of renewable feedstocks* and *Use of renewable feedstocks: The recovery of high value molecules from waste of renewable feedstocks: soybean hull* are based on principle 7 of green chemistry. They mention the importance of adding value to agricultural residues and waste, due to the high value molecules that these materials contain. In the chapter *Reduce Derivatives*, there are some strategies to reduce the formation of waste during the synthesis by analyzing alternative approaches. It is possible to obtain compounds with high yields, cleaner and faster reactions with the use of catalysts, which is described in the chapter on *Catalysis*, besides, the importance of the stereochemistry of the catalysts in the organic reactions is explained in detail.

On the other hand, environmental pollution due to use of pesticides is shown in the chapter *Design for Degradation*, where it is explained that this type of contamination is remedied by different methods, however the method of biodegradation through the use of microorganisms, which eliminate pesticides from the environment is an effective method compared to others. In addition, it has been proposed to produce less hazardous materials and reduce waste as a solution to pollution prevention as established in the chapter *Real-Time Analysis for Pollution Prevention*, making it clear that real-time monitoring of processes can reduce waste.

Finally, the chapter, *Green Precipitation with Polysaccharide as a Tool for Enzyme Recovery*, refers to principle 12 "Inherently safer chemistry for accident prevention" shows that the isolation and purification of enzymes extracted from natural sources by precipitation with polysaccharide, will reflect economic savings and benefits to the environment.

Chapter 2

Atom Economy

Kunnambeth M. Thulasi[1], Sindhu Thalappan Manikkoth[1],
Manjacheri Kuppadakkath Ranjusha[1],
Padinjare Veetil Salija[1], Vattakkoval Nisha[1],
Shajesh Palantavida[2] and
Baiju Kizhakkekilikoodayil Vijayan[1]*

[1]Department of Chemistry/Nanoscience, Kannur University,
Swami Anandha Theertha Campus, Payyannur, Edat P.O.,
Kerala-670 327, India.
[2]Centre for Nano and Materials Science, Jain University,
Jakkasandra, Ramanagaram, Karnataka, India.

INTRODUCTION

The concept of atom economy was developed by Barry M. Trost in 1991. Organic synthesis requires multiple reagents, facilitating agents and solvents to obtain the desired product. At the end of the reaction everything except for the desired product and reagents that can be recycled, like solvents and catalysts, will end up as wastes, mostly hazardous wastes. Conceptually, if the desired product contains all the atoms making up the reagents there will be no waste generated. The concept of atom economy can be used to identify synthetic methodologies that will retain the maximum number of atoms from the reactants in the final product and thereby reduce wastage. The atom economy concept allows quantification of the efficiency of a reaction with respect to the number of atoms transferred from the reactants to the final desired product (Trost, 1995; Trost, 2002). The concept of atom economy can be applied to every synthesis and be used to define new pollution prevention benchmarks (Cann and Dickneider, 2004; Song et al., 2004). Atom economy calculation, broadly presents a measure of the greenness of a chemical reaction.

*For Correspondence: baijuvijayan@kannuruniv.ac.in; baijuvijayan@gmail.com

The commonly used indicator for efficiency of a reaction in organic synthesis is the percentage yield, which neglects mass flow (Cann and Dickneider, 2004). A synthetic chemist records the yield of a particular reaction as percentage yield. The percentage yield can be calculated as

$$\% \text{yield} = \frac{\text{Actual yield of the product}}{\text{Theoretical yield of the product}} \times 100$$

A yield above 90% is considered good.

From the above calculations, it can be understood that if one mole of the reactant produces one mole of the product we get 100% yield. However, the amount of waste produced is not factored in the above calculation. So a synthesis cannot be characterized in terms of greenness, even if it has a 100% yield (Eissen et al., 2004). An example is the Grignard reaction (Scheme 2.1). It cannot be considered green, even though it gives a 100% yield. A large amount of by-products is formed in the reaction.

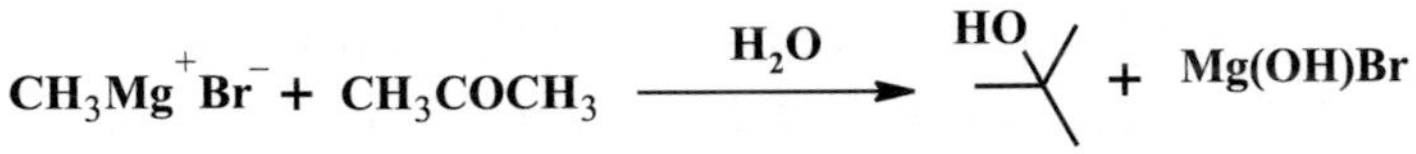

$$CH_3Mg^+Br^- + CH_3COCH_3 \xrightarrow{H_2O} \overset{HO}{\underset{}{\diagup\!\!\!\diagup}} + Mg(OH)Br$$

Scheme 2.1   Grignard reaction.

Here one mole of acetone gives one mole of tert-butyl alcohol. But one mole of Mg(OH)Br is formed as a by-product which is not taken into account while calculating the yield.

Here lies the need for an accurate gauge for the greenness of a chemical reaction and leads to the concept of atom economy. If there are two possible schemes for a synthesis, the one, more consistent with green principles should be chosen. Such a synthetic route, will mainly incorporate the maximum number of atoms from the reagents into the desired final product (Anastas et al., 2002). Hence the concept of atom economy concentrates on developing synthetic schemes that will reduce waste by maximizing the incorporation of atoms from the reactants in the desired product. By following the atom economy one can design greener reactions.

The quantification of atom economy concept was conducted by Roger A Sheldon (Kidwai and Mohan, 2005), a professor at Delft University in the Netherlands. He introduced the percentage atom utilization as a measure of atom economy and is given as

$$\% \text{ Atom utilization} = \frac{\text{Molecular weight of desired product}}{\text{Molecular weight of formed product} + \text{Waste product}} \times 100$$

Percentage atom utilization is the percentage of atoms in the reactants that are incorporated into the final product. If more than one product of interest is formed in a reaction, percentage atom utilization equation is modified as

$$\% \text{ Atom utilization} = \frac{(\Sigma N_p W_p)}{[(\Sigma N_p W_p) + (\Sigma N_r W_r)]} \times 100$$

where,

N_p = Number of molecules of the products
N_r = Number of molecules of the residues
W_p = Formula weight of products
W_r = Formula weight of residues

An example of a 100% atom utilization reaction is the reaction between sodium sulfite and sulfur powder to form sodium thiosulfate, which is given in Scheme 2.2.

$$Na_2SO_3 + Sulfur \longrightarrow Na_2S_2O_3$$

Scheme 2.2   Formation of sodiumthiosulfate.

In several reactions, it is difficult to determine the identities of waste products and the concept of percentage atom utilization cannot be used. The law of conservation of mass provides a better platform for understanding the maximum inclusion of all the atoms of the reactants in the final products and is applied for calculating percentage atom economy. It is given by:

$$\% \text{ Atom economy} = \frac{\text{formula weight of atoms utilized}}{\text{total formula weight of all the reagents used in this reaction}} \times 100$$

ATOM ECONOMY IN DIFFERENT TYPES OF REACTIONS IN ORGANIC SYNTHESIS

The concept of atom economy can be best understood by considering examples of different kinds of reactions. There are four main types of reactions in organic synthesis; rearrangement, addition, substitution and elimination reactions. By calculating the percentage atom economy, we can find out which one is more atom economical.

Rearrangement Reactions

All rearrangement reactions are carried with 100% atom economy as it involves only the rearrangements of the atoms of a molecule. As an illustration, consider the rearrangement reaction of propan-1-ol to propan-2-ol (Scheme 2.3).

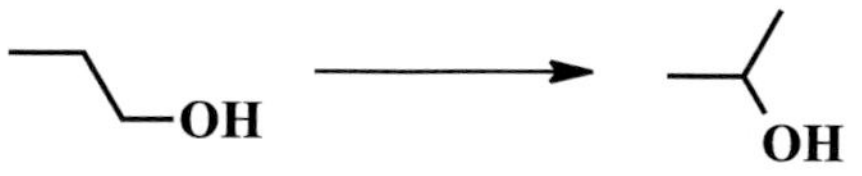

Scheme 2.3   Rearrangement reaction of propan-1-ol to propan-2-ol.

Here all the atoms in the reactant molecule are utilized in the final desired product.

$$\% \text{ Atom economy} = \frac{\text{formula weight of atoms utilized}}{\text{total formula weight of all the reagents used in this reaction}} \times 100$$

$$= \frac{60}{60} \times 100 = 100\%$$

Consider the rearrangement of 3, 3-dimethyl-1-butene to 2, 3-dimethyl-2-butene in presence of acid catalyst (Scheme 2.4).

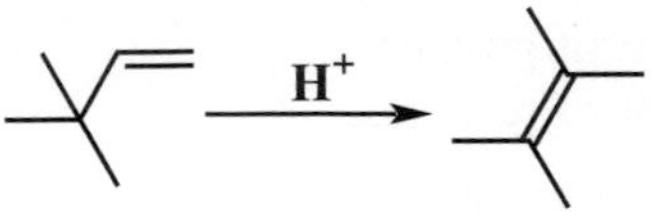

Scheme 2.4 Rearrangement of 3, 3-dimethyl-1-butene to 2, 3-dimethyl-2-butene in presence of acid catalyst.

$$\% \text{ Atom economy} = \frac{60}{60} \times 100 = 100\%$$

Here also all the atoms in the reactant are rearranged to form the final product and atom economy is found to be 100%. The acid used in the reaction is not included in the calculation because it is only required in catalytic amounts and does not form part of the product. The Claisen rearrangement is another classical example of 100% atom efficiency reaction, which involves rearrangement of aromatic allyl ethers producing ortho-substituted product, is illustrated in Scheme 2.5.

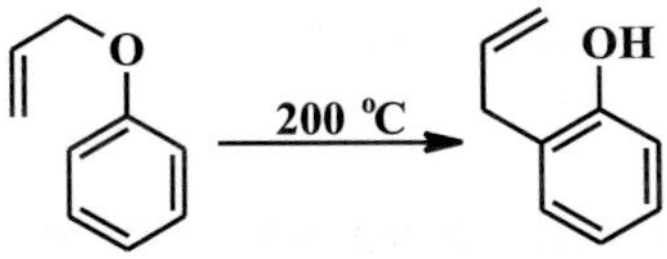

Scheme 2.5 Claisen rearrangement.

$$\% \text{ Atom economy} = \frac{134.75}{134.75} \times 100 = 100\%$$

An important rearrangement involving catalysts is the beckmann reagent where a cyclohexanone oxime converted to caprolactum in the presence of oleum (Scheme 2.6), a major precursor for the synthesis of nylon 6. Fries rearrangement of phenolic esters also go through a catalytic route involving lewis acid such as $AlCl_3$ as catalyst, which is given in Scheme 2.7. However the use of a large amount of catalyst increases the aluminum waste thereby reducing atom efficiency. In order to make it atom economic, photo-fries rearrangement is carried out where the reaction takes place through an intermolecular free radical route.

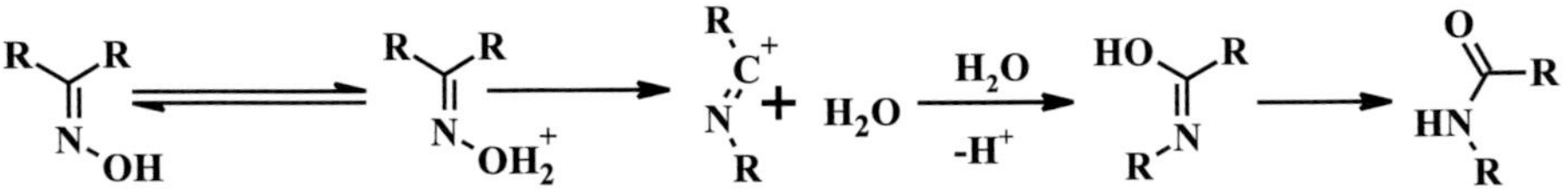

Scheme 2.6 Beckmann rearrangement.

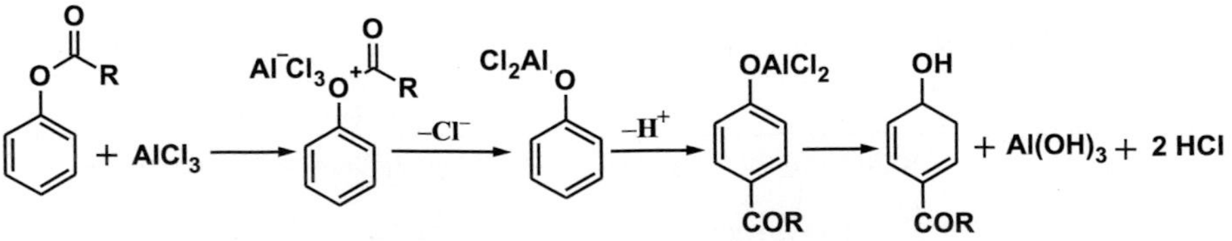

Scheme 2.7 Fries rearrangement.

Even though rearrangement reactions are considered to show 100% atom economy, there are few exceptions. Some rearrangement reactions involve elimination of water reducing the theoretical atom economy. Similarly, rearrangements involving intermediates along the reaction pathway show less atom economy.

Addition Reactions

Addition reactions are reactions in which there occur an addition of molecules or groups across a double or triple bond. These reactions are 100% atom economical (Trost, 2002), since the reactants combine to form the final product, without any elimination of side products. We can consider an example for addition reaction.

Considering the hydrogenation of ethene (Scheme 2.8), all the atoms of ethene and hydrogen molecule are utilized in forming the final product ethane. Here, nickel is only required in catalytic amounts and will not influence the calculation of atom economy.

$$H_2C{=}CH_2 + H_2 \xrightarrow{\text{Ni}} H_3C{-}CH_3$$

Scheme 2.8 Hydrogenation of ethene.

In this reaction, the total formula weight of reagents used is 30 g/mole and the total formula weight of atoms utilized is also 30 g/mole and thus the percentage atom economy is 100%.

Other examples of simple addition reactions include carbonylation, bromination, chlorination, etc. which also proceeds with 100% atom economy. Geier et al. in 2010 reported metal free reductions of N- heterocycles in the presence of a catalytic amount of borane, $B(C_6F_5)_3$ and hydrogen.

Substitution Reactions

Substitution reactions take place by the replacement of one or a group of atoms by another atom or group of atoms. These reactions are usually poor in terms of atom economy.

Consider the example of acid promoted nucleophilic substitution reaction, given in Scheme 2.9.

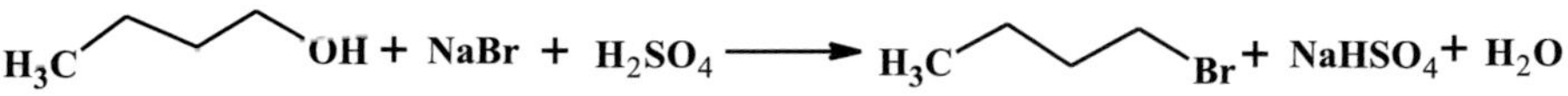

$$H_3C\diagup\diagdown\diagup OH + NaBr + H_2SO_4 \longrightarrow H_3C\diagup\diagdown\diagup Br + NaHSO_4 + H_2O$$

Scheme 2.9 Bromination of butanol.

Here, some atoms of the reactants are wasted and some are incorporated in the desired products. We can see that half of the mass of the reactants converts to the desired products and the rest is wasted as unwanted materials.

Total formula weight of reagents used = 275 g/mole

Total formula weight of atoms utilized = 138 g/mole

$$\% \text{ Atom economy} = \frac{\text{formula weight of atoms utilized}}{\text{total formula weight of all the reagents used in this reaction}} \times 100$$

$$= \frac{138}{275} \times 100 = 50\%$$

i.e., for the above reaction % atom economy is only 50%.

In the Mitsunobu reaction of conversion of secondary alcohols to ester, diethyl azodicarboxylate and triphenyl phosphine are used as reagents. Diethyl azocarboxylate is unstable and explosive in nature. The byproducts of the reactions also have considerable mass. Therefore the reaction is not ideal and has low atom economy. Instead of these reagents, Barret in 1998 used imidate esters (Barrett et al., 1998), produced from DMF and oxalyl chloride. This, on reaction with potassium benzoate leads to SN_2 substitution reaction of secondary alcohols. This reaction, given in Scheme 2.10 minimizes the quantity of nucleophile which improves atom economy. Moreover, the side products (DMF and KCl) can be removed easily.

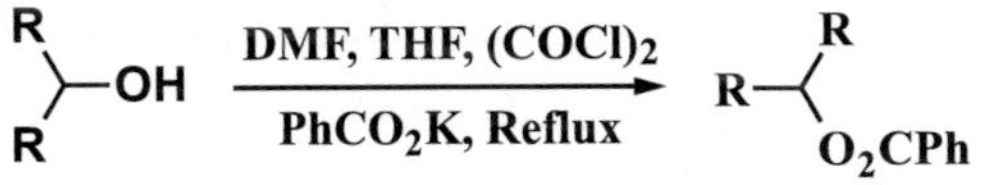

Scheme 2.10 Mitsunobu reaction using imidate esters.

Generally, bromination of alkenes is carried out with liquid bromine in non-reacting or acidic solvents like CH_2Cl_2, CCl_4 and glacial acetic acid. This easy method can be completed in 10 minutes. But the chemicals used are harmful, carcinogenic and a skin irritant. Mckenzie et al. in 2005, utilized pyridiniumtribromide as a bromine source to make this reaction greener. This reaction used ethanol as a solvent instead of glacial acetic acid. Reaction was carried out at room temperature. But a disadvantage was that, the reaction was less atom economical than the classical reaction. Instead of the above mentioned reactants, they used hydrobromic acid in the presence of hydrogen peroxide as a bromine source and high atom economy could be achieved. Consider bromination of Stilbene given in Scheme 2.11.

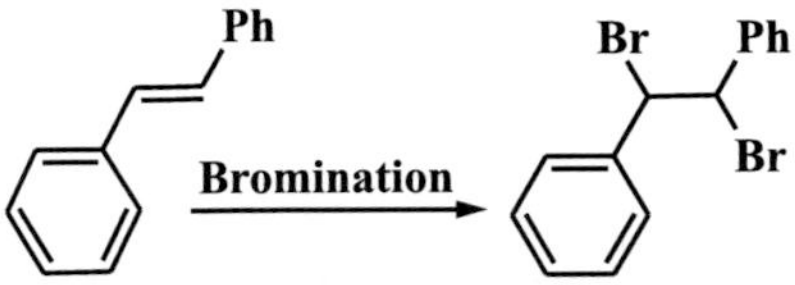

Scheme 2.11 Bromination of stilbene.

In the process stilbene is dissolved in ethanol and heated to reflux. HBr and 30% H_2O_2 are then added to the mix. Orange color appears in the solution, indicating the formation of bromine in the system. The reaction mixture becomes colorless as the product precipitates out. The atom economies of bromination of stilbene in classical and green methods are given in Scheme 2.12.

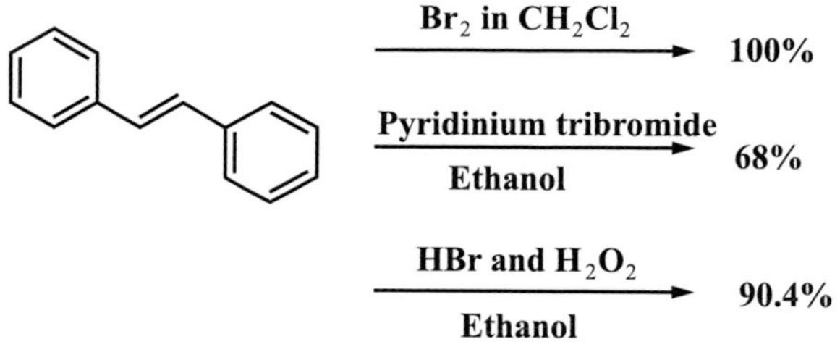

Scheme 2.12   Atom economies of bromination of stilbene.

Fishback in 2016 compared the aromatic electrophilic substitution (bromination and nitration) reactions of toluene (Fishback et al., 2016). Bromination using N-bromosuccinimide and nitration with nitric acid, illustrated in Scheme 2.13, were chosen for their study.

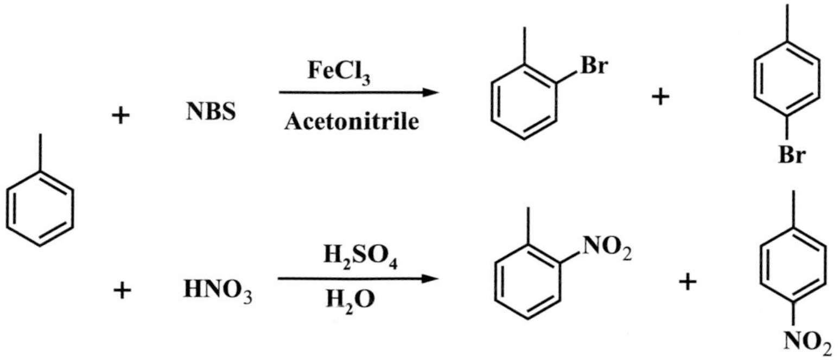

Scheme 2.13   Bromination and nitration of toluene.

Atom economy of nitration reaction is 85% and that of bromination is 63%. Functionalization by nitration reaction is greener than bromine substitution to obtain monosubstituted toluene. Even though nitration is greener from the perspective of atom economy and green principles, it is more difficult to execute than bromination.

Elimination reactions

In the case of elimination reactions, two atoms or group of atoms are lost from the reactant. Therefore, the reaction can never be atom economical. These are considered the worst atom economical reactions among the four.

Consider preparation of methyl propane from 2-bromo 2-methyl propane, given in Scheme 2.14.

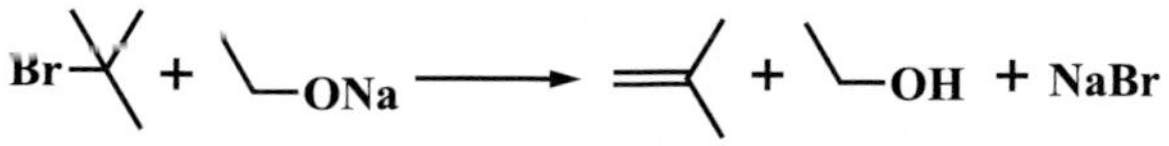

Scheme 2.14   Preparation of methyl propane.

Total formula weight of reagents used = 205 g/mole
Total formula weight of atoms utilized = 56 g/mole

$$\% \text{ Atom economy} = \frac{\text{formula weight of atoms utilized}}{\text{total formula weight of all the reagents used in this reaction}} \times 100$$

$$= \frac{56}{205} \times 100 = 27\%$$

i.e., for the above reaction percentage atom economy is only 27.

There are several other types of reactions, such as pericyclic reactions, carbon-carbon coupling reactions, non carbon-carbon coupling reactions, condensation reactions, redox type reactions, multicomponent reactions, isomerizations, etc. The atom economy of these reactions can also be calculated.

Pericyclic Reactions

Pericyclic reactions are concerted reactions in which reorganization of electrons occur via a single cyclic transition state. There are different types of pericyclic reactions, such as cycloaddition, electrocyclic reaction, sigmatropic rearrangement, group transfer reaction, chelotropic reaction and diatropic rearrangement. Pericyclic reactions are 100% atom economical, like simple addition and rearrangement.

Consider the cycloaddition of butadiene and ethene (Diels-Alder reaction), the product obtained is cyclohexene and takes place in presence of heat, illustrated in Scheme 2.15.

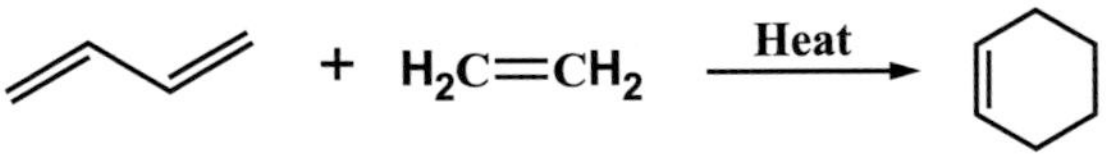

Scheme 2.15   Diels – Alder reaction.

Here the total formula weight of reagents used is 82 g/mole and the total formula weight of atoms utilized is also 82 g/mole giving a percentage atom economy of 100%.

Coupling Reactions

In organic chemistry, the term coupling reaction encompasses a variety of reactions, where 2 reaction fragments join each other by forming new bonds between atoms, with the help of a catalyst. Fischer indole synthesis, Danishefsky synthesis, Meerweinarylation, Kolbe synthesis, Zincke-Suhl synthesis, Robinson annulation, Gattermann-Koch reaction, Henry reaction, Heck reaction, Nazarov cyclization, Stoltz aerobic etherification, Buchwald-Hartwig cross coupling, Stahl aerobic amination, Polonovski reaction, etc. are some named organic reactions belonging to this category.

Consider the Heck reaction between methyl chloride and propene in presence of a base and palladium catalyst to form 2-butene, given in Scheme 2.16.

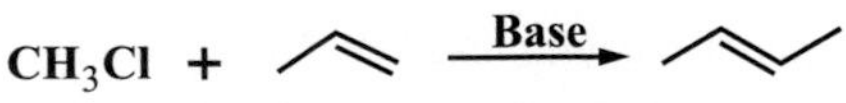

Scheme 2.16   Heck reaction.

Formula weight of all the reactants used = 92.5 g/mole
Formula weight of atoms utilized = 56 g/mole

$$\% \text{ Atom economy} = \frac{56}{92.5} \times 100 = 60.5\%$$

Here the percentage atom economy is only 60.5%.

Fishback et al. in 2016 compared various reactions through a green approach. According to him, the Barbier reaction which is an aqueous version of the Grignard reaction, is less toxic and greener than the Grignard reaction, illustrated in Scheme 2.17. Unlike the Barbier reaction, the Grignard synthesis has to be performed in anhydrous condition in organic solvents. Atom economy of the Barbier reaction is 65%.

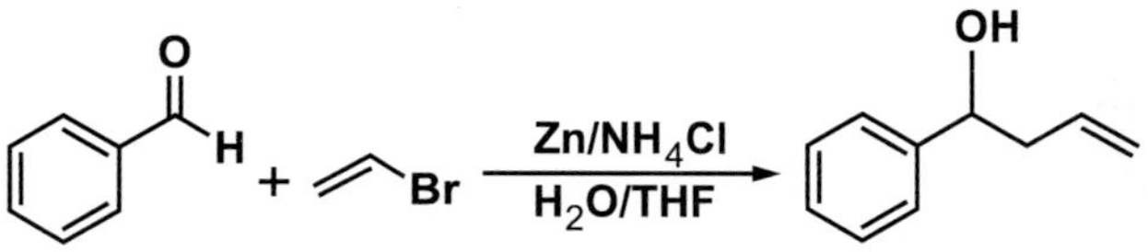

Scheme 2.17   Barbier reaction.

Classical Wittig reaction using organic solvents and waste generation is very high. For making it greener, Fishback et al. 2016 proposed a grinding method to induce contact between the reactants in the absence of any solvent (Scheme 2.18). This greener method has an improved atom economy of 45%.

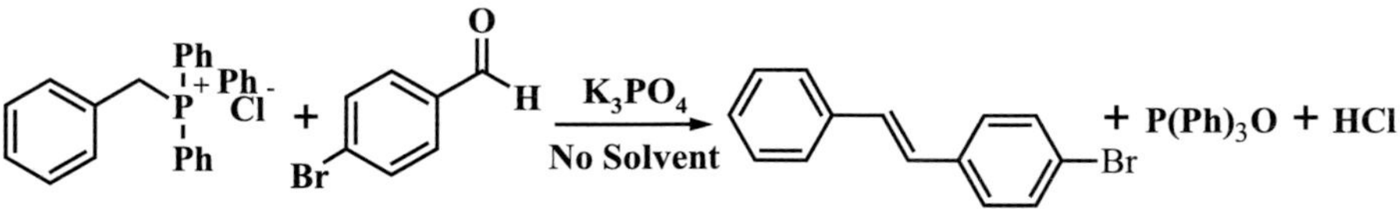

Scheme 2.18   Green wittig reaction.

Multicomponent Reactions

These are a special type of addition reactions in which at least three substrates react together in one reaction vessel either sequentially or all at once (Posner, 1986; Ugi et al., 1994; Dömling, 2002). Here the order of addition of substrates determines the final product of the reaction. Strecker synthesis of R-cyanoamines, Hantzschdihydropyridine synthesis, Radziszewski imidazole synthesis reaction, Riehmquinoline synthesis, Doebner reaction, Pinner triazine synthesis, Hantzsch synthesis, Biginelli reaction, Thiele reaction, etc. belong to the category of multicomponent reactions.

Consider the Doebner reaction of aniline with acetaldehyde and pyruvic acid forming quinolone-4-carboxylic acid (Scheme 2.19).

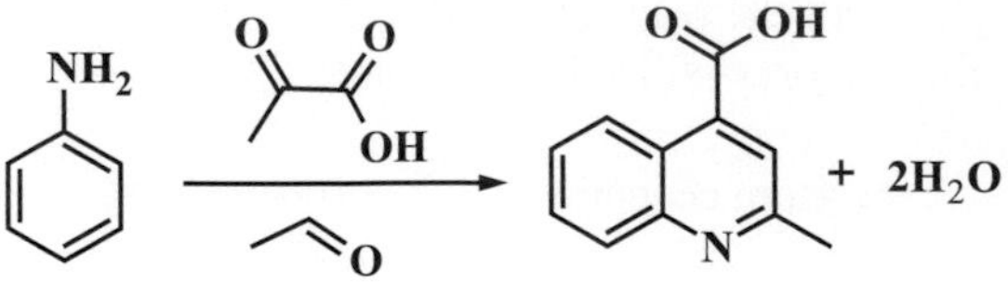

Scheme 2.19   Doebner reaction.

Formula weight of all the reactants used = 225 g/mole
Formula weight of atoms utilized = 187 g/mole

$$\% \text{ Atom economy} = \frac{187}{225} \times 100 = 83.1\%$$

From the calculation here, the percentage atom economy is 83.1%.

Condensation Reactions

Condensations are reactions in which carbon-carbon coupling takes place by straight-chain additions or cyclizations, producing hydroxylic molecules like water or alcohol as a byproduct. Dieckmann condensation, Stobbe condensation, Nenitzes-cuindole synthesis, Mukaiyamaaldol condensation and acyloin condensation are some examples.

Consider the example of Claisen condensation between methyl propanoate and methyl methanoate, in the presence of the base sodium ethoxide, forming the ketone derivative, with the elimination of ethanol as given in Scheme 2.20.

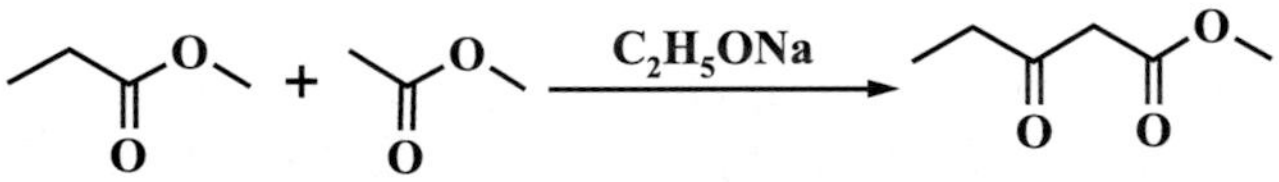

Scheme 2.20   Claisen condensation.

Formula weight of all the reactants used = 162 g/mole
Formula weight of atoms utilized = 130 g/mole

$$\% \text{ Atom economy} = \frac{130}{162} \times 100 = 80.2\%$$

The percentage atom economy is 80.2%.

Amide formation reaction is the most important topic in the area of pharmaceutics. For the preparation of amide bonds, the two classical methods used are acid chloride route and coupling agent method. The first method is the reaction of amine with an acid halide, an acid anhydride or an ester. In the second method, amine reacts with carboxylic acid in the presence of coupling reagents like dicyclohexylcarbodiimide. These methods have the disadvantage of being less safe and having a low atom economy. Fennie in 2016 proposed a catalytic method for the formation of amides involving generation of water as the byproduct, a greener alternative (Fennine and Roth, 2016).

As an example, consider the synthesis of N-benzyl-3-phenylpropanamide using the above three methods, given in Scheme 2.21.

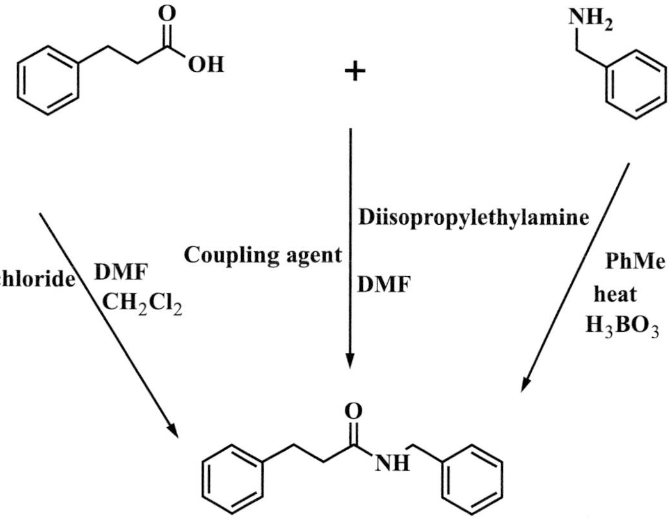

Scheme 2.21   Synthesis of N-benzyl-3-phenylpropanamide through various methods.

A comparison of the atom economy of these three reactions show that the acid chloride route has 49% atom economy, the coupling reagent method has only 31% atom economy and the catalytic method using boric acid catalyst has 93% atom economy. This affirms the greenness of the process along with the low cost.

Redox-type Reactions

Redox-type reactions involve oxidation or reduction of reactants in the presence of oxidizing or reducing agents respectively. These reactions usually produce significant by product wastes mostly from the oxidation or reduction of the reducing and oxidizing reagents (Andaros, 2005a, b; Sipos et al., 2013). Redox type reactions require the presence of another redox couple for converting byproducts back to the initial oxidizing or reducing reagents (Sato, 1998). These reactions are not green in terms of atom economy.

Consider the oxidation of 2- butanol to 2- butanone using the Jone's reagent (Scheme 2.22).

Scheme 2.22   Oxidation using Jone's reagent.

Formula Weight of all the reactants used = 820 g/mole
Formula weight of atoms utilized = 216 g/mole

$$\% \text{ Atom economy} = \frac{820}{216} \times 100 = 26.3\%$$

Cascade redox reactions, resulting in water as the ultimate waste product has recently been developed (Wardman, 1989). Molecular oxygen from air and hydrogen peroxide are the optimal oxidizing agents which produce water as the byproduct (Choudary et al., 2003) and hydrogen gas is the optimal reducing agent which produces no byproducts. Catalytic oxidations are always superior to the above stoichiometric oxidations in the calculation of atom economy.

Consider the above example of oxidation of 2-butanol to 2-butanone, in the presence of oxygen and a catalyst, illustrated in Scheme 2.23.

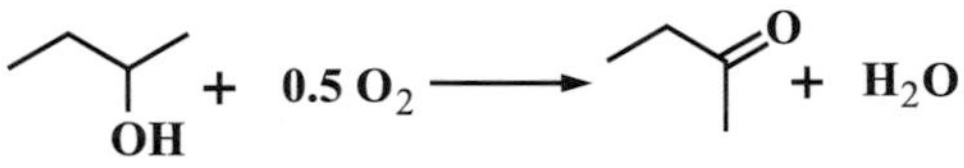

Scheme 2.23   Catalytic oxidation.

Formula weight of all the reactants used = 90 g/mole
Formula weight of atoms utilized = 72 g/mole

$$\% \text{ Atom economy} = \frac{72}{90} \times 100 = 80\%$$

Percentage atom economy is almost three times compared to stoichiometric oxidation.

Fishback et al. in 2016, studied epoxidation of trans-Anethole using potassium peroxymonosulfate and meta- Chloroperoxy benzoic acid as oxidizing agents to produce trans - Anethole oxide, illustrated in Scheme 2.24.

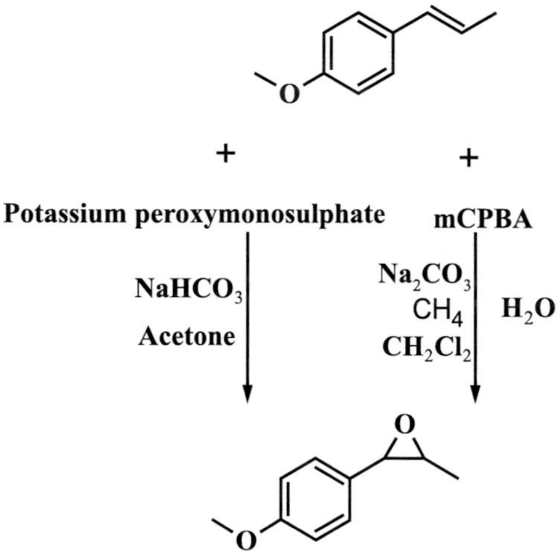

Scheme 2.24   Epoxidation of trans-anethole.

The atom economy of epoxidation reaction using mCPBA (51%) is higher as compared to that involving potassium peroxymonosulphate (36%). Epoxidation using mCPBA is also cheaper. Therefore, it is the greener one from the perspective of atom economy and cost consideration.

Nguyen and Retailleau in 2017, reported for the first time, a redox-neutral, catalyst free, completely atom-economical synthesis of sultams. Sultams are cyclic derivatives of sulfonamides and are used as important scaffolds in medicinal chemistry. Their uses also include antibacterial activity, lipoxygenase inhibition, drugs for heart disease, etc. Redox and non-redox approaches were widely used for the preparation of sulfonamides and sultams. In the redox method, the widely used strategy is the oxidation of sulfonamide derivatives such as sulfenylamides and sufinylamides and in non-redox method, the condensation of amines with activated sulfonyl group is utilized. In both strategies, the atom economy can never attain 100%. In the new method, elemental sulfur was used as a versatile synthetic tool in sulfonamide synthesis without any byproducts. Nguyen presented a total atom economical approach for the synthesis of sultams (Scheme 2.25) which involves heating of 2-nitrochalcones with elemental sulfur in the presence of N-methylmorpholine or 3-picoline without any catalyst.

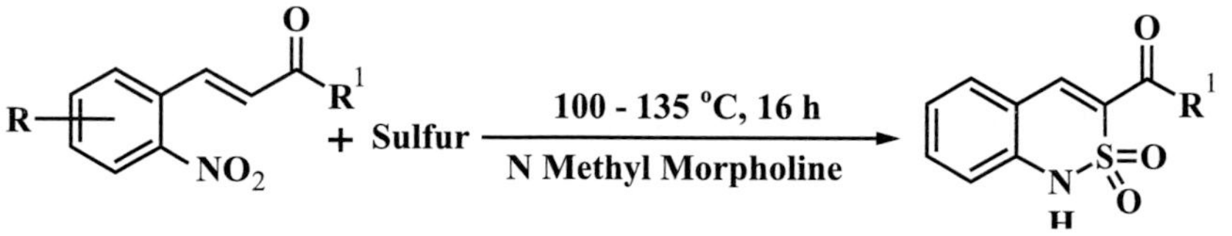

Scheme 2.25   Synthesis of sultams from 2-nitrochalcone.

Isomerizations

Synthesis of conjugated carbonyl compounds by the isomerization of propargyl alcohol represents an atom economical reaction. The isomerization of propargyl alcohols into enones occur via a traditional two step stoichiometric redox reaction in presence of a Ruthenium catalyst. Total synthesis of adociacetylene B through an enantioselective route is shown in Scheme 2.26. Conversion of alkynylvinylcyclopropene to a fused ring structure with ring size 5 or 7 also shows ideal atom economy (Li and Trost, 2008), which is shown in Scheme 2.27.

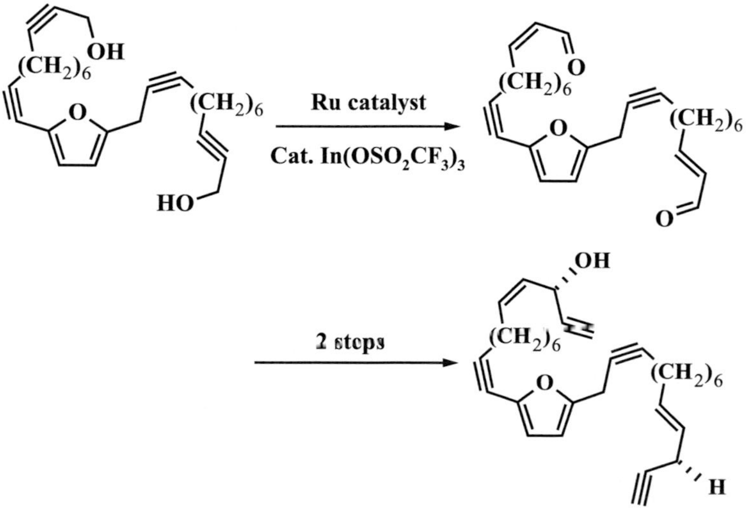

Scheme 2.26   Isomerization of propargyl alcohol.

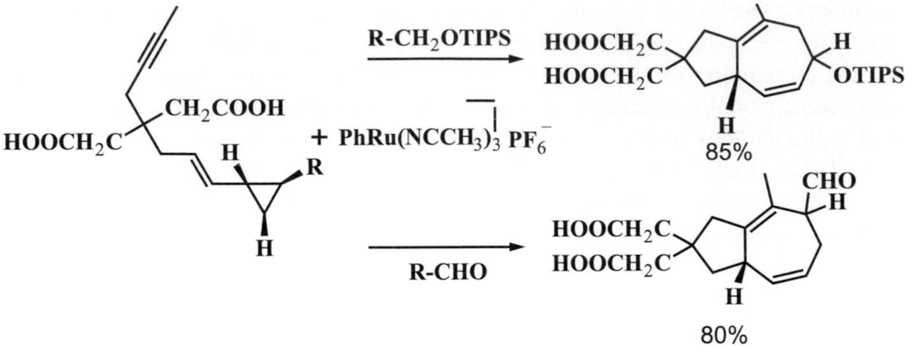

Scheme 2.27 Conversion of alkynylvinylcyclopropene.

ATOM ECONOMY IN CATALYSIS

Catalysis offers atom efficient reactions through an alternative reaction pathway by lowering the activation energy and thereby enhancing the rate of chemical processes. Catalysts can be excluded from the atom economy calculation, as they are not used in the reaction. The use of homogenous, heterogenous or biocatalysts helps to reduce reaction steps, stoichiometric components and energy thus giving higher atom economies (Anastas et al., 2001).

Heterogenous Catalysis

Atom economy of some reactions can be increased by using a suitable heterogenous catalyst. Here the reactant in one phase and the catalyst in a different phase interact through adsorption. As an illustration, consider the case of nickel-catalyzed hydrogenation ofnitrobenzene. The atom economy of aniline production is found to be 35% without nickel catalyst and hikes to 70% with this robust, cost-effective, easily recyclable heterogeneous catalyst, Scheme 2.28 (Grant et al., 2005).

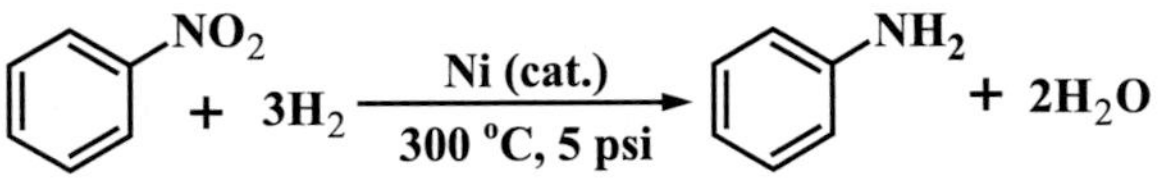

Scheme 2.28 Nickel catalyzed hydrogenation process.

$$\% \text{ Atom economy} = \frac{93.13}{129.17} \times 100 = 72\%$$

In the case of amide bond formation, there are many inefficient coupling reagents such as phosphonium/uranium salts and carbodiimides, which not only reduce the atom economy but also forms toxic or corrosive side products (Valeur and Bradley, 2008; Constable et al., 2007). Many methods have been proposed to overcome this problem. Among them, the most significant approach is the use of readily available and affordable, thermally activated K60 Silica as a heterogenous catalyst (Comeford et al., 2009) (Scheme 2.29).

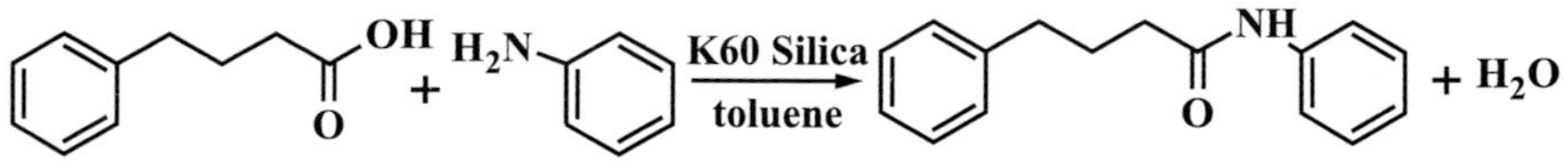

4-phenylbutanoic acid

N,4-diphenylbutanamide

Scheme 2.29 Synthesis of N, 4-diphenylacetamide from 4-phenyl butanoic acid in presence of K60 silica catalyst.

$$\% \text{ Atom economy} = \frac{239.32}{257.33} \times 100 = 93\%$$

The above reaction of 4-phenylbutanoic acid with aniline have 74% yield with 93% atom economy although they were carried out at high temperature to inhibit the trapping of product into the silica pores. In addition, catalytic recyclability and completion of reaction were established through continuous flow experiments for industrial applications.

An important form of heterogenous catalysis is phase transfer catalysis. The transfer of a reactant from one phase into another where reaction occurs is facilitated by using phase transfer catalysts. Studies reveal that benzoin condensation in aqueous media using quaternary ammonium salt (Q^+X^-) as pseudo phase transfer catalysts (Scheme 2.30) tend to be 100% atom economical (Yadav and Kadam, 2012).

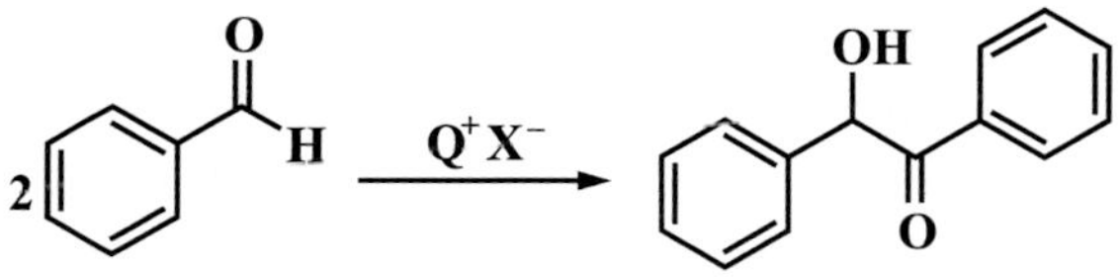

Scheme 2.30 Quaternary ammonium salt catalyzed Benzoin condensation.

$$\% \text{ Atom economy} = \frac{212.225}{212.24} \times 100 = 100\%$$

Homogenous Catalysis

Homogenous catalysis is gaining attention despite the fact that 90% of industrial processing involves heterogeneous catalysis. An important completely atom economic industrial process is the synthesis of adiponitrile using nickel-tetrakis (phosphite) complex as catalyst, given in Scheme 2.31.

$$\% \text{ Atom economy} = \frac{108.14}{108.15} \times 100 = 100\%$$

Research has progressed to improve the efficiency of catalytic reactions. Allen and Crabtree proposed such systems in the reference (Allen and Crabtree, 2010). The enhancement of atom economy in β-alkylation of alcohols is usually a three step process (Scheme 2.32) with minimum atom economy which can be improved by the use of an alkali metal based catalyst, which is shown in Scheme 2.33.

The suggested mechanism begins with oxidation followed by aldol condensation involving base catalyst and subsequent reduction using aluminium alkoxide catalysts. This method presents an energy efficient and cheaper process using less toxic metals, apart from the atom efficiency consideration.

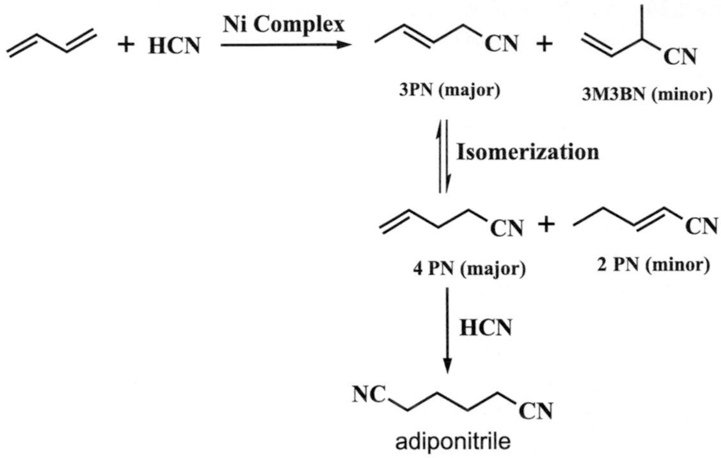

Scheme 2.31   Ni catalyzed synthesis of adiponitrile.

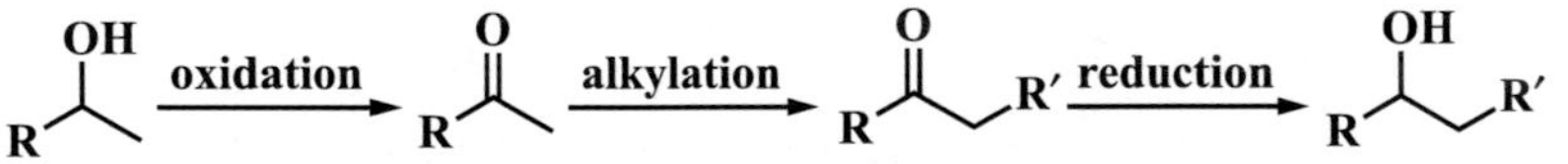

Scheme 2.32   Three-step beta alkylation of alcohols.

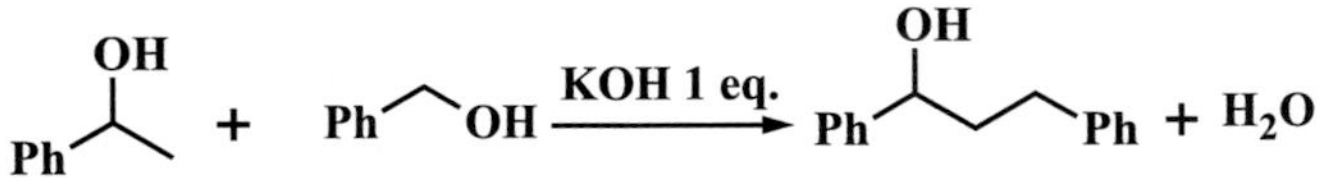

Scheme 2.33   Base catalyzed beta alkylation of secondary alcohols.

$$\% \text{ Atom economy} = \frac{212.29}{286.42} \times 100 = 74\%$$

Hydrogen borrowing is another significant approach in homogenous catalysis (Berliner et al., 2011). This process combines one-pot oxidation-reaction-reduction sequence using alcohols or amines as alkylating agents with inherently high atom economy. Conventionally second and third row transition metals particularly Ru, Os, Rh and Ir were used. In need of sustainable methodologies, finding efficient alternatives becomes a major challenge. In recent times first row transition metals, which are very abundant on earth, have been established for hydrogen borrowing catalysis. Iron catalyzed methylation is the best example for the hydrogen borrowing approach using a Knölker-type (cyclopentadienone) iron carbonyl complex as a catalyst, illustrated in Scheme 2.34. A variety of amines, indoles, ketones, sulfonamides and oxindoles undergo mono- or dimethylation in excellent yields following this method (Polidano et al., 2018).

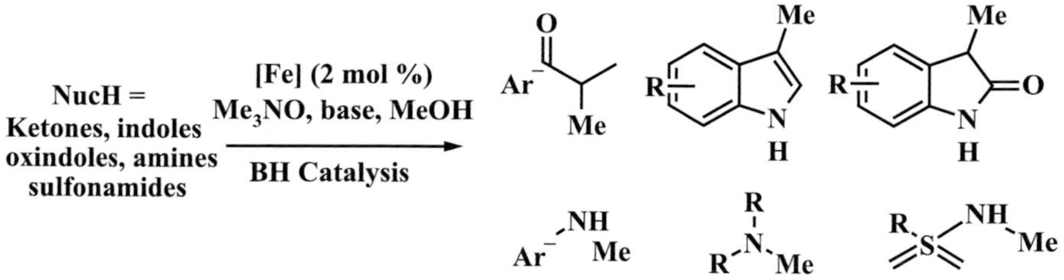

Scheme 2.34 Iron catalyzed methylation.

Biocatalysis

In biocatalytic processes natural catalysts such as enzymes accomplish desired transformations (Straathof et al., 2002). Modern technologies including recombinant DNA technology, immobilization methods and protein engineering providenumerous ways for realizing these reactions (Parmar et al., 2000; Powell et al., 2001). Extremely high reaction rates, high selectivity and biodegradability make enzymes a good candidate for such reactions. However high substrate specificity causes difficulties when wide substrate applicability is needed. Concerning atom economy, synthesis of 6- aminopencillanic acid from pencilin G can be considered (Scheme 2.35). Usually 6-aminopencillanic acid is synthesized through a four step deacylation process (Weissenburger and van der Hoeven, 2010).

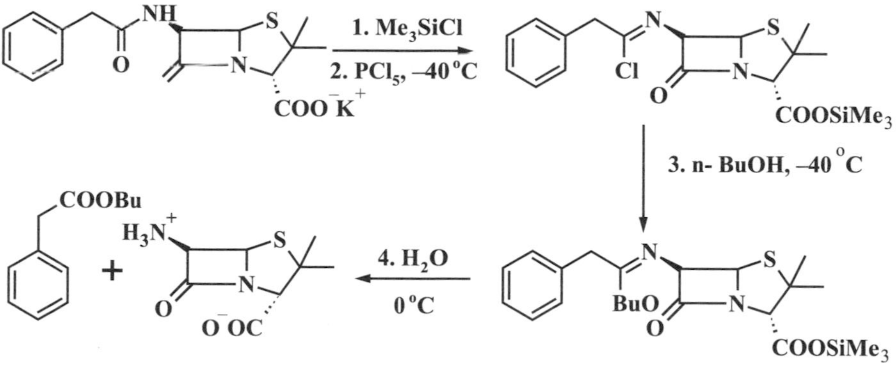

Scheme 2.35 Synthesis of 6- aminopencillanic acid.

The mechanism includes protecting carboxyl group of penicillin G by silyl, conversion of amide into imine chloride and enol ether formation followed by hydrolysis. The atom economy of the reaction is only 28%. Later Wegman et al., 2001 demonstrated 58% atom economy by employing stable enzyme penicillin G acylase (Wegman et al., 2001), given as Scheme 2.36. Similar studies have been done extensively by completely replacing the conventional deacylation process.

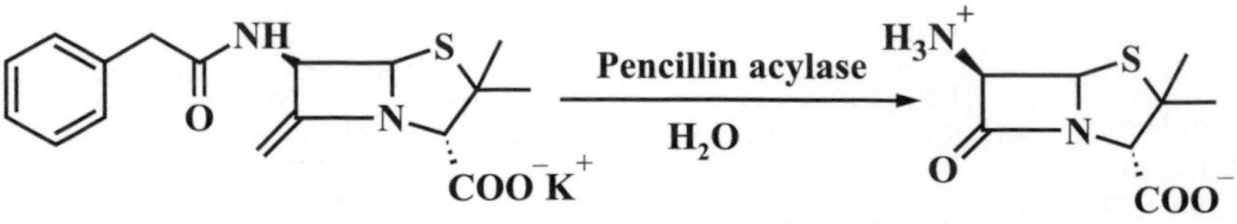

Scheme 2.36   Penicillineacylase catalyzed production of 6-amino penicillanic acid.

Research is progressing by taking atom economy as a valuable tool to design the sustainable process of pharmaceutical chemicals. Recently, Oeggl and co-workers used biocatalysts in cascades producing pharmaceutical chemicals with enhanced productivity and selectivity (Oeggl et al., 2018). In this work, all four isomers of anti-inflammatory drug 4-methoxyphenyl-12-propanediol were synthesized via two step cascade reactions using NADPH-dependent alcohol dehydrogenases. The reaction takes place with 99.9% atom economy.

Novel Examples for Catalytic Synthesis Involving 100% Atom Economy

Kang and co-workers, in 2013 reported a 100% atom economical catalytic redox-neutral method for synthesis of amides from an alcohol and a nitrile (Kang et al., 2013). The earlier method of amide production is given in Scheme 2.37, which are not atom economical. Many approaches, including direct oxidation of primary alcohols and amines have been attempted for a 100 % atom economical amide synthesis (Zweifel et al., 2009; Gunanathan et al., 2007,). Ru catalyzed hydrogenation of nitriles yielding primary amines without side products had been reported (Enthaler et al., 2008; Takemoto et al., 2002). Kang et al., 2013 inspired by this, used a Ru catalyst, $RuH_2(CO)(PPh_3)_3$ for transfer of hydrogen from alcohol to nitrile, which was followed by C–N bond formation between α-carbon of the primary alcohol and nitrogen of the nitrile and yielded amides without any byproduct. The completely atom-economical redox neutral amide synthesis method is given in Scheme 2.38.

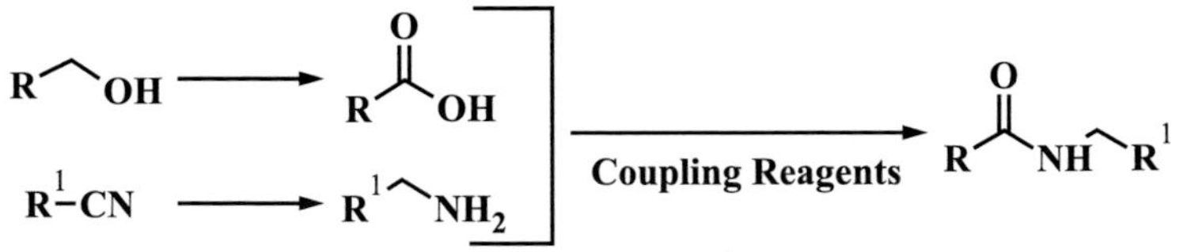

Scheme 2.37   Earlier method of amide production.

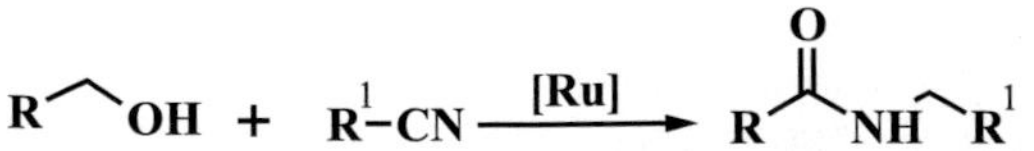

Scheme 2.38   Amide formation from alcohols and cyanides in presence of Ru catalyst.

Nitrogen-containing polycyclic compounds are essential for the agrochemical, pharmaceutical and fine chemical industries. Thus the improvement of existing methods and development of new methodologies for synthesizing nitrogen-containing polycyclic compounds are very important. Qiu et al. in 2017 reported a 100% atom economic method for synthesizing indanone fused pyrrolidine from enynals and propargylamines using CuCl/Et$_3$N as the catalyst, which is given in Scheme 2.39. Here in this reaction, two rings and four bonds are formed in a single step with high atom economy.

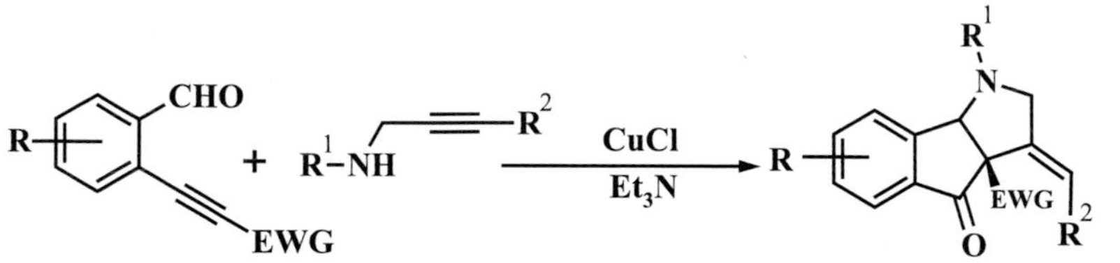

Scheme 2.39 Indanone fused pyrrolidine synthesis in presence of CuCl/Et$_3$N catalyst.

Large varieties of natural bioactive compounds, such as lignans, terpenes, steroids, nucleosides, etc. contain 2-methylene tetrahydrofurans and g-lactones as their core structures. Also the polycyclic structures such as euchroquinol, tricholomalide, parthenolide, dehydrocostus lactone, solanacol, etc. are formed by fusing these skeletons with other cyclic units. But all these syntheses require multisteps with low atom and step economy. Thus development of an efficient method for the synthesis of polycyclic compounds containing 2-methylene tetrahydrofuran and g-lactone units is of great importance. Liang in 2016 reported a 100% atom economic one pot rapid process for synthesizing indanone-fused 2-methylene tetrahydrofurans from enynals and propynols, which involves hydrolysis, followed by Knoevenagel condensation, Michael addition and Conia-ene reactions (Liang et al., 2016), which is illustrated in Scheme 2.40. Here in this method two rings and four bonds are formed with complete atom-economy.

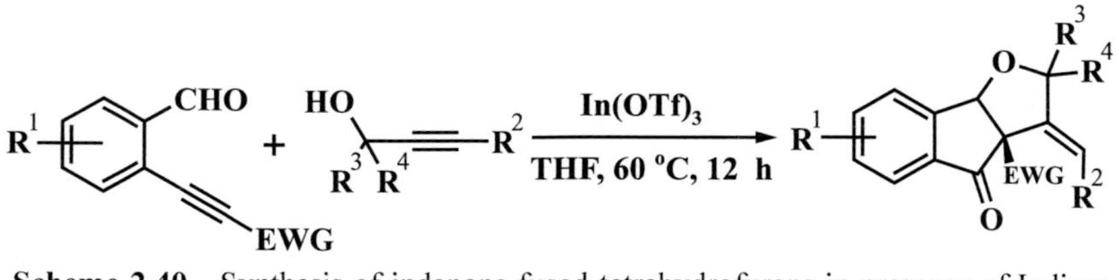

Scheme 2.40 Synthesis of indanone-fused tetrahydrofurans in presence of Indium catalyst.

Tri aryl methanes find various applications in medicinal chemistry and the dye industry. They also find various applications in material science, mostly as photochromic agents and fluorescent probes. Reddy and Vijaya in 2015 reported a highly efficient, 100% atom economic, one pot methodology (Scheme 2.41) for the synthesis of unsymmetrical diarylindolyl methane derivatives by a Pd catalyzed addition of o-alkynyl anilines to p-quinonemethides (Reddy and Vijaya, 2015). The main features of this methodology are its 100% atom economy and its broad substrate scope. Unlike the reported methodologies so far, this method does not need the protection of the amino group.

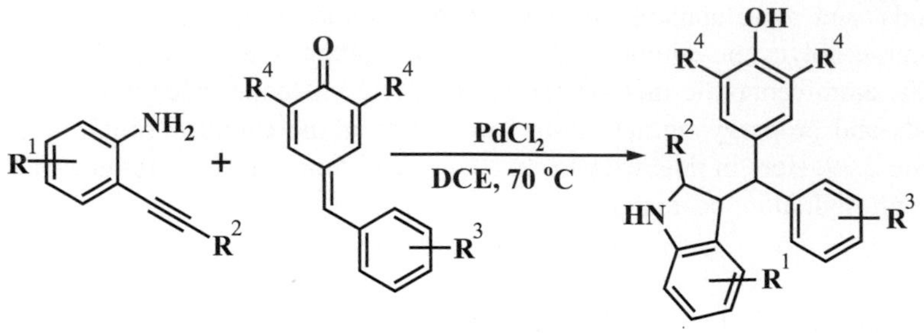

Scheme 2.41   Pd catalyzed synthesis of unsymmetrical diarylindolyl methane derivatives.

Many natural products and bioactive molecules contain spiro[4,5]cyclohexadienones as their skeletons. Hence they are considered as dominant intermediates in many natural product syntheses. Traditional methods for the synthesis of spiro[4,5]cyclohexadienones are intra molecular dearomatization reaction of phenol derivatives. These reactions need the use of heavy metals and hard reaction conditions. Yuan in 2017 reported a highly efficient, 100% atom economic method for the synthesis of spiro[4,5]cyclohexadienones from para-quinonemethides and propargylmalonates through a silver-catalyzed 1,6-addition reaction (Yuan et al., 2017), explained in Scheme 2.42.

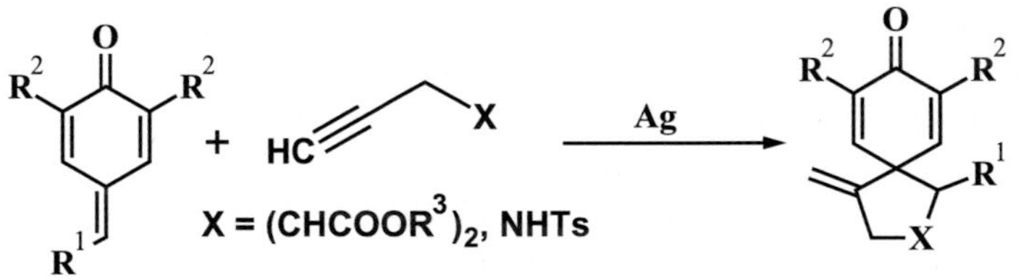

Scheme 2.42   Silver catalyzed synthesis of spiro[4,5]cyclohexadienones.

CHEMISTRY OF NANOSIZED METAL PARTICLES VIA ATOM ECONOMY CONCEPT

In the modern era, advanced technologies for synthesis are gaining wide attention. In case of metal nanoparticle synthesis, a long standing dilemma exists between the classical top-down and bottom-up approach. Although each method is significant in some way, size control is a crucial problem with physical methods. Even though the colloidal approach can address the problem, low yields hamper the advantage. Moreover, large scale production of monodisperse nanoparticles cannot be achieved using physical methods. An atom economical approach could be more feasible in this context. Selecting reagents with multiple capability and reducing reagents eliminates unwanted products and purification processes. Wostek et al. investigated nanoparticles of Ru, Co and Rh through decomposition of organo metallic precursors (Wostek et al., 2005). Greckler and co-workers synthesized Au nanoparticle using poly(vinylpyrrolidone) as the protecting agent and $NaBH_4$ as the reducing agent in the solid state using high speed vibration milling (Debnath et al., 2009).

In 2010, Kalidindi and co-workers suggested a facile green route (scheme) for synthesizing air stable Cu, Ag and Au nanoparticles. The method used the mechanical stirring of mixtures of a metal salt and ammonia borane. Ammonia borane plays both the role of a reducing agent and the precursor for the stabilizing agent. Here ammonia borane is maintained at 60°C in an oil bath and metal salt is added in small batches under an inert atmosphere. The resulting powders are desired metal nanoparticles of corresponding metal salts.

The metal particles formed can be identified from the color of the final product. For example, $CuCl_2$ has a brown color, while the copper nanoparticle has a reddish brown appearance.

OTHER CONCEPTS RELATED TO ATOM ECONOMY

There are various other measures, which can be used for calculating the greenness of a chemical reaction. Selectivity, E-factor, effective mass yield, reaction mass efficiency, etc. are some concepts related to atom economy utilised in green chemistry (Sheldon, 2018; Andaros and Sayed, 2007; Hudson et al., 2016). Selectivity is a widely accepted concept for measuring the greenness of a reaction and is described in detail below.

Selectivity

Selectivity is one of the important concepts in green chemistry and arises when we have to control the reaction to get the desired product over side products. Selectivity is related to yield an atom economy. A number of compounds are prepared through selective reactions. Selectivity is influenced by steric and electronic effects. Depending on the nature of selection chemoselectivity (functional group reactivity), regioselectivity (orientation) and stereoselectivity (spatial arrangement) can be defined (Trost, 1983).

Chemoselectivity

According to IUPAC, Chemoselectivity is defined as "the preferential reaction of a chemical reagent with one of two or more different functional groups" (Shenvi et al., 2009). In the absence of chemoselective reagents the reaction is indiscriminate and all possible products are formed, decreasing the atom economy of the process. Chemoselectivity becomes important when complex organic molecules with more than one reactive site or functional groups are involved in the reaction. In this case, the reagent has to be directed to the preferred site for the desired outcome. The discrimination ability of a reagent, the ability of the reagent to select sites to react with it, is called chemoselectivity (Trost, 1983).

For example, $SnCl_2$/HCl does not react with –CHO group, when both the nitro and aldehyde groups are present in the same molecule. Thus only the nitro group will be reduced as shown in the Scheme 2.43.

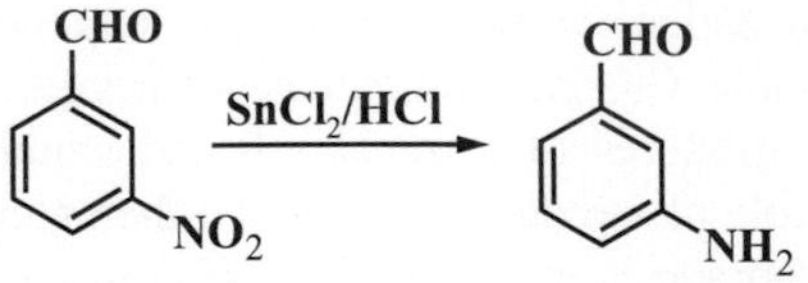

Scheme 2.43   Reduction using SnCl$_2$/HCl.

Another example is the reduction of –CHO group to –CH$_2$OH in presence of another substituent –COOEt group, by NaBH$_4$ as reducing agent, given in Scheme 2.44.

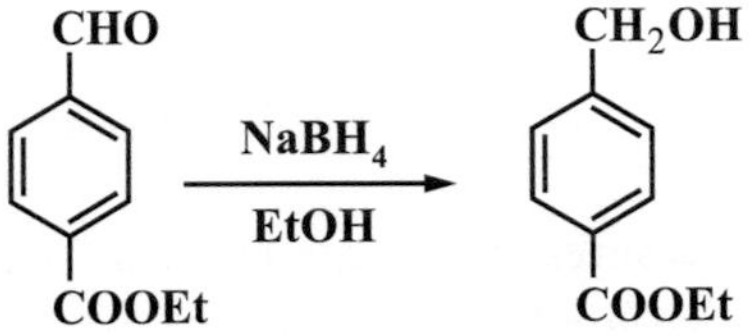

Scheme 2.44   Reduction using NaBH$_4$.

Most of the industrial synthesis of dyes, medicinal supplies, agricultural chemicals, etc. is accompanied by reduction of nitroarenes to aromatic amines (Tsukinoki and Tsuzuki, 2001). The conventional procedure for reduction of aromatic nitro compounds requires organic solvents under drastic conditions and irritant reagents like NH$_3$, conc.HCl or 20% aq. NaOH, so this reaction is not environmentally benign. To make the reaction green Zinc metal and NH$_4$Cl in water at 80°C can be used as reducing agents. The sterically hindered 2,6-dimethylnitrobenzene can be reduced selectively by these reagents, which is shown in Scheme 2.45. The reaction is chemoselective for nitrogroups only; other substituents such as amide, ester and halide on aromatic rings are rarely affected.

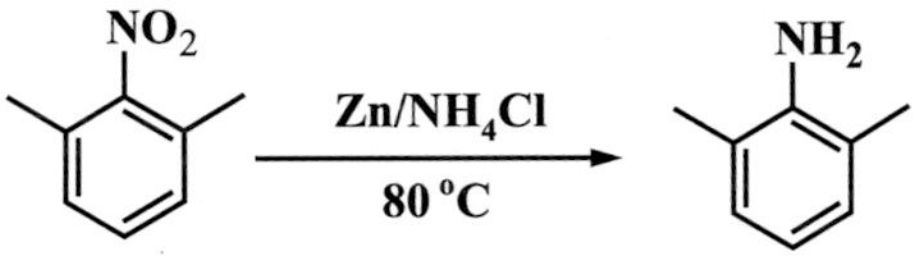

Scheme 2.45   Reduction using Zn/NH$_4$Cl.

Regioselectivity

In regioselective reactions, one of the positional isomer is preferentially formed. The question of regioselectivity arises when a reagent can approach a particular reactive site through different regions or positions. A regioselective reagent is required to obtain the desired product. A simple example is addition of HX over an unsymmetrical olefin.

Example: Reaction of Propene with HBr

(a) In the absence of hydrogenperoxide - Markovnikov's addition: The positive end goes to the carbon having more number of hydrogen atoms (Scheme 2.46).

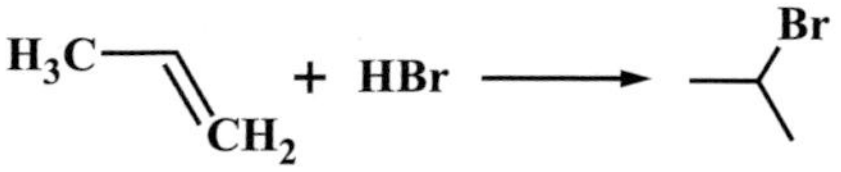

Scheme 2.46 Markovnikov's addition.

(b) In the presence of hydrogenperoxide—AntiMarkovnikov's addition: The positive end goes to carbon having lesser number of hydrogen atoms (Scheme 2.47).

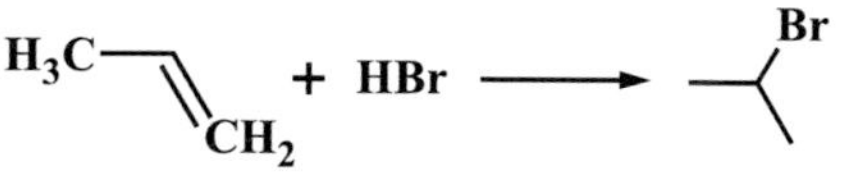

Scheme 2.47 AntiMarkovnikov's addition.

Traditional procedure for the nitration of phenol is the use of conc. nitric acid and sulfuric acid. The products obtained here are ortho- nitrophenol, para-nitrophenol and dinitro phenol. i.e. the nitration of phenol is not selective.

Regio selective ortho nitration of phenol can be done by using $NaNO_2/KHSO_4$ in CH_3CN (Heravi et al., 2007), Scheme 2.48. The reagent used is also inexpensive, easy to handle and ecofriendly compared to the traditional nitration reagents.

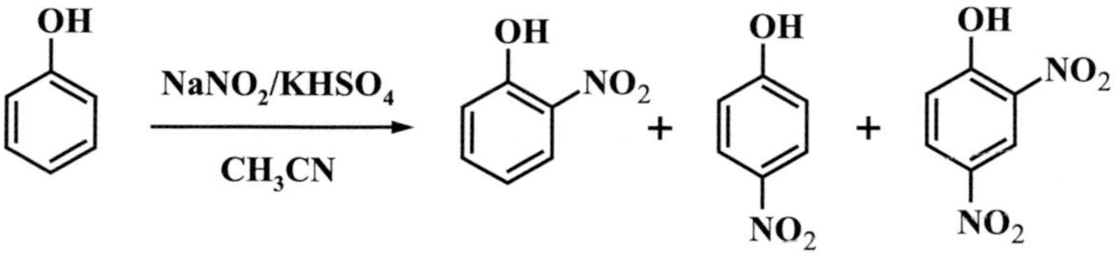

63.282%   12.131%   6.891%

Scheme 2.48 Regio selective orthonitration of phenol.

Another regioselective method for the nitration of phenol and substituted phenols is the use of dilute HNO_3 in the presence of a phase transfer catalyst tetrabutylammonium bromide (TBAB) using ethylene dichloride (EDC) as a solvent (Joshi et al., 2003), given in Scheme 2.49. As a consequence, reaction rate increases, yield increases and production of waste reduce. The nitration data of phenol and substituted phenols using different mol% of phase transfer catalyst reagent is provided in Table 2.1.

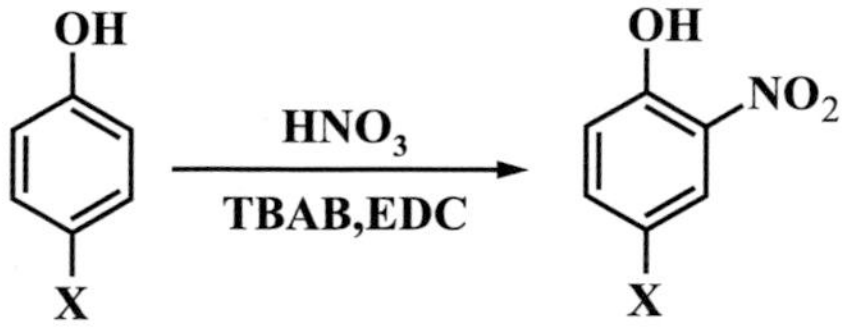

where X = CH_3, Cl or F

Scheme 2.49 Nitration using HNO_3 in presence of TBAB.

Table 2.1   Nitration with phenol and substituted phenols (Joshi et al., 2003)

Sl no.	Reactant	TBAB (mol%)	Time (hours)	% Yield
1	Phenol	0	4	0.4
2	Phenol	5	4	90
3	4-Cresol	10	5	97
4	4-Chlorophenol	10	6	97
5	4-Fluorophenol	10	6	97

Stereoselectivity

Stereoselectivity in a reaction allows preferential formation of one stereoisomer in the reaction. Stereoisomers differ only in the spatial orientation of constituent atoms in space. Stereoisomers can be optically active or inactive. Stereoisomers that are mirror images are called enantiomers and those that are not mirror images are called diastereomers. Stereoselectivity have been achieved in many cases between enantiomers and diastereomers. They are considered separately below.

a) Enantioselectivity

In enentioselective reactions, one enantiomer is formed preferentially over the other. Enantiomers have similar physical and chemical properties, but significantly differ in biological activity. Enantiomers are optically active and contain chiral reaction centers. Hence chiral features in substrate, reagent or catalyst are required for an enantioselective reaction to occur. Non-chiral reagents produce racemic mixtures as products. Enantioselectivity is of great importance in pharmaceutical chemistry (Federsel, 2005) and agrochemistry, particularly in cases where one enantiomer has good efficiency and is non toxicity, while the other one has no activity and is toxic. Consider the case of ibuprofen, a pain killer. The S-form of the molecule is an effective analgesic while the R-form is inactive. As separation of two forms is difficult, the drug is supplied as a racemic mixture. Enantioselective synthesis of ibuprofen has been addressed in may research reports (Acemoglu and Williams, 2003; Tanaka et al., 2001).

     An example of enantioselective reaction is the reduction of ketone using the enzyme Thermoanaerobiumbrockii. Lower ketones like 2-butanone produce R-alcohol and higher ketones like 2-hexanone produce S-alcohol (Scheme 2.50).

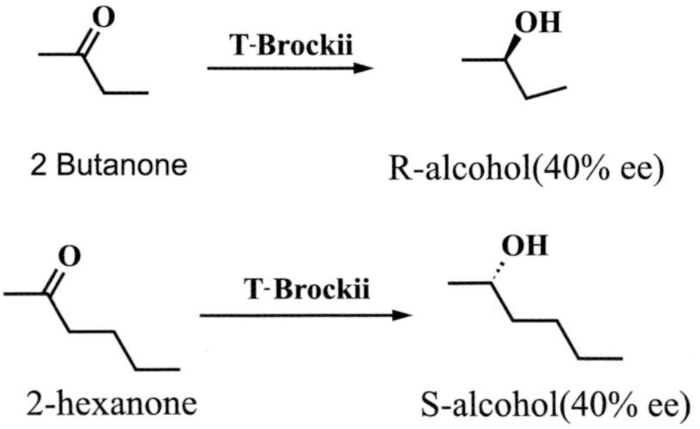

Scheme 2.50   Enantio selective reduction of ketone with Thermoanaerobiumbrockii.

'ee' is the enantiomeric excess, which is the differencein mole fractions of enantiomers in the mixture, used to measure degree of selectivity. Since a racemic mixture contains equal amount of enantiomers, the enantiomeric excess is zero. If a mixture contains the (+) enantiomer in four times than (−) enantiomer, the difference is 0.8–0.2 = 0.6 or 60% ee. A pure enantiomer has 100% ee.

b) Diastereoselectivity

Consider the hydrogenation of substituted cyclohexene given in Scheme 2.51.

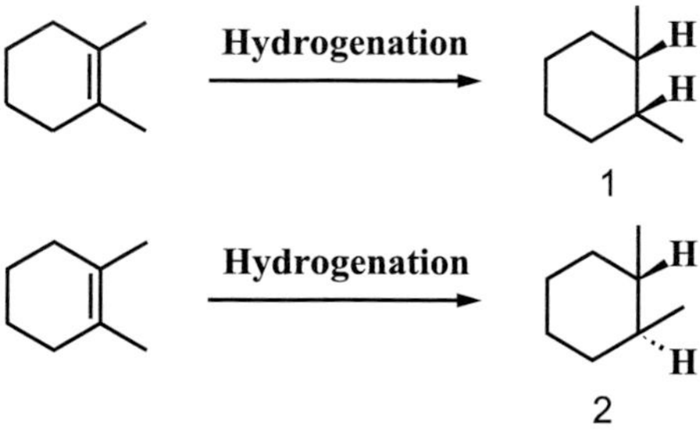

Scheme 2.51   Hydrogenation of substituted cyclohexene.

Addition of the two hydrogen atoms through the same side of the plane of the double bond, results in product I and through opposite sides results in product II. Catalytic hydrogenation usually results in product I rather than II since the hydrogen atoms are delivered from the catalyst surface, on one side. The two products are diastereomers.

This preferential formation of one diastereomer over the other is called diastereoselectivity. Diastereomers contain more than one chiral center. Consider the reaction between anthrone and R(+)-N-α-methyl benzyl maleimide in the presence of triethyl amine as achiral base in water-ethanol medium, given in Scheme 2.52. The reaction has excellent yields, but is not diastereoselective. Instead of the achiral base, chiral Brownsted bases like cinchonine, quinine, etc. produce diastereoselective products.

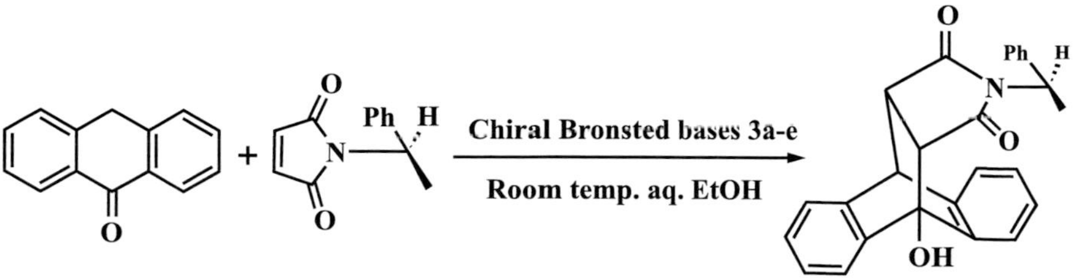

Scheme 2.52   Reaction between anthrone and R-(+)-N-α-methyl benzyl maleimide in the presence of triethyl amine.

E-Factor

Waste production is an unavoidable consequence of industrial production. Managing the waste problem is of prime importance in green and sustainable chemistry. One aspect of minimizing waste production is recycling reagents and solvents. If catalysts and reagents are in the solid phase, they can be filtered off and reused. Substrate materials are continuously added to yield the product which is simultaneously removed, while heterogeneous reagents and catalysts are kept stationary. Environmental factor (E-factor) was introduced by Roger Sheldon to determine the mass efficiency of such type of chemical reactions (Andaros, 2009). E-factor can be effectively used as a tool to measure how much waste is generated. It gives a perspective on the environmental impact of a particular reaction.

$$E\text{-}factor = \frac{\text{Mass of produced wastes}}{\text{Mass of desired product}}$$

Since estimation of the exact amount of wastes is difficult, an easier way of calculating the E-factor is given below.

$$E\text{-}factor = \frac{\text{Mass of raw materials} - \text{Mass of product}}{\text{Mass of product}}$$

For almost all petroleum refining processes E-factors have been estimated to be close to zero, an ideal production with zero waste generation. But in the case of pharmaceutical and fine chemical industries, E-factors have been estimated to be very high, in the range 5–100, indicating production of a large amount of waste. The E-factors of various industries is given in Table 2.2.

E-factor and other concepts do not discuss the toxicity of wastes. For comparison of alternate processes, we need to consider the toxicity aspect , along with the amount of waste production. The term Environmental Quotient (EQ) was proposed and is calculated as the product of E-factor and a correction factor Q (unfriendliness quotient). The unfriendliness quotient, Q is the measure of toxic substances. If Q-factor is 1, the waste has no environmental impact. The waste can be recycled, if the E-factor is less than 1 and the wastes produced are toxic, hazardous and cannot be recycled, if it is greater than 1.

Table 2.2   E-factors for different sectors of chemical industry (Sheldon, 2008)

Industry sector	Annual production (tones)	E-factor
Oil refining	10^6–10^8	Less than 0.1
Bulk chemicals	10^4–10^6	≤ 1
Fine chemicals	10^2–10^4	5–50
Pharmaceuticals	10–10^3	25–100

We can compare E-factor of various chemical industries. Production of bulk chemicals is highly efficient compared to the production of fine chemicals and pharmaceuticals. Use of inorganic reagents such as metals (Zn, Fe), metal hydrides ($NaBH_4$, $LiAlH_4$), etc. in reduction steps and $KMnO_4$, $K_2Cr_2O_7$ etc. in oxidation steps helps in decreasing the E-factor. Many reactions are accelerated by Lewis acids ($AlCl_3$, BF_3, $ZnCl_2$) or mineral acids (H_2SO_4, H_3PO_4, HF). Substituting outdated stoichiometric approaches with suitable catalytic alternatives is a way to manage the waste. A catalytic process in which hydrogen acts as a reducing agent is more atom efficient than a hydride reducing agent. Similarly in comparison of E-factors of different oxidants, oxygen has less E-factor than other oxidants and is more reliable.

Effective mass yield (EMY)

Another concept related to atom economy is the Effective Mass Yield (EMY) proposed by Hudlicky

$$EMY = \frac{\text{Mass of desired product}}{\text{Mass of non-benign reagents}} \times 100$$

A non benign substance is defined as those by-products, solvents or surplus reagents being associated with environmental threats. The difficulty to find out whether the reagent is non benign or not is the major drawback of this concept. The scarcity of toxicology data and measuring of non-benign nature makes this parameter a less useful one.

Reaction Mass Efficiency (RME)

We cannot ensure that a reaction with high atom economy will give high yields or vice versa. So a need for a better quantitative measure arises, which accounts for both yield and atom economy. Reaction Mass Efficiency (RME), also called real atom economy is a more realistic metric for describing the greenness of a chemical reaction, put forward by Constable ct al. at Glaxo Smith Kline in 2002. RME comprises yield, stoichiometry and atom economy and thus can be considered as a superior tool among all.

$$RME = \frac{\text{Mass of desired product}}{\text{Total mass of reactants}} \times 100\%$$

It can be expressed in another way as,

$$RME = \text{Yield} \times \text{Atom economy} \times \frac{1}{\text{Stoichiometric factor}}$$

Consider the esterification reaction of n- butanol with acetic acid to form n-butyl acetate ester, as an example for illustration (Table 2.3).

Table 2.3 Number of moles calculation for esterification reaction of n- butanol with acetic acid

Reactant/Product	Mol. weight (g)	Mass (g)	No. of moles
n-Butanol	74	37	0.5
Acetic acid	60	60	1
Acetate ester	116	40	0.34

$$\text{Yield} = \frac{0.34}{0.5} \times 100 = 68\%$$

$$\text{Atom economy} = \frac{116}{74 + 60} \times 100 = 87\%$$

From the calculation, it is clear that, this esterification reaction is green in atom economical context, but in terms of yield, it is only moderately green. Here we can calculate the RME, which is a measure encompassing both these aspects.

$$\text{RME} = \frac{40}{97} \times 100 = 41\%$$

As one can imagine multiple factors need to be taken into account to access the greenness of a process along with atom economy. Recently multi-metric analysis including acidification potential, smog formation potential, global warming potential, ozone depletion potential, abiotic resource depletion potential, persistence, bioaccumulation, ingestion toxicity, inhalation toxicity, etc. of different synthetic routes of aniline from benzene have been attempted to identify the greener process (Mercer et al., 2012). Similar assessments are required to completely characterize a process as being completely green.

CONCLUSION

Use of green chemistry and green principles are going to be pivotal if sustainable life is to be established on earth. It is high time that our synthetic methodologies and industrial production are revamped to align with green principles that outline safe, sustainable and waste minimizing synthetic methods. Among the 12 principles of green chemistry, the second one, concept of atom economy allows a quantitative estimation of greenness of a synthetic process. The process with higher atom economy is expected to result in low wastage. But one needs to keep in mind that it does not encapsulate product yield, extend of waste generation or nature of waste. Definitions with wider scope that can reflect toxicity, nature and management of waste, general usefulness and environmental impact would be required to precisely quantify the environmental and sustainability impact of a chemical process.

REFERENCES

Acemoglu, L., Williams, J.M.J. Palladium-catalysed enantioselective synthesis of ibuprofen. *Journal of Molecular Catalysis A: Chemical,* 2003, 196(1–2), 3–11.

Allen, L.J., Crabtree, R.H. Green alcohol couplings without transition metal catalysts: base-mediated β-alkylation of alcohols in aerobic conditions. *Green Chemistry,* 2010, 12(8), 1362–1364.

Anastas, P.T., Kirchhoff, M.M., Williamson, T.C. Catalysis as a foundational pillar of green chemistry. *Applied Catalysis A: General,* 2001, 221(1–2), 3–13.

Anastas, P.T., Kirchhoff, M.M. Origins, current status, and future challenges of green chemistry. *Journal of Chemical Education,* 2002, 35(9), 686–694.

Andraos, J. Unification of reaction metrics for green chemistry: Applications to reaction analysis. *Organic Process Research and Development,* 2005a, 9(2), 149–163.

Andraos, J. Unification of reaction metrics for green chemistry ii: Evaluation of named organic reactions and application to reaction discovery. *Organic Process Research and Development*, 2005b, 9(4), 404–431.

Andraos, J., Sayed, M. On the use of "Green" metrics in the undergraduate organic chemistry lecture and lab to assess the mass efficiency of organic reactions. *Journal of Chemical Education*, 2007, 84(6), 1004.

Andraos, J. Global green chemistry metrics analysis algorithm and spreadsheets: Evaluation of the material efficiency performances of synthesis plans for oseltamivir phosphate (Tamiflu) as a test case. *Organic Process Research and Development*, 2009, 13(2), 161–185.

Barrett, A.G.M., Braddock, D.C., James, R.A., Koike, N., Procopiou, P.A. Nucleophilic substitution reactions of (Alkoxymethylene) dimethylammonium chloride. *Journal of Organic Chemistry*, 1998, 63(18), 6273–6280.

Berliner, M.A., Dubant, S.P.A., Makowski, T., Ng, K., Sitter, B., Wager, C., Zhang, Y. Use of an iridium-catalyzed redox-neutral alcohol-amine coupling on kilogram scale for the synthesis of a GlyT1 inhibitor. *Organic Process Research and Development*, 2011, 15(5), 1052–1062.

Cann, M.C., Dickneider, T.A. Infusing the chemistry curriculum with green chemistry using real-world examples, web modules, and atom economy in organic chemistry courses. *Journal of Chemical Education*, 2004, 81(7), 977–980.

Choudary, B.M., Chowdari, N.S., Madhi, S., Kantam, M.L. A trifunctional catalyst for one-pot synthesis of chiral diols via heck Coupling–N-Oxidation–Asymmetric dihydroxylation: Application for the synthesis of diltiazem and taxol side chain. *Journal of Organic Chemistry*, 2003, 68(5), 1736–1746.

Comerford, J.W., Clark, J.H., Macquarrie, D.J., Breeden, S.W. Clean, reusable and low cost heterogeneous catalyst for amide synthesis. *Chemical Communications*, 2009, 18, 2562–2564.

Constable, D.J.C., Curzons, A.D., Cunningham, V.L. Metrics to 'Green' chemistry—which are the best? *Green Chemistry*, 2002, 4(6), 521–527.

Constable, D.J.C., Dunn, P.J., Hayler, J.D., Humphrey, G.R., Leazer, J.L. Jr., Linderman, R.J., Lorenz, K., Manley, J., Pearlman, B.A., Wells, A., Zaks, A., Zhang, T.Y. Key green chemistry research areas—A perspective from pharmaceutical manufacturers. *Green Chemistry*, 2007, 9(5), 411–420.

Debnath, D., Kim, S.H., Geckeler, K.E. The first solid-phase route to fabricate and size-tune gold nanoparticles at room temperature. *Journal of Material Chemistry*, 2009, 19(46), 8810.

Dömling, A. Recent advances in isocyanide-based multicomponent chemistry. *Current Opinion in Chemical Biology*, 2002, 6(3), 306–313.

Eissen, M., Mazur, R., Quebbemann, H.G., Pennemann, K.H. Atom economy and yield of synthesis sequences. *Helvetica Chimica Acta*, 2004, 87(2), 524–535.

Enthaler, S., Junge, K., Addis, D., Erre, G., Beller, M. A practical and benign synthesis of primary amines through ruthenium-catalyzed reduction of nitriles. *ChemSusChem*, 2008, 1(12), 1006–1010.

Federsel, H.J. Asymmetry on large scale: the roadmap to stereoselective processes. *Nature Reviews Drug Discovery*, 2005, 4, 685.

Fennie, M.W., Roth, J.M. Comparing amide-forming reactions using green chemistry metrics in an undergraduate organic laboratory. *Journal of Chemical Education*, 2016, 93(10), 1788–1793.

Fishback, V., Reid, B., Schildkret, A. Three modules incorporating cost analysis, green principles, and metrics for a sophomore organic chemistry laboratory. *ACS Symposium Series,* 2016, 1233, 33–53.

Geier, S.J., Chase, P.A., Stephan, D.W. Metal-free reductions of n-heterocycles via lewis acid catalyzed hydrogenation. *Chemical Communications,* 2010, 46(27), 4884–4886.

Grant, S., Freer, A.A., Winfield, J.M., Gray, C., Lennon, D. Introducing undergraduates to green chemistry: An interactive teaching exercise. *Green Chemistry,* 2005, 7(3), 121–128.

Gunanathan, C., Ben-David, Y., Milstein, D. Direct synthesis of amides from alcohols and amines with liberation of H_2. *Science,* 2007, 317(5839), 790–792.

Heravi, M.M., Oskooie, H.A., Baghernejad, B. Facile, Regioselective and green synthesis of ortho-nitrophenoles using $NaNO_2$, $KHSO_4$. *Journal of the Chinese Chemical Society,* 2007, 54(3), 767–770.

Hudson, R., Leaman, D., Kawamura, K.E., Esdale, K.N., Glaisher, S., Bishop, A., Katz, J.L. Exploring green chemistry metrics with interlocking building block molecular models. *Journal of Chemical Education,* 2016, 93(4), 691–694.

Joshi, A.V., Baidoosi, M., Mukhopadhyay, S., Sasson, Y. Nitration of phenol and substituted phenols with dilute nitric acid using phase-transfer catalysts. *Organic Process Research and Development,* 2003, 7(1), 95–97.

Kalidindi, S.B., Sanyal, U., Jagirdar, B.R. Metal nanoparticles via the atom-economy green approach. *Inorganic Chemistry,* 2010, 49(9), 3965–3967.

Kang, B., Fu, Z., Hong, S.H. Ruthenium-catalyzed redox-neutral and single-step amide synthesis from alcohol and nitrile with complete atom economy. *Journal of American Chemical Society,* 2013, 135(32), 11704–11707.

Kidwai, M., Mohan, R. Green chemistry: An innovative technology. *Foundations of Chemistry,* 2005, 7(3), 269–287.

Li, C.J., Trost, B.M. Green chemistry for chemical synthesis. *Proceedings of the National Academy of Sciences,* 2008, 105(36), 13197–13202.

Liang, R., Chen, K., Zhang, Q., Zhang, J., Jiang, H., Zhu, S. Rapid access to 2-methylene tetrahydrofurans and γ-lactones: A tandem four-step process. *Angewandte Chemie International Edition,* 2016, 55(7), 2587–2591.

McKenzie, L.C., Huffman, L.M., Hutchison, J.E. The evolution of a green chemistry laboratory experiment: greener brominations of stilbene. *Journal of Chemical Education,* 2005, 82(2), 306–310.

Mercer, S.M., Andraos, J., Jessop, P.G. Choosing the greenest synthesis: a multivariate metric green chemistry exercise. *Journal of Chemical Education,* 2012, 89(2), 215–220.

Nguyen, T.B., Retailleau, P. Redox-neutral access to sultams from 2-Nitrochalcones and sulfur with complete atom economy. *Organic Letters,* 2017, 19(14), 3879–3882.

Oeggl, R., Maßmann, T., Jupke, A., Rother, D. Four atom efficient enzyme cascades for all 4-Methoxyphenyl-1,2-Propanediol isomers including product crystallization targeting high product concentrations and excellent E-factors. *ACS Sustainable Chemistry and Engineering,* 2018, 6(9), 11819–11826.

Parmar, A., Kumar, H., Marwaha, S., Kennedy, J. Advances in enzymatic transformation of penicillins to 6-Aminopenicillanic acid (6-APA). *Biotechnology Advances,* 2000, 18(4), 289–301.

Polidano, K., Allen, B.D.W., Williams, J.M.J., Morrill, L.C. Iron-catalyzed methylation using the borrowing hydrogen approach. *ACS Catalysis,* 2018, 8(7), 6440–6445.

Posner, G.H. Multicomponent one-pot annulations forming 3 to 6 bonds. *Chemical Reviews,* 1986, 86(5), 831–844.

Powell, K.A., Ramer, S.W., del Cardayré, S.B., Stemmer, W.P.C., Tobin, M.B., Longchamp, P.F., Huisman, G.W. Directed evolution and biocatalysis. *Angewandte Chemie International Edition,* 2001, 40(21), 3948–3959.

Qiu, S., Chen, L., Jiang, H., Zhu, S. CuCl/Et$_3$ N-catalyzed synthesis of indanone-fused 2-Methylene pyrrolidines from enynals and propargylamines. *Organic Letters,* 2017, 19(17), 4540–4543.

Reddy, V., Vijaya Anand, R. Expedient access to unsymmetrical diarylindolylmethanes through palladium-catalyzed domino electrophilic cyclization–extended conjugate addition approach. *Organic Letters,* 2015, 17(14), 3390–3393.

Sato, K.A. "Green" Route to adipic acid: Direct oxidation of cyclohexenes with 30 percent hydrogen peroxide. *Science,* 1998, 281(5383), 1646–1647.

Sheldon, R.A. E factors, green chemistry and catalysis: An odyssey. *Chemical Communications,* 2008, 29, 3352–3365.

Sheldon, R.A. Metrics of green chemistry and sustainability: Past, present, and future. *ACS Sustainable Chemistry and Engineering,* 2018, 6(1), 32–48.

Shenvi, R.A., O'Malley, D.P., Baran, P.S. Chemoselectivity: The mother of invention in total synthesis. *Accounts of Chemical Research,* 2009, 42(4), 530–541.

Sipos, G., Gyollai, V., Sipőcz, T., Dormán, G., Kocsis, L., Jones, R.V., Darvas, F. Important industrial procedures revisited in flow: Very efficient oxidation and *N*-Alkylation reactions with high atom-economy. *Journal of Flow Chemistry,* 2013, 3(2), 51–58.

Song, Y., Wang, Y., Geng, Z. Some exercises reflecting green chemistry concepts. *Journal of Chemical Education,* 2004, 81(5), 691–692.

Straathof, A.J., Panke, S., Schmid, A. The production of fine chemicals by biotransformations. *Current Opinion in Biotechnology,* 2002, 13(6), 548–556.

Takemoto, S., Kawamura, H., Yamada, Y., Okada, T., Ono, A., Yoshikawa, E., Mizobe, Y., Hidai, M. Ruthenium complexes containing bis(diarylamido)/Thioether ligands: synthesis and their catalysis for the hydrogenation of benzonitrile. *Organometallics,* 2002, 21(19), 3897–3904.

Tanaka, J.I., Oda, S., Ohta, H. Synthesis of (S)-Ibuprofen via enantioselective degradation of racemic ibuprofen with an isolated yeast, trichosporon cutaneum KPY 30802, in an interface bioreactor. *Journal of Bioscience and Bioengineering,* 2001, 91(3), 314–315.

Trost, B.M. Selectivity: A key to synthetic efficiency. *Science,* 1983, 219(4582), 245–250.

Trost, B.M. Atom economy-a challenge for organic synthesis: Homogeneous catalysis leads the way. *Angewandte Chemie International Edition in English,* 1995, 34(3), 259–281.

Trost, B.M. On inventing reactions for atom economy. *Accounts of Chemical Research,* 2002, 35(9), 695–705.

Tsukinoki, T., Tsuzuki, H. Organic reaction in water. Part 5.1 novel synthesis of anilines by zinc metal-mediated chemoselective reduction of nitroarenes. *Green Chemistry,* 2001, 3(1), 37–38.

Ugi, I., Dömling, A., Hörl, W. Multicomponent reactions in organic chemistry. *Endeavour,* 1994, 18(3), 115–122.

Valeur, E., Bradley, M. Amide bond formation: Beyond the myth of coupling reagents. *Chemical Society Reviews,* 2008, 38(2), 606–631.

Wardman, P. Reduction potentials of one-electron couples involving free radicals in aqueous solution. *Journal of Physical Chemistry Reference Data,* 1989, 18(4), 1637–1755.

Wegman, M.A., Janssen, M.H.A., van Rantwijk, F., Sheldon, R.A. Towards biocatalytic synthesis of β-lactam antibiotics. *Advanced Synthesis and Catalysis,* 2001, 343(6–7), 559–576.

Weissenburger, H.W.O., van der Hoeven, M.G. An efficient nonenzymatic conversion of benzylpenicillin to 6-aminopenicillanic acid. *Recueil des Travaux Chimiquesdes Pays-Bas,* 2010, 89(10), 1081–1084.

Wostek-Wojciechowska, D., Jeszka, J.K., Amiens, C., Chaudret, B., Lecante, P. The solid-state synthesis of metal nanoparticles from organometallic precursors. *Journal of Colloid and Interface Science,* 2005, 287(1), 107–113.

Yadav, G.D., Kadam, A.A. Atom-efficient benzoin condensation in liquid–liquid system using quaternary ammonium salts: pseudo-phase transfer catalysis. *Organic Process Research and Development,* 2012, 16(5), 755–763.

Yuan, Z., Liu, L., Pan, R., Yao, H., Lin, A. Silver-catalyzed cascade 1,6-addition/cyclization of *para* -quinone methides with propargyl malonates: An approach to spiro[4.5]deca-6,9-dien-8-ones. *Journal of Organic Chemistry,* 2017, 82(16), 8743–8751.

Zweifel, T., Naubron, J.V., Grützmacher, H. Catalyzed dehydrogenative coupling of primary alcohols with water, methanol, or amines. *Angewandte Chemie,* 2009, 121(3), 567–571.

3
Chapter

Prevention

Maria Isabel Martinez Espinoza*

Department of Chemistry, Materials, and Chemical Engineering "Giulio
Natta", Polytechnic University of Milan, Milan, Italy,
Via Luigi Mancinelli, 7, 20131, Milan Italy

INTRODUCTION

Continuous scientific and technological advances have allowed human beings to
extend and improve life conditions. But on the other hand, these advancements
have brought some of the short and long-term negative effects –that these days–
can not be ignored.

Over the past few decades, awareness has grown that the quality of air,
water, soil, and food affects human health and the environment. The causes of the
various environmental diseases are numerous and include pollution generated by
transport, agricultural activity, industrial processes, domestic effluents and waste
management. Addressing environment problems means fighting on several fronts
and in most cases, the better strategy is prevented.

It was in the 90's when green chemistry was introduced to avoid chemical
hazards that organic and inorganic compounds had on global health and to offer
innovative and sustainable strategies to reduce pollution, (Anastas and Kirchhoff,
2002) and how to prevent, reduce and eliminate the most dangerous risks. These
strategies are explained by the 12 principles of the green chemistry, (Ivankovic
et al., 2017). The first principle is dedicated to prevention which is described by:
"It is better to prevent waste than to treat or clean up waste after it is formed"
(Ivanković et al., 2017). It is the most important because it is considered the basis
of all the principles, while the others are the strategies to achieve it (Fig. 3.1).

*For Correspondence: q.isabel09@gmail.com

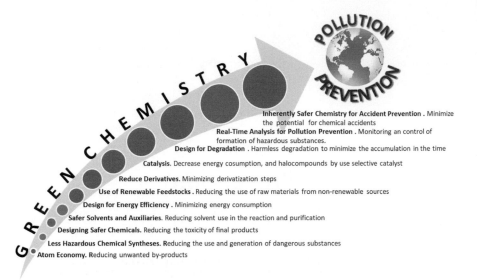

Inherently Safer Chemistry for Accident Prevention . Minimize the potential for chemical accidents
Real-Time Analysis for Pollution Prevention . Monitoring an control of formation of hazardous substances.
Design for Degradation . Harmless degradation to minimize the accumulation in the time
Catalysis. Decrease energy cosumption, and halocompounds by use selective catalyst
Reduce Derivatives. Minimizing derivatization steps
Use of Renewable Feedstocks . Reducing the use of raw materials from non-renewable sources
Design for Energy Efficiency . Minimizing energy consumption
Safer Solvents and Auxiliaries. Reducing solvent use in the reaction and purification
Designing Safer Chemicals. Reducing the toxicity of final products
Less Hazardous Chemical Syntheses. Reducing the use and generation of dangerous substances
Atom Economy. Reducing unwanted by-products

Figure 3.1   Prevention as the basis of all principles of green chemistry.

As shown in the Fig. 3.1, the 12 principles focus on prevention; prevent different points such as hazardous waste, solvents and other substances that can affect the environment.

It is important that a scientist takes into consideration that any substance used or generated can directly or indirectly influence health and the surrounding environment. For this reason, before designing an experiment, it is necessary to consider the possible waste generated and the best options for their disposal.

In reality, these points must be one of the main lessons in the professional training of chemists, engineers, biologists, and other professionals who are directed to generate new processes and products. Teaching new generations to prevent the generation of waste of all kinds, through i) the principles of green chemistry and their application in laboratory practices, ii) respect for the medium environment, iii) through the awareness that science can bring many benefits but also generate so much serious damage when it is not done properly and iv) that green chemistry is not a philosophy of the future, but of the present, are the key to the conservation of the planet and all that it offers us.

PREVENTION WASTE

Waste can be defined as generated or unwanted material of any type that does not have any value and especially what is left after useful substances have been removed. Waste can be represented in many forms, and the environment impact depending on its chemical nature, toxicity, quantity and the way it was disposed (Chhipa et al., 2013; U.S. EPA, 2018). According to the Resource Conservation and Recovery Act (RCRA) over the course of 2001–2015, the quantity of hazardous waste generated in the U.S. ranged from 20.3 to 29.1 million tons, and over 31%

corresponded to the chemical sector which produces waste such as solvents, acids, salts, and other organic chemicals (U.S. EPA, 2018).

To control and monitor the regeneration of waste, Roger Sheldon introduced the E-Factor, which relates the weight of waste co-produced to the weight of the desired product (Sheldon, 2007; 2017). This metric helps to quantify the amount of waste generated per kilogram of product. More recently, the ACS Green Chemistry Institute Pharmaceutical Roundtable (ACS GCIPR) has modified the concept of the E-Factor involving the ratio of the weights of all materials (water, organic solvents, materials, reagents, process aids, catalyst, sub-products) used to the weight of theactive drug ingredient (API) produced (Sheldon, 2007; 2017; Betts, 2018). This fact has the origin in the historically large amount of waste coproduced during drug production. Currently, many companies apply green chemistry principles to the design of the API process and fourtunately, dramatic reductions in waste are often achieved, sometimes as much as ten-fold (Betts, 2018). These impressive results obtained by the application of the prevention principle to all parts of the chemical industry, from the pharmaceuticals to others where synthetic chemistry is used to produce their products can be considered a first success of the green philosophy and demostrates that it is possible design alternative process more friendly for the enviorment (Betts, 2018; Pfizer Inc, Waste and Recycling, 2018; Novartis Materials & Waste, 2018; Janssen Global Services, Green Chemistry: Innovation in Chemistry Development, 2018; Boehringer Ingelheim International GmbH., 2018; L'Oreal group, Green chemistry for new active ingredients, 2018; Merck group, Corporate Responsibility Report, 2017; 2018; DuPont. DuPont Bio-Based Polymer Platform Earns Green Chemistry Award, 2018).

Today, some companies such as Pfizer, Janssen, Boehringer Ingelheim, Novartis, L'Oréal, Merck group, DuPont and many others, have implemented a team dedicated to the research and development of new sustainable and green technologies that offers a "greener" product and safeguards the environment. The paint and varnish industry is already free-solvents. The detergent industry has already eliminated all phosphorus-containing detergents. Asbestos is no longer used in practice. Along with finding new drugs, one of the biggest challenges of the pharmaceutical industry is to apply procedures that have a minimal environmental impact to the manufacturing process. This mainly means reducing the amount of waste generated in production processes and the disposal of expired drugs.

E-FACTOR

In laboratories, reactions are carried out on a scale of grams and milliliters, which means that the waste (subproducts and solvents) are also produced on the same scale. But at an industrial level these quantities are multiplied more than 1000 times, using a quantity of reagents and solvents that vary from kilograms to tons and liters of solvent. Obviously, the large-scale production of a product consequently produces the generation of waste even on the same scale.

In the industrial sectors, the most common basic reactions of organic synthesis include the halogenation, oxidation, alkylation, nitration, and sulfonation (Brown and Bostro, 2016). Unfortunately, they also are known as "dirty reaction" due to generating unwanted products and requiring dangerous and toxic reagents and solvent. Today companies want to demonstrate that their chemical reactions, processes, and final products are green.

Green chemistry have proposed several metrics to increase the awareness of generated waste sources from the reaction and to identify opportunities for further improvement (Table 3.1) (Papadakis et al., 2017; Andraos, 2005; Roschangar et al., 2015). These parameters indicate how much a process or product is "green".

Related to waste prevention, and as was mentioned before, the E-Factor (eq. shown in Table 3.1) is used to measure the ratio of waste generation and final product and can be applied to all types of reaction. Some years ago, the typical E-Factor for chemical industry sectors was published showing that the pharmaceuticals have higher E-Factor of 25 to > 100 (Phan et al., 2015; Sheldon, 1992). In other words, the pharmaceutical industry typically generates between 25 and 100 kilograms of waste per kilogram of product. The waste includes the solvents, acids and bases, and sub-products that do not have any further use.

However, one fact is true, as Sanderson published in her title article: Chemistry: It's not easy being green, (Sanderson, 2011) The chemical sectors have devoted all its efforts to improve production processes, following (as much as possible) the principles of green chemistry.

Table 3.1   List of metrics that have been proposed for "green" chemistry

Metric	Description	Equation
Effective Mass yield (EM)	The percentage of the mass of product over the overall mass of non-benign compounds used during the synthesis.	$EM\ (\%) = \dfrac{\text{Mass of products (kg)}}{\text{Mass of non-benign reagents (kg)}} \times 100$
E-Factor	The mass of total waste produced for a given amount of produced product.	$E\text{-factor} = \dfrac{\text{Total waste (kg)}}{\text{kg of product}}$
Atom Economy (AE)	How much of the reactants remain in the product.	$AE\ (\%) = \dfrac{\text{Molecular weight of products}}{\Sigma(\text{MW of } A,B,C,D,E,F,G,H,I)} \times 100$ *where $A, B, C, D, E, F, G, H, I$: molecular weight of reactants*
Mass Intensity (MI)	Total mass used to produce the product.	$MI = \dfrac{\text{Total mass used in a process or process step (kg)}}{\text{Mass of product (kg)}}$
Carbon efficiency (CE)	Percentage of carbon of the reactants that remain in the final product.	$CE\ (\%) = \dfrac{\text{Amount of carbon in product}}{\text{Total carbon present in the reactants}} \times 100$
Reaction mass efficiency (RME)	Mass of reactants remaining in the product.	$RME\ (\%) = \dfrac{\text{Mass of product (kg)}}{\text{Mass of product reactants (kg)}} \times 10$

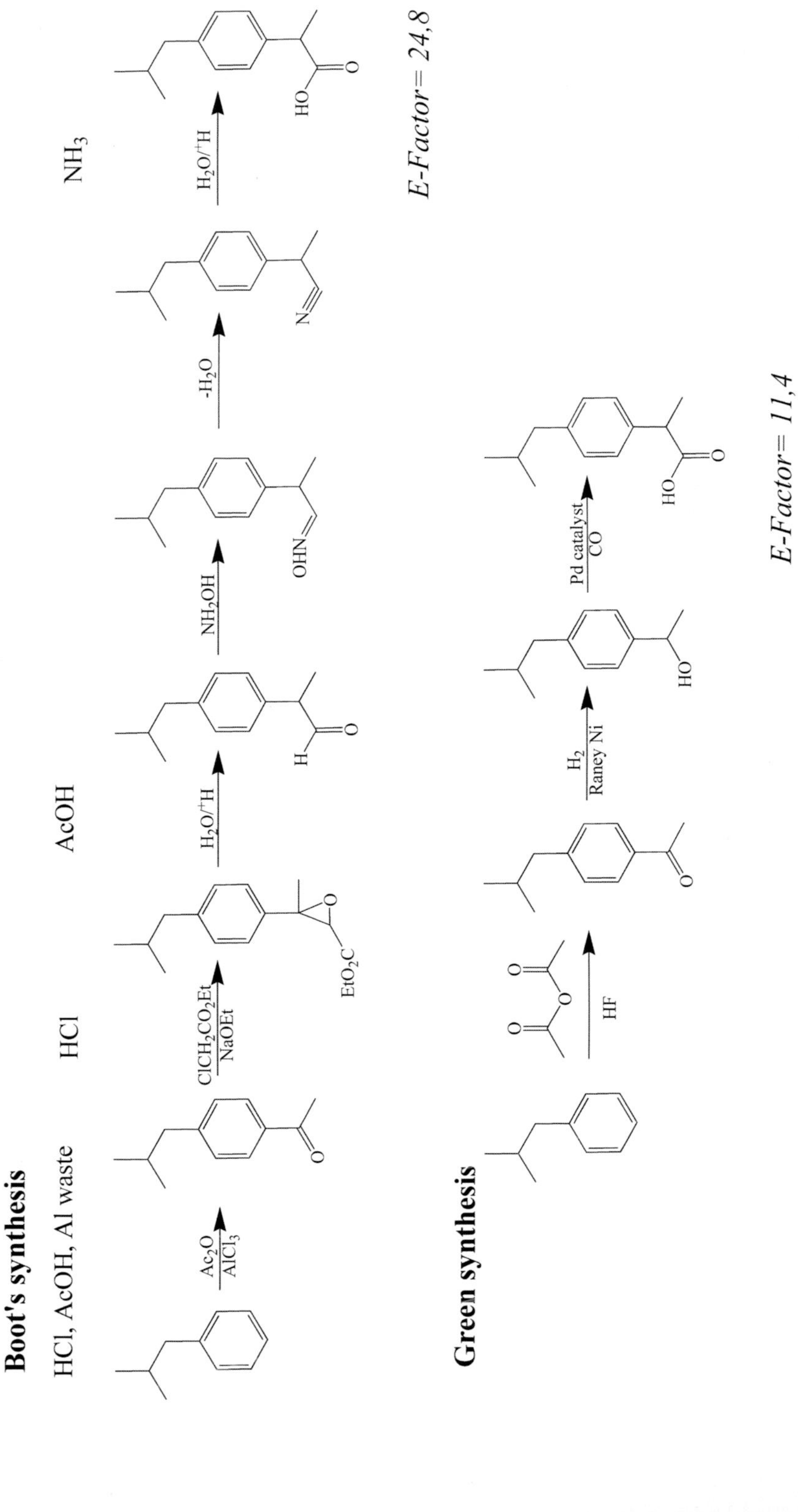

Boot's synthesis

HCl, AcOH, Al waste   HCl   AcOH   NH$_3$

E-Factor= 24,8

Green synthesis

E-Factor= 11,4

Scheme 3.1 Conventional and greener synthesis of Ibuprofen

A few examples of this effort are the new synthesis of ibuprofen (Nagendrappa, 2005; Murphy, 2018; Andraos, 2016; Bogdan et al., 2009; Cleij et al., 1999). Ibuprofen, known for its use as an efficient anti-inflammatory, can be obtained by green synthesis in three steps whereas the Boot's procedure consists of six steps with the consequent production of waste to be disposed of and other by-products that need separate treatment.

Other strategies for the synthesis of Ibuprofen were proposed changing the starting materials, reagents and numbers of steps, and in all cases the plan was and evaluated according to the E-Factor. The summary of these studies are presented in the Table 3.2 (Murphy, 2018; Andraos, 2016; Bogdan et al., 2009; Cleij et al., 1999).

Table 3.2   Summary of E-factor for the synthesis Ibuprofen

Autor	*Year*	*E-Factor*
Upjohn	1977	10.7
DD113889	1975	11.4
Hoechst-Celanese	1988	23.6
Boots	1968	24.8
Ruchardt	1991	29.8
duPont	1985	44.6
Pinhey	1984	144.3
RajanBabu	2009	394.2
Furstoss	1999	477.1
McQuade	2009	8,718

Pfizer, one of the most biggest pharmaceutical companies, has optimized the synthesis of the anti-impotence drug "Viagra". A Pfizer team has reviewed and reorganized each synthesis phase at the Sandwich facility in the United Kingdom. It replaced chlorinated solvents with other less toxic solvents, introduced processes to recover and reuse these solvents, and eliminated the need to use hydrogen peroxide. Finally, it eliminated the need for Oxalyl Chloride, a reagent that produces carbon monoxide during the reactions, reducing the E-Factor at 50.3 (Pfizer Inc, Waste and Recycling, 2018; Roschangar et al., 2015). Also Pfizer has reduced the E-Factor of other drugs such as the anticonvulsant "lyrica", the antidepressant "Sertalina" and in the anti-inflammatories "Celecobix". This success has enabled Pfizer to eliminate more than half a million tonnes of chemical waste (Anderson et al., 2009; Sheldon, 2010). The savings in raw materials and in the use of energy, the costs of disposing of waste and any penalties due to pollution, are excellent incentives for industries that choose green chemistry.

THE WASTE HIERARCHY

Effective waste generation prevention is practically impossible due to the fact that no raw material can be used 100%. Another point to consider is that the final product after use is considered a waste to be disposed of, and consequently the treatment of this waste can cause a high environmental impact. Therefore, it is necessary to design the products in such a way that it is possible to prevent the generation of waste and if it is not possible, generate the method that the waste generated by the product itself, can be reused in the best possible way.

The "waste hierarchy" ranks waste management options according to what is best for the environment. The principle of waste hierarchy introduced by the European Union provides for the implementation of a series of waste management initiatives that in the short term - in the "transition" phase can be considered complementary, but which in the long term should be considered as alternatives (Fig. 3.2) (European commission, 2018; Williams, 2005).

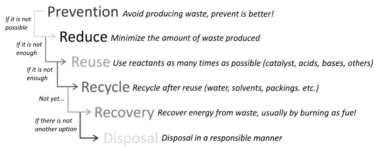

Figure 3.2   Waste hierarchy.

Prevention.   In this case, one option is biodegradable waste, that in the time it is not accumulated. A good example is the extensive research in the area of biodegradable plastics.

Reduction of waste.   It is possible thanks to a new and the increasingly large attention to design, optimization of packaging, the choice of materials used in the production process.

Reuse.   In the plastic industry, the re-extrusion (primary reusing) is considered a very efficient solution to onsite process scrap, re-introduced into the production loop by various heat cycles to produce a number of plastic products.

Recycling.   The Pfizer industry, polymers companies and others implement a recycling proces of the materials that make up the goods transformed into waste as raw materials, thanks to the organization of a supply chain of the recycling and separation and separate collection phases. ETH Zurich does its part by recycling solvents through re-destillation while minimizing discharges.

Energy recovery.   Waste can be converted in an energy font from the disposal of non-reusable waste or recyclable, through technological systems such as biodigesters, waste-to-energy plants and new pyrolytic processes. Energy from waste has been accepted as a sustainable practice for waste management systems, especially when compared with landfilling. With the development of waste management assessment methods it is now easier to compare material recovery (namely recycling) with energy recovery but also evaluating the environmental burdens associated with a product, process or activity, by identifying and quantifying energy and materials used and waste released to the environment.

Landfill disposal. It is the last option, destined to disappear in the future or, at least, to undergo a sharp reduction.

CONCLUSION

In general, the prevention principle together with the others ones give a guide to plan ahead a chemical reaction/process and prevent waste at every step, reducing the formation of sub-products and considering the toxicity of the all chemicals used, at the same time, to explore alternative chemicals to reduce the toxicity of the final product and to evaluate the reaction conditions to prevent the generation of solvent waste, the enviormental and economy impacts and prevent the use of additional reagents avoiding unnecessary steps. Also, the prevention principle can help to prevent release of hazardous and polluting substances and prevent acidents minimizing the risk and reducing or eliminating the exposure to hazardous products or reagents.

REFERENCES

Anastas,P.T., Kirchhoff, M.M. Origins, Current Status, and Future Challenges of Green Chemistry, *Accounts of Chemical Research*, 2002, 35, 686–694.

Anderson, A., O'Brien, K., Larson, A. *ACS* 2009, 12/20/2018, Available from https://www.acs.org/content/dam/acsorg/greenchemistry/industriainnovation/Pfizer-business-case-study.pdf.

Andraos, J. Unification of reaction metrics for green chemistry: Applications to Reaction Analysis. *Organic Process Research & Development*, 2005, 9, 149–163.

Andraos, J., Designing a green organic chemistry lecture course. *In*: Andrew, P.D. (ed.), Green Organic Chemistry in Lecture and Laboratory. CRC Press Taylor & Francis Group; 2016; pp. 29–68.

Betts, K., How industrial applications in green chemistry are changing our world, 2018/10/12, Available from: https://www.acs.org/content/dam/acsorg/membership/acs/benefits/extra-insights/green-chemistry-applications.pdf

Boehringer Ingelheim International GmbH. Boehringer Ingelheim taking the lead in green chemistry, 2018/11/02 Available from: https://www.boehringer-ingelheim.com/corporate-profile/sustainability/green-chemistry.

Bogdan, A.R., Poe, S.L., Kubis, D.C., Broadwater, S.J., McQuade, D.T. The continuous-flow synthesis of Ibuprofen. *Angewandte Chemie International Edition*, 2009, 48, 8547–50.

Brown, D.G., Bostro, J. Analysis of past and present synthetic methodologies on medicinal chemistry: where have all the new reactions gone? *Journal of Medicine Chemistry*, 2016, 59, 4443–4458.

Chhipa, N.M.R., Jatakiya, V.P., Gediya, P.A., Patel, S.M., Jyoti Sen, D. Green chemistry: An unique relationship between waste and recycling. *International Journal of Advances in Pharmaceutical Research*, 2013, 4, 2000–2008.

Cleij, M., Archelas, A., Furstoss, R. Microbiological transformations 43. Epoxide hydrolases as tools for the synthesis of enantiopure α-methylstyrene oxides: A new and efficient synthesis of (S)-Ibuprofen. *The Journal of Organic Chemistry*, 1999, 64, 5029–5035.

DuPont. DuPont Bio-Based Polymer Platform Earns Green Chemistry Award, 2018/10/01, Available from: http://www.dupont.com/corporate-functions/our-approach/innovation-excellence/science/scientists-engineers/awards-and-recognition/articles/green-chemistry-award.html

European commission. Directive 2008/98/EC on waste (Waste Framework Directive), 12/25/2018, Available from: http://ec.europa.eu/environment/waste/hazardous_index.html

Ivanković, A., Dronjić, A., Martinović Bevanda, A., Talic, S. Review of 12 principles of green chemistry in practice. *International Journal of Sustainable and Green Energy*, 2017, 6, 39–48.

Janssen Global Services, Green chemistry: Innovation in chemical development, 2018/10/10, Available from https://www.janssen.com/sustainability/green-chemistry

L'Oreal group, Green chemistry for new active ingredients, 2018/10/24, Available from: https://www.loreal.com/research-and-innovation/people-behind-our-science/green-chemistry

Merck group, Corporate responsibility report 2017, 2018/10/24 Available from: http://reports.merckgroup.com/2017/cr-report/products/sustainable-products/sustainable-product-design.html

Murphy, M.A. Early industrial roots of green chemistry and the history of the BHC ibuprofen process invention and its quality connection. *Foundations of Chemistry*, 2018, 20, 121–165

Nagendrappa, G. Green chemistry 2005, 12/20/2018, Available from http://www.chemvista.org/greenchemistryarticlepage6.html

Novartis, A.G., Materials & Waste, 2018/12/02, Available from https://www.novartis.com/our-company/corporate-responsibility/environmental-sustainability/materials-waste

Papadakis, E., Anantpinijwatna, A., Woodley, J.M., Gani, R. A reaction database for small molecule pharmaceutical processes integrated with process information. *Processes*, 2017, 5, 58–82.

Pfizer Inc, Waste and Recycling, 2018/12/12 Available from: https://www.pfizer.com/purpose/workplace-responsibility/green-journey/waste-and-recycling

Phan, T.V.T., Gallardo, C., Mane, J. Green Motion: A new and easy to use green chemistry metric from laboratories to industry. *Green Chemistry*, 2015, 17, 2846–2852.

Roschangar, F., Sheldon, R.A., Senanayake, C.H. Overcoming barriers to green chemistry in the pharmaceutical industry—the Green Aspiration Level™ concept. *Green Chemistry*, 2015, 17, 752–768.

Sanderson, K. It's not easy being green. *Nature*, 2011, 469, 18–20.

Sheldon, R., Introduction to green chemistry, organic synthesis and pharmaceuticals. *In*: Dunn, P.J., Wells, A.S., Williams, M.T. (eds), Green Chemistry in the Pharmaceutical Industry. KGaA; WILEY-VCH Verlag GmbH & Co.; 2010; pp. 1–20.

Sheldon, R.A. Organic synthesis; past, present and future, *Chemistry & Industry*, 1992, 7, 903–906.

Sheldon, R.A. The E Factor: Fifteen years on. *Green Chemistry*, 2007, 9, 1273–1283.

Sheldon, R.A. The E factor 25 years on: The rise of green chemistry and sustainability. *Green Chemistry*, 2017, 19, 18–43.

U.S. EPA. U.S. Environmental Protection Agency, BR data files for analyses- Report on the Environment, 2018/11/05, Available from URL:https://www.epa.gov/roe/

Williams, P.T. Waste Treatment and Disposal, 2nd Ed., John Wiley & Sons, 2005.

Chapter 4

Less Hazardous Chemical Synthesis from Palm Oil Biomass

Raja Safazliana Raja Sulong*,
Seri Elyanie Zulkifli, Fatimatul Zaharah Abas,
Muhammad Fakhrul Syukri Abd Aziz
and Zainul Akmar Zakaria*

Institute of Bioproduct and Development, School of Chemical and
Energy Engineering, Faculty of Engineering,
Universiti Teknologi Malaysia, 81310 Johor Bahru, Johor, Malaysia.

INTRODUCTION

The oil palm industry forms the economic backbone of Malaysia and contributes to the high export earnings in the agriculture sector. The Malaysian palm oil industry is the fourth largest contributor to the Malaysian Gross National Income (GNI) with RM 52.7 billion and was set to increase to RM 178.0 billion in 2020 (ETP, n.d). Malaysia had contributed 21.25 million metric tonnes to the world's oil palm production in 2014 compared to previous year with 20.2 million metric tones (Aditiya et al., 2016). Oil palm plantations are expanding across both Peninsular and East Malaysia with the total area of 5.64 million hectares (ha) in 2015 compared to 5.40 million ha in 2014 (MPOB, 2015; MPOB, 2014). According to the Malaysian Palm Oil Board (MPOB), in 2015, Sabah had the largest area of oil palm cultivation at 1.54 million ha followed by Sarawak (1.44 million ha) and Johore was third with 0.74 million ha. Malaysia has approximately 450 palm oil mills with 5.81 million ha total area planted in January 2017, processing about 110 million tons of oil palm fresh fruit bunch (MPOB, 2017). According to Ng et al. (2012), about four kg of dry biomass is produced for every kg of palm oil produced.

*For Correspondence: Zainul Akmar Zakaria: zainulakmar@utm.my;
Raja Safazliana Raja Sulong: rsafazliana@gmail.com

During harvesting of fresh fruit bunches and pruning of palm trees, oil palm frond are regularly cut and during oil palm trees replanting, the oil palm trunk is unable to be utilized and these biomasses are usually available throughout the year in the plantation. For the Empty Fruit Bunch (EFB), Mesocarp Fiber (MF) and Palm Kernel Shell (PKS), these biomasses can be found in the oil palm mill. After removal of the palm fruit from the fruit bunches, the EFB remains. Meanwhile, the MF and PKS are recovered during the crude palm oil and palm kernel oil extraction, respectively. Palm Oil Mill Effluent (POME) is another waste in liquid form that accumulates at the mills during oil palm milling (AIM, 2013).

Oil palm produces two types of oils from the fibrous mesocarp and palm kernel sections, respectively. The mesocarp fiber is a thick fleshy part of oil palm fruit which is rich in oil (80% dry mass, known as crude palm oil) while the palm kernel is the white-hard innermost part of the oil palm fruit (producing palm kernel oil). Oil Palm Fiber (OPF) is the residue that is obtained after pressing the mesocarp of palm fruitlets to obtain the palm oil. Figure 4.1, shows the typical cross section of mature oil palm fruits. The estimated OPF and PKS in 2015 is more than 13.5 million tones/year and 5 million tonnes/year and the amount is expected to increase continuously (Aditiya et al., 2016).

(a) OPF   (b) PKS

Figure 4.1   Cross section of mature oil palm fruit (Cookwithkathy, 2013).

The physical and chemical properties of OPF and PKS normally varied according to the plant species. The lignocellulosic content, proximate and ultimate analysis of OPF and PKS are as shown in Table 4.1.

Table 4.1   Lignocellulosic, proximate and ultimate analysis of OPF and PKS

Reference	OPF Chen and Lin (2016)	PKS Onochie et al. (2017) Abnisa et al. (2013)
Lignocellulosic content (%)		
Cellulose	46.63	23.7
Hemicellulose	19.98	30.5
Lignin	10.92	27.3
Proximate analysis		
Moisture content	3.27	10.23
Volatile matter	71.49	85.11
Fixed Carbon	19.18	1.42
Ash	6.06	3.24
Ultimate analysis		
Carbon	42.70	47.85
Oxygen	50.75	42.70
Nitrogen	1.10	0.92
Sulfur	–	0.10
Hydrogen	5.45	5.13

OPF has higher amounts of cellulose compared to hemicellulose and lignin, while the PKS has a higher amount of hemicellulose and lignin compared to cellulose. OPF and PKS have low moisture contents which is desirable for the pre-processing stage of biomass notably on the short drying time required. Drying of biomass prior to the conversion process increase the energy efficiency of the process and improves the quality of the products (Kan et al., 2016). The fixed carbon is the carbon determined after volatile matters are evaporated from the biomass. OPF and PKS contain high carbon contents due to its woody characteristic (Vassilev et al., 2010).

THERMO-CHEMICAL CONVERSION OF BIOMASS

Thermo-chemical conversion is a process commonly used to convert biomass waste into bio-fuels in the forms of solid (e.g., charcoal), liquid (e.g., bio-oils, methanol and ethanol), and gas (e.g., methane and hydrogen), which can be used further for heat and power generation (Zhang et al., 2010). This process can be divided into four major conversion routes such as combustion, gasification, liquefaction and pyrolysis processes as shown in Fig. 4.2.

Combustion is widely used to burn biomass waste into a charcoal product with the complete presence of oxygen. This process is inconvenient and causes serious air pollution because of extensive smoke formation. The powdery charcoal produced from combustion process can be converted into high energy-concentrated fuel pellets orother different geometric forms (Kibwage et al., 2006). The gasification process is an efficient and environmentally friendly way to produce energy (Hanne et al., 2011). This process is well known in a conversion of biomass fuel into gaseous fuel. This whole process is completed at the elevated temperature range of 800–1300°C (McKendry, 2002). The gaseous fuel generated from gasification is used as an energy in power generation cycles. The liquefaction process operates at a low-temperature and high pressure to break down the biomass waste into fragments of small molecules in water or another suitable solvent (Zhang et al., 2010). Liquefaction has some similarity with pyrolysis where the preferred products is the liquid product. However, in order to operate the liquefaction process, catalysts are essential. Compared with pyrolysis, liquefaction technology is more challenging as it requires more complex and expensive reactors and fuel feeding systems (Demirbas, 2001).

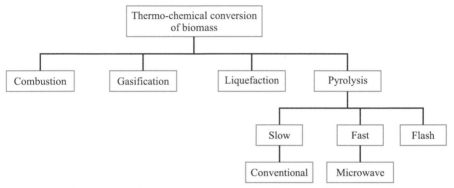

Figure 4.2. Thermo chemical conversion route of biomass (Zhang et al., 2010).

Pyrolysis has received special attention as it can convert biomass directly into gaseous, oil-like liquid and carbon rich solid residue products by thermal decomposition of biomass in the absence of oxygen (Goyal et al., 2008; Balat et al., 2009). Besides, using pyrolysis the emission of millions tonnes of smoke to the atmosphere can be avoided and the products produced from this process have a low environmental impact and are also good for organic agriculture. Pyrolysis can be divided into three main categories involving slow pyrolysis, fast pyrolysis and flash pyrolysis. These categories can be differentiated depending on solid residence time, heating rate and temperature as shown in Table 4.2.

Table 4.2 Operating parameters for pyrolysis processes (Balat et al., 2009)

Pyrolysis process	Residence time (s)	Heating rate	Temperature (°C)	Prefer product
Slow	450–550	Low	550–950	Char, oil, gas
Fast	0.5–5	Very high	850–1250	Bio-oil
Flash	< 1	High	1050–1300	Bio-oil

Slow pyrolysis has been used for charcoal production at low temperature with low heating rates and long residence time. Fast pyrolysis described by high reaction temperature, fast heating rate and short residence time. Fast pyrolysis is considered as modern pyrolysis technology which is carefully controlled, hence provides fast decomposition of biomass. Fast pyrolysis is the condition of choice when the primary product of interest is bio-oil (Kan et al., 2016).

PYROLYSIS PROCESS

The products yield produced after pyrolysis process (Fig. 4.3) depends on the final temperature, heating rate, holding time, gas flow rate, biomass feed size and lignocellulose composition as well as the characteristic of biomass (Ertas and Alma, 2010; Heo et al., 2010). According to Collard and Blin (2014), the pyrolysis process of biomass can be separated into primary mechanisms and secondary mechanisms. Primary mechanisms can be described by three main pathways which are char formation, depolymerization, and fragmentation. Primary mechanism involves the breaking of different chemical bonds within the polymers which results in the release of volatile compounds and rearrangement reactions within the matrix of the residue.

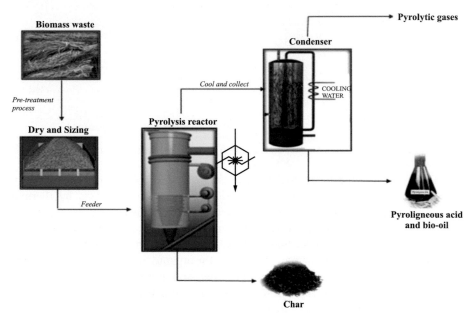

Figure 4.3 Flow diagram of biomass pyrolysis process (Ertas and Alma, 2010; Heo et al., 2010).

The char formation pathway occurs with the formation of the benzene ring and combination of these rings in polycyclic structure, accompanied by the release of water and incondensable gas. Depolymerization involves breaking of the bonds between the monomers unit of the polymers to produce volatile molecules. This pathway generally occurs at temperatures of 250 to 500°C. The molecules are condensable at ambient temperature and frequently found in liquid fraction (e.g., guaiacol, catechol, creasol and phenol). At temperatures ranging from 450 to 550°C, fragmentation occurs with the linkage of many covalent bonds of the polymers resulting in the formation of incondensable gas and small chain organic compounds that are condensable at ambient temperatures. Secondary mechanism occurs when the released volatile compounds are not stable and this mechanism involves two reactions namely cracking and recombination. At temperature >600°C, cracking of the chemical bond within the volatile compounds results in the formation of lower molecular weight molecules. Recombination of volatie compounds occur at temperature ≥800°C to give higher molecular weight molecules and this reaction leads to the secondary char formation and the appearance of Polycyclic Aromatic Hydrocarbons (PAHs).

Pyroligneous Acid (PA) known as pyrolysis liquid or wood vinegar can be obtained by condensing the smoke produced during pyrolysis heating process of plant biomass in the absence of oxygen. It generally consists of 80–90% of water and 10–20% organic compounds with a distinctive smoky odor, reddish brown appearance, and acidic pH. PA comprises of complex mixtures of compounds such as phenolics compound (guaiacol, phenol, catechol, syringol), organic acids, aldehyde, ketone, furan, pyran and nitrogen compounds (Wanderley et al., 2012). Various biomass feedstocks from agricultural waste have been reported for the production of PA which includes Oil Palm Fiber (OPF) (Abas et al., 2018), Palm Kernel Shell (PKS) (Ariffin et al., 2017), Citrus Plant (CP) (Bacanli et al., 2015), Coconut Shell (CS) (Wititsiri, 2011), Walnut Shell (WS) (Zhai et al., 2015), Bamboo (B) (Harada et al., 2013), Pineapple Waste (PW) (Mathew et al., 2015), Cotton Stalk (CS) (Wu et al., 2015) just to mention a few.

Various pyrolysis techniques can be applied for the production of PA including via conventional thermal heating by using charcoal kiln and pyrolysis reactor (Mathew et al., 2015; Mungkunkamchao et al., 2013) as well as a microwave-assisted heating system. The production of PA via a conventional heating system has been well reported by earlier studies. However, this technique has a few drawbacks such as heat transfer resistance, heat losses to the surroundings, damage to reactor walls due to continuous electric heating and utilization of a portion of heat supplied to biomass materials (Salema and Ani, 2011). This can markedly deteriorates the quality and the yield of the product required. The most significant and unique features for microwave pyrolysis heating is involving the energy conversion compared to transfer of energy for conventional heating. In addition, microwave heating normally has in-core volumetric heating, also known as internal heating that involves superficial heating via conduction and convection.

Pyroligneous acid has been described to have a pleasant smoky aroma with yellowish or reddish brown color. The physical properties of PA may slightly differ based on the nature of biomass involved during the production of PA. The density of PA is around 1.2 kg/dm^3 at 15°C while a different range of PA acidity has been reported by earlier studies such as 4.4 with density of 1.003 g/cm^3 (Mathew and Zakaria, 2015b), 1.5–5.5 (Montazeri et al., 2013), 2.5–3.5 (Oasmaa and Peacocke, 2001). The flash point and pour point of PA are in the range of 50–70°C and 20°C respectively, and it starts to boil below 100°C. Meanwhile the vapor pressure of PA is similar to water and the auto ignition temperature of PA is approximately above 500°C (Oasmaa and Peacocke, 2001).

In general, various kinds of chemical compounds have been observed in PA namely phenolics, organic acids, aldehyde, ketone, flavonoids, alcohols, derivatives of furan and pyran and nitrogen compounds, of which the major ones are organic acids and phenolics (Loo, 2008). According to Yang et al. (2016), these chemical constituent of PA can be analyzed and identified by using GC-MS analysis. Table 4.3 demonstrates the chemical compounds available in PA from PKS and OPF. Fitriady et al. (2016) and Abas et al. (2018) found that PA from PKS and OPF has the highest fraction of phenol and its derivatives which are respectively around 43.5% (PKS) and 68.25% (OPF).

Table 4.3   Chemical constituents of PA from different agricultural biomass

Chemical Compounds	PKS (Fitriady et al., 2016)	OPF (Abas et al., 2018)
Ketones (%)	5.88	1.17
Benzene (%)	–	1.41
Organic acid (%)	38.24	–
Phenol and derivatives (%)	43.50	68.25
Aldehydes (%)	–	–
Esters (%)	–	–
Amide (%)	–	3.11
Furan and Pyran derivatives (%)	6.78	4.51
Toluene	–	4.12
Alkyl and aryl ether (%)	–	–
Alkane		0.87
Nitrile		5.52
Alcohols (%)	2.77	1.72
Nitrogenated derivatives (%)	–	–
Sugar derivatives (%)	–	–
Heterocyclic organic compound	1.32	–
Others	1.51	2.12

APPLICATION OF PYROLIGNEOUS ACID

Solid biomass can be converted into PA through pyrolysis, making it easier to transport, store and upgrade (Demirbas, 2011). PA, also known as bio-oil and wood vinegar is a dark brown color aqueous phase fraction produced by condensing the vapor released from pyrolysis process (Abnisa et al., 2011). PA has a sour and smoky odor, ususally characterized as acidic with pH ranging from 1.8 to 3.8 (Wei et al., 2010a). It consists of mainly a complex mixture of water and organic compunds which are 80–90% and 10–20% respectively. PA contains a huge amount of oxygenated molecules which made it unstable, viscous and corrosive but it depends on biomass feedstock and the pyrolysis condition (Demirbas, 2011). Table 4.4 shows the pH and yield of PA produced from different plant sources reviewed by Mathew and Zakaria (2015a).

Currently, there is much research on biomass focusing on how to make it another power generation source. There are many biomass and agricultural residues that can benefit us to produce energy and value-added products. In some earlier studies, production of PA from oil palm biomass concluded that the PA had the potential as a fuel source and chemical feedstocks (Sukiran et al., 2009,

Khor et al., 2009, Abnisa et al., 2013). The generated PA may be directly used for a boiler or furnace and also can be considered to be used for vehicles after the refinement process (Khor et al., 2009).

Table 4.4 pH and yield of pyroligneous acid produced from different plant sources

Sources	pH	Yield (%)	References
Rubber wood	2.9–3.8	23–29	Ratanapisit et al., 2009
Camellia oleifera shell	3.5	30	Xu et al., 2013
Eucalyptus	–	45.5	Souza et al., 2012
Birch heartwood	1.8–2.9	38–43	Fagernäs et al., 2012
Walnut shell	2.98	38.79	Wei et al., 2010b
Walnut tree branch	3.32	31.4	Wei et al., 2010a

Pyroligneous acid can been used to stimulate growth in various plants. A study on Japanese pears cultivars by Kadota et al. (2002) showed that PA gave a positive effect on the development of roots in the shoot culture of the plant. In another finding by Kadota and Niimi (2004) PA mixed with charcoal could increase plants survival rate and shorten the number of days of flowering for plants depending on the species. Jain and Van Staden (2006) proved that PA helped promote seedling development and seed germination for tomato growth. Additionally, PA also has potential for improvement of rockmelon such as increasing the fruit yield and weight, improving the productivity, and enhancing the fruit size and sweetness (Zulkarami et al., 2011). The okra vegetable was improved in terms of root length, height, fruit weight, leaf numbers and diameter by applying fertilizers mixed with different concentrations of PA (Mahmud et al., 2016).

Yatagai et al. (2002) found that PA exhibited high termiticidal activities against Japanese termite, Reticulitermes speratus. Later more research was also conducted to study the utilization of PA as an anti-termite (Lee et al., 2010; Lee et al., 2011b). Tiilikkala et al. (2010) reviewed the application of PA as herbicides and pesticides. In addition, PA combined with insecticides showed improvement on the agronomic effect of the soybean crop as it can control caterpillars (Petter et al., 2013) and the use of herbicide can be reduced by mixing it with PA to control weed, such as barnyardgrass (Acenas et al., 2013).

Pyroligneous acid has been used traditionally in Asia as a folk remedy to treat inflammation diseases inspite of poor scientific research and evaluation on their efficacy and toxicity on their application. Kimura et al. (2002) studied the carcinogenicity of commercialized PA from bamboo and found that the PA had no carcinogenic effect and was unable to promote tumors. PA obtained from oak tree charcoal production has been used to alleviate cellulitis, athlete's foot and atopic dermatitis symptoms (Lee et al., 2011a). In Japan, PA obtained from bamboo and broad-leaf trees wood burners has been used for long time ago as an alternative treatment for eczema, scabies, atopic dermatitis and other skin diseases. Ho et al. (2013) successfully demonstrated the anti-inflammatory activity of PA from bamboo both in vitro and in vivo. Their study showed that the PA from bamboo with concentration upto 2%, decreases the expression of inflammatory mediator in lipopolysaccharide (LPS)-activated murine macrophages cell line.

The presence of organic acids, phenolic compounds and carbonyls in PA contributed a bigger potential as an antimicrobial agent. PA prevented an activity of Listeria innocua and L. monocytogenes that generally contaminated cold-smoked salmon due to higher acidity of PA and the presence of smoke flavor compounds such as phenols and carbonyls such as formaldehyde inhibited the bacterial growth (Vitt et al., 2001). PA was also able to prevent growth of Escherichia coli, Corynebacterium agropyri, and Alternaria mali (Mahmud et al., 2016; Jung, 2007). The studies concluded that the PA can be a good alternative for replacing agrochemical fungicides.

PA from walnut shell is a good antioxidant agent. This was proved by Wei et al. (2010b) and Ma et al. (2011) where their study indicated that phenols are the functional components to reduce oxidants or to scavenge free radicals. They identified that 2,6-dimethoxyphenol and 1,2-benzenediol and 2,6-dimethoxyphenol accounted as the major compounds of phenol and derivatives present in PA from the walnut shell respectively. A study on the antioxidant activity of phenolic compounds in PA was also conducted by Rungruang and Junyapoon (2010) who found that antioxidant activity of PA from eucalyptus wood heightened with the increased concentration of phenolic compounds thus indicating that the PA produced from this specie was able to be a source of natural antioxidants. Moreover, phenols were found responsible for the antioxidant activity with 1S-α-Pinene and eucalyptol being the main composition of rosemary (Rosmarinus officialis) leaves PA (Ma et al., 2013).

PA has been used traditionally in Japan as a detoxifier to remove toxins from the body. Moreover, PA was also used as smoke flavoring due to its organoleptic properties (Guillén and Manzanos, 1999). PA also has been used for odor removal of landfill site leachates by reducing the odor constituent namely H_2S, NH_3, NH_2 and CO_2 (Hur et al., 1999). PA has become an alternative over traditional flue gas smoking for smoke flavoring that has been applied for various food products such as meat, fish, and cheese (Hattula et al., 2001).

Study by Abnisa et al. (2013) found that the oxygenated contents of PA from pyrolysis process were quite high (71.40%). Higher content of oxygenated compounds makes PA less attractive for transport fuels production due to decrement of high heating values. Thus, the oxygenated contents and water content need to be removed first before using it in transportation. Methods to remove those contents include hydrotreating and the hydrocracking process. These processes could increase the heating value and reduce the corrosiveness of PA (Abnisa et al., 2011). Elliott and Neuenschwander (1996) carried out the study on hydrotreatment of PA aimed to reduce oxygen content. The experiment was performed using two fixed bed reactors with Co-Mo and Ni-Mo as hydrotreating catalysts in 21MPa hydrogen environment. It was found that Liquid Hourly Spaced Velocity (LHSV) affected the deoxygenation rate and complete deoxygenation was achieved at LHSV of ca. 0.1. Another study of upgrading PA was performed by Zheng and Wei (2011) by distilling it and it was found that the distilled PA has lower content of oxygen (9.2 wt%), higher heating value, lower corrosivity and more stablility compared to PA from fast pyrolysis.

REFERENCES

Abas, F.Z., Zakaria, Z.A., Ani, F.N. Antimicrobial properties of optimized microwave-assisted pyroligneous acid from oil palm fiber. *Journal of Applied Pharmaceutical Science,* 2018, 8(07), 065–071.

Abnisa, F., Daud, W.M.A.W., Husin, W.N.W. Sahu, J.N. Utilization possibilities of palm shell as a source of biomass energy in Malaysia by producing bio-oil in pyrolysis process. *Biomass and Bioenergy,* 2011, 35, 1863–1872.

Abnisa, F., Arami-Niya, A., Daud, W.W., Sahu, J.N. Characterization of bio-oil and bio-char from pyrolysis of palm oil wastes. *BioEnergy Research,* 2013, 6(2), 830–840.

Acenas, X.S., Nuñez, J.P.P., Seo, P.D., Ultra Jr, V.U., Lee, S.C. Mixing pyroligneous acids with herbicides to control barnyardgrass (*Echinochloa crus-galli*). *Weed & Turfgrass Science,* 2013, 2(2), 164–169.

Aditiya, H.B., Chong, W.T., Mahlia, T.M.I., Sebayang, A.H., Berawi, M.A., Nur, H. Second generation bioethanol potential from selected Malaysia's biodiversity biomasses: A review. *Waste Management,* 2016, 47, 46–61.

Agensi Inovasi Malaysia (AIM). National biomass strategy 2020: new wealth creation for malaysia's palm oil industry. *Agensi Inovasi Malaysia,* Kuala Lumpur; 2013.

Ariffin, S.J., Yahayu, M., El-Enshasy, H., Malek, R.A., Aziz, A.A., Hashim, N.M., Zakaria, Z.A. Optimization of pyroligneous acid production from palm kernel shell and its potential antibacterial and antibiofilm activities. *Indian Journal of Experimental Biology,* 2017, 55, 427–435.

Bacanli, M., Başaran, A.A., Başaran, N. The antioxidant and antigenotoxic properties of citrus phenolics limonene and naringin. *Food and Chemical Toxicology,* 2015, 81, 160–170.

Balat, M., Balat, M., Kırtay, E., Balat, H. Main routes for the thermo-conversion of biomass into fuels and chemicals. Part 1. Pyrolysis systems. *Energy Convers Manage,* 2009, 50, 3147–3157.

Chen, W.H., Lin, B.J. Characteristics of products from the pyrolysis of oil palm fiber and its pellets in nitrogen and carbon dioxide atmospheres. *Energy,* 2016, 94, 569–578.

Collard, F.X., Blin, J. A review on pyrolysis of biomass constituents: Mechanisms and composition of the products obtained from the conversion of cellulose, hemicelluloses and lignin. *Renewable and Sustainable Energy Reviews,* 2014, 38, 594–608.

Cookwithkathy, (2013) Retrieved on 23/1/2017 from http://cookwithkathy.wordpress.com/2013/08/02/what-is-the-difference-between-palm-oil-and-coconut-oil/

Demirbas, A. Biomass resource facilities and biomass conversion processing for fuels and chemicals. *Energy Conservation Management,* 2001, 42, 1357–1378.

Demirbas, A. Competitive liquid biofuels from biomass. *Applied Energy,* 2011, 88, 17-28.

Economic Transformation Programme (ETP) (n.d). Retrieved on October 17, 2014, from http://www.mpob.gov.my/images/stories/pdf/NKEA_-Chapter_9_Palm_Oil.pdf

Elliott, D.C., Neuenschwander, G.G. Developments in thermochemical biomassconversion. London: Blackie Academic and Professional, 1996.

Ertaş, M , Alma, M.H. Pyrolysis of laurel (*Laurel nobilis* L.) extraction residues in a fixed-bed reactor: Characterization of bio-oil and bio-char. *Journal of Analytical and Applied Pyrolysis,* 2010, 88(1), 22–29.

Fagernäs, L., Kuoppala, E., Tiilikkala, K., Oasmaa, A. Chemical composition of birch wood slow pyrolysis products. *Energy & Fuels,* 2012, 26(2), 1275–1283.

Fitriady, M.A., Mansur, D., Simanungkalit, S.P. Utilization of natural iron ore for catalytic reaction of (Pyroligneous Acid) derived from palm kernel shells. *Widyariset*, 2016, 2(2), 118–130.

Goyal, H.B., Seal, D., Saxena, R.C. Bio-fuels from thermochemical conversion of renewable resources: A review. *Renewable & Sustainable Energy Reviews*, 2008, 12, 504–517.

Guillén, M.D., Manzanos, M.J. Smoke and liquid smoke. Study of an aqueous smoke flavouring from the aromatic plant *Thymus vulgaris* L. *Journal of the Science of Food and Agriculture*, 1999, 79(10), 1267–1274.

Hanne, R., Kristina, K., Alexander, K., Arunas, B., Pekka, S., Matti, R., Outi, K., Marita, N. Thermal plasma-sprayed nickel catalysts in the clean-up of biomass gasification gas. *Fuel*, 2011, 90, 1076–1089.

Harada, K., Iguchi, A., Yamada, M., Hasegawa, K., Nakata, T. Hikasa, Y. Determination of maximum inhibitory dilutions of bamboo pyroligneous acid against pathogenic bacteria from companion animals: An in vitro study. *Journal of Veterinary Advance*, 2013, 3(11), 300–305.

Hattula, T., Elfving, K., Mroueh, U.M., Luoma, T. Use of liquid smoke flavouring as an alternative to traditional flue gas smoking of rainbow trout fillets (*Oncorhynchus mykiss*). *LWT-Food Science and Technology*, 2001, 34(8), 521–525.

Heo, H.S., Park, H.J., Yim, J.H., Sohn, J.M., Park, J., Kim, S.S., Ryu, C., Jeon, J.K., Park, Y.K. Influence of operation variables on fast pyrolysis of *Miscanthus sinensis var. purpurascens*. *Bioresource Technology*, 2010, 101(10), 3672–3677.

Ho, C.L., Lin, C.Y., Ka, S.M., Chen, A., Tasi, Y.L., Liu, M.L., Chiu, Y.C., Hua, K.F. Bamboo vinegar decreases inflammatory mediator expression and NLRP3 inflammasome activation by inhibiting reactive oxygen species generation and protein kinase C-α/δ activation. *PloS one*, 2013, 8(10), 1–11.

Hur, K.S., Jeong, E.D., Paek, U.H. A study on odor removal of landfill site leachate by pyroligneous liquid. *Journal of Korean Environmental Sciences Society*, 1999, 8, 607–610.

Jain, N., Van Staden, J. A smoke-derived butenolide improves early growth of tomato seedlings. *Plant Growth Regulation*, 2006, 50(2-3), 139–148.

Jung, I.S., Kim, Y.J., Gal, S.W., Choi, Y.J. Antimicrobial and antioxidant activities and inhibition of nitric oxide synthesis of oak wood vinegar. *Journal of Life Science*, 2007, 17(1), 105–109.

Jung, K.H. Growth inhibition effect of pyroligneous acid on pathogenic fungus, *Alternaria mali*, the agent of alternaria blotch of apple. *Biotechnology and Bioprocess Engineering*, 2007, 12(3), 318–322.

Kadota, M., Hirano, T., Imizu, K., Niimi, Y. Pyroligneous acid improves *in vitro* rooting of Japanese pear cultivars. *HortScience*, 2002, 37(1), 194–195.

Kadota, M., Niimi, Y. Effects of charcoal with pyroligneous acid and barnyard manure on bedding plants. *Scientia Horticulturae*, 2004, 101(3), 327–332.

Kan, T., Strezov, V., Evans, T.J. Lignocellulosic biomass pyrolysis: A review of product properties and effects of pyrolysis parameters. *Renewable and Sustainable Energy Reviews*, 2016, 57, 1126–1140.

Khor, K.H., Lim, K.O., Zainal, Z.A. Characterization of bio-oil: A by-product from slow pyrolysis of oil palm empty fruit bunches. *American Journal of Applied Sciences*, 2009, 6(9), 1647.

Kibwage, J.K., Munywe, S.N., Mutonyi, J., Savala, C.N., Wanyonyi, E. Organicresource management in Kenya: perspectives and guidelines. Recycling waste into fuelbriquettes. Organic Resource Management in Kenya – Chapt 15. 2006. http://www.formatkenya.org/ormbook/Chapters/chapter15.htm

Kimura, Y., Suto, S., Tatsuka, M. Evaluation of carcinogenic/co-carcinogenic activity of chikusaku-eki, a bamboo charcoal by-product used as a folk remedy, in BALB/c 3T3 cells. *Biological and Pharmaceutical Bulletin*, 2002, 25(8), 1026–1029.

Lee, S.H., H'ng, P.S., Lee, A.N., Sajap, A.S., Tey, B.T., Salmiah, U. Production of pyroligneous acid from lignocellulosic biomass and their effectiveness against biological attacks. *Journal of Applied Sciences*, 2010, 10(20), 2440–2446.

Lee, C.S., Yi, E.H., Kim, H.R., Huh, S.R., Sung, S.H., Chung, M.H., Ye, S.K. Anti-dermatitis effects of oak wood vinegar on the DNCB-induced contact hypersensitivity via STAT3 suppression. *Journal of Ethnopharmacology*, 2011a, 135(3), 747–753.

Lee, S.H., H'ng, P.S., Chow, M.J., Sajap, A.S., Tey, B.T., Salmiah, U., Sun, Y.L. Effectiveness of pyroligneous acids from vapour released in charcoal industry against biodegradable agent under laboratory condition. *Journal of Applied Sciences*, 2011b, 11(24), 3848–3853.

Loo. A.Y. Isolation and Characterization of Antioxidants Compound from Pyroligneous Acid of *Rhizophora apiculata*. Doctor of Philosophy Thesis. UniversitiSains Malaysia; 2008.

Luque, R., Menendez, J.A., Arenillas, A., Cot, J. Microwave-assisted pyrolysis of biomass feedstocks: The way forward?, *Energy & Environmental Science*, 2012, 5, 5481–5488.

Ma, X., Wei, Q., Zhang, S., Shi, L., Zhao, Z. Isolation and bioactivities of organic acids and phenols from walnut shell pyroligneous acid. *Journal of Analytical and Applied Pyrolysis*, 2011, 91(2), 338–343.

Ma, C., Song, K., Yu, J., Yang, L., Zhao, C., Wang, W., Zu, G., Zu, Y. Pyrolysis process and antioxidant activity of pyroligneous acid from *Rosmarinus officinalis* leaves. *Journal of Analytical and Applied Pyrolysis*, 2013, 104, 38–47.

Mahmud, K.N., Yahayu, M., Md Sarip, S.H., Rizan, N.H., Min, C.B., Mustafa, N.F., Ngadiran, S., Ujang, S., Zakaria, Z.A. Evaluation on efficiency of pyroligneous acid from palm kernel shell as antifungal and solid pineapple biomass as antibacterial and plant growth promoter. *Sains Malaysiana*, 2016, 45(10), 1423–1434.

Malaysia Palm Oil Board (MPOB) 2014. Oil palm planted area by state as at December 2014 (hectares). Retrieved on February 01, 2017, from http://bepi.mpob.gov.my/images/area/2015/Area_summary.pdf

Malaysia Palm Oil Board (MPOB) 2015. Oil palm planted area by state as at December 2015 (hectares). Retrieved on February 01, 2017, from http://bepi.mpob.gov.my/images/area/2014/Area_summary.pdf

Malaysia Palm Oil Board (MPOB) 2017. Number and capacities of palm oil sectors January 2017 (Tonnes/Year). Retrieved on February 01, 2017, from http://bepi.mpob.gov.my/index.php/en/statistics/sectoral-status/179-sectoralstatus2017/803-number-a-capacities-of-palm-oil-sectors-2017.html

Mathew, S., Zakaria, Z.A. Pyroligneous acid—The smoky acidic liquid from plant biomass. *Applied Microbiology and Biotechnology*, 2015a, 99(2), 611–622.

Mathew, S., Zakaria, Z.A., Musa, N.F. Antioxidant Property and Chemical Profile of Pyroligneous Acid from Pineapple Plant Waste Biomass. *Process Biochemistry*, 2015b, 50(11), 1985–1992.

McKendry, P. Energy production from biomass (Part 2): Conversion technologies. *Bioresource Technology*, 2002, 83, 47–54.

Montazeri, N., Oliveira, A., Himelbloom, B.H., Leigh, M.B., Crapo, C.A. Chemical characterization of commercial liquid smoke products. *Food Science and Nutrition*, 2013, 1(1), 102–115.

Mungkunkamchao, T., Kesmala, T., Pimratch, S., Toomsan, B., Jothityangkoon, D. Wood vinegar and fermented bioextracts: Natural products to enhance growth and yield of tomato (*Solanum Lycopersicum* L.). *Scientia Horticulturae*, 2013, 154, 66–72.

Ng, W.P.Q., Lam, H.L., Ng, F.Y., Kamal, M., Lim, J.H.E. Waste-towealth: Green potential from palm biomass in Malaysia. *Journal of Cleaner Production*, 2012, 34, 57–65.

Oasmaa, A., Peacocke, C. A Guide to Physical Property Characterisation of Biomass-Derived Fast Pyrolysis Liquids. (No. 450). Technical Research Centre of Finland Espoo: ValtionTeknillinenTutkimuskeskus; 2001.

Onochie, O.P., Obanor, A.I., Aliu, S.A., Ighodaro, O.O. Proximate and ultimate analysis of fuel pellets from oil palm residues. *Nigerian Journal of Technology*, 2017, 36(3), 987–990.

Petter, F.A., Silva, L.B., Sousa, I.J., Magionni, K., Pacheco, L.P., Almeida, F.A., Pavan, B.E. Adaptation of the use of pyroligneous acid in control of caterpillars and agronomic performance of the soybean crop. *Journal of Agricultural Science*, 2013, 5(8), 27–36.

Ratanapisit, J., Apiraksakul, S., Rerngnarong, A., Chungsiriporn, J., Bunyakarn, C. Preliminary evaluation of production and characterization of wood vinegar from rubber wood. *Songklanakarin Journal of Science and Technology*, 2009, 31(343349), 9.

Rungruang, P., Junyapoon, S. Antioxidative activity of phenolic compounds in pyroligneous acid produced from Eucalyptus wood. *In*: Proceedings of the 8th International Symposium on Biocontrol and Biotechnology. Pattaya, Thailand, 4–6 October 2010. Khon Kaen, Thailand; Institute of Technology Ladkrabang and Khon Kaen University; 2010; pp.102–106.

Salema, A.A., Ani, F.N. Microwave induced pyrolysis of oil palm biomass. *Bioresource Technology*, 2011, 102(3), 3388–3395.

Souza, J.B.G., Re-Poppi, N., Raposa Jr., J.L. Characterization of pyroligneous acid used in agriculture by gas chromatography-mass spectometry. *Journal of the Brazilian Chemical Society*, 2012, 23(4), 610–617.

Sukiran, M.A., Chin, C.M., Bakar, N.K. Bio-oils from pyrolysis of oil palm empty fruit bunches. *American Journal of Applied Sciences*, 2009, 6(5), 869–875.

Tiilikkala, K., Fagernäs, L., Tiilikkala, J. History and use of wood pyrolysis liquids as biocide and plant protection product. *The Open Agriculture Journal*, 2010, 4, 111–118.

Vassilev, S.V., Baxter, D., Andersen, L.K., Vassileva, C.G. An overview of the chemical composition of biomass. *Fuel*, 2010, 89(5), 913–933.

Vitt, S.M., Himelbloom, B.H., Crapo, C.A. Inhibition of *Listeria inocula* and *Listeria monocytogenes* in a laboratory medium and cold-smoked salmon containing liquid smoke. *Journal of Food Safety*, 2001, 21(2), 111–125.

Wanderley, C.S., Faria, R.T., Ventura, M.U. Chemical Fertilization, Organic Fertilization and Pyroligneous Extract in the Development of Seedlings of Areca Bamboo Palm (*Dypsislutescens*). *Acta Scientiarum. Agronomy*, 2012, 34, 163–167.

Wei, Q., Ma, X., Dong, J. Preparation, chemical constituents and antimicrobial activity of pyroligneous acids from walnut tree branches. *Journal of Analytical and Applied Pyrolysis*, 2010a, 87(1), 24–28.

Wei, Q., Ma, X., Zhao, Z., Zhang, S., Liu, S. Antioxidant activities and chemical profiles of pyroligneous acids from walnut shell. *Journal of Analytical and Applied Pyrolysis*, 2010b, 88(2), 149–154.

Wititsiri, S. Production of wood vinegars from coconut shells and additional materials for control of termite workers, *Odontotermes* Sp. and striped mealy bugs, *Ferrisia Virgata. Journal of Science and Technology*, 2011, 33(3), 349–354.

Wu, Q., Zhang, S., Hou, B., Zheng, H., Deng, W., Liu, D., Tang, W. Study on the preparation of wood vinegar from biomass residues by carbonization process. *Bioresource Technology*, 2015, 179, 98–103.

Xu, X., Jiang, E., Li, B., Wang, M., Wang, G., Ma, Q., Shi, D., Guo, X. Hydrogen production from wood vinegar of camellia oleifera shell by Ni/M/γ-Al$_2$O$_3$ catalyst. *Catalysis Communications*, 2013, 39, 106–114.

Yang, J.F., Yang, C.H., Liang, M.T., Gao, Z.J., Wu, Y.W., Chuang, L.Y. Chemical composition, antioxidant, and antibacterial activity of wood vinegar from litchi chinensis. *Molecules,* 2016, 21(9), 1150, 1–10.

Yatagai, M., Nishimoto, M., Hori, K., Ohira, T., Shibata, A. Termiticidal activity of wood vinegar, its components and their homologues. *Journal of Wood Science*, 2002, 48(4), 338–342.

Zhai, M., Shi, G., Wang, Y., Mao, G., Wang, D., Wang, Z. Chemical compositions and biological activities of pyroligneous acids from walnut shell. *BioResources*, 2015, 10(1), 1715–1729.

Zhang, L., Xu, C., Champagne, P. Overview of recent advances in thermo-chemical conversion of biomass. *Energy Conversion and Management*, 2010, 51(5), 969–982.

Zheng, J.L., Wei, Q. Improving the quality of fast pyrolysis bio-oil by reducedpressure distillation. *Biomass and Bioenergy*, 2011, 35, 1804–1810

Zulkarami, B., Ashrafuzzaman, M., Husni, M.O., Ismail, M.R. Effect of pyroligneous acid on growth, yield and quality improvement of rockmelon in soilless culture. *Australian Journal of Crop Science*, 2011, 5(12), 1508–1514.

Chapter 5

Designing Safer Chemicals

Cesar Garcias Morales[*1], Armando Ariza Castolo[2]
and Mario Alejandro Rodriguez[3]

[1]Departamento de Química Orgánica, Facultad de Ciencias Químicas,
Universidad Autónoma de Coahuila, Saltillo Coahuila México.
[2]Departamento de Quimica, Centro de Investigación y de Estudios Avanzados
del Instituto Politecnico Nacional, Ciudad de Mexico, Mexico.
[3]Grupo de Propiedades Opticas de la Materia, Centro de
Investigaciones en Optica, Leon, Guanajuato, México.

INTRODUCTION

The fourth principle of green chemistry states that "chemical products should be designed to effect their desired function while minimizing their toxicity hazards" (Anastas and Kirchhoff, 2002; Erythropel et al., 2018). Among the 12 principles of green chemistry it is the least developed since the introduction of the green chemistry concept. One of the main reasons for its slow development is because it involves different areas such as: organic chemistry, toxicology, biology, environment, among others. Despite more than 20 years since the principle was introduced, some guide or design criteria have been established for the synthesis of safe chemicals that allow the reduction of compounds toxicity in a rapid manner, efficient and reliable, though it is not enough (Melnikov et al., 2017; Anastas et al., 1999).

Less dangerous chemical compounds, not only refers to those which have an adverse effect on human health, but to the effects on the environment (Beach, 2007).

*For Correspondence: Departamento de Química Orgánica, Facultad de Ciencias Químicas, Universidad Autónoma de Coahuila, Blvd. V. Carranza e Ing. José Cárdenas Valdez C.P. 25280, Saltillo Coahuila México. Email: cgarcias@uadec.edu.mx

Before continuing with the design of "Safer Chemicals" it is necesary to establish what is considered a safe chemical and what characteristics should be collected. DeVito lists some characteristics that can help answer these questions (Devito, 2016; Garrett, 1996).

I. A Safer chemical needs to have good potency: In the pharmaceutical field the potency is a measure of the amount (dose) required of the drug to obtain the same response in the same degree through the same pharmacological process in relation to another substance. When a less amount of active ingredient is used, it is said that the chemical compound is more powerful. Designing more powerful drugs has environmental advantages since it requires a lower amount of active principle, this leads to reduction of raw materials, solvents and other chemicals required for its manufacture, resulting in a reduction of waste produced.

II. Good efficacy: It refers to the ability of one drug to induce the desired pharmacological effect without causing adverse effects. Table 5.1 shows both potency and efficacy of a drug for hypoglycemia treatment. In 1955, Tolbutamide was introduced, when the patient under treatment required three doses per day of 500 mg because the pharmacological effect is from 6 to 12 hours, in 1984, Glipizide was introduced when the patient under treatment required 1 dose a day of 5 mg in this case the pharmacological effect is up to 24 hours.

Table 5.1   Examples of potency and efficacy in pharmaceutical products of the same structure class and pharmacologic mechanism.

Drug product generic name	Structure	Typical oral dose and dosing frequency	Biological half-life	Duration of effect
Sulfonylurea hypoglycemics				
Tolbutamide		500 mg three times a day	5 h	6–12 h
Tolazamide		100 mg twice a day	7 h	12–14 h
Glyburide		2.5 mg once daily	10 h	Up to 24 h
Glipizide		5 mg once daily	4 h	Up to 24 h

III. Can be manufactured easily, efficiently, inexpensively and is green: This property is obvious because if chemical production requires the use of many chemical compounds for drug preparation and in addition several reaction steps, accordingly a large amounts of byproducts and waste are produced, in addition to raising the cost of manufacturing. The reduction of reaction steps, elimination of solvents and the increase of yields must be implemented.

IV. A safe chemical has minimal hazards. The chemical compounds must be designed to have low toxicity, meaning that it does not cause adverse effects both to human health and the environment. Especially those chemicals that may cause effects on prenatal development or postnatal development in humans are very restricted by regulatory agencies such as FDA. A well-known example is the use of Thalidomide which was marketed in 1957 as an over-the-counter "safe" sedative/ tranquilizer. Later it was marketed to calm morning sickness during pregnancy without knowing the teratogenic effect of the drug. Severe congenital anomalies were soon observed in babies born from pregnant women who had used this drug. In 1960, a variety of deformities, especially amelia (lack of limb) or phocomelia (seal limb), were reported in children exposed to thalidomide. Deformities of the heart, kidney and eyes were also reported; external ears absent or abnormal; cleft lip or palate, defects of the spinal cord; and gastrointestinal tract disorders, 12,000 babies were affected worldwide (Matthews and McCoy, 2003).

V. Minimal physical and global hazard: A safe chemical must not have explosive or flammable properties and a minimum global risk which means that it does not have adverse effects to the environment after a long period of exposure (e.g., it does not cause deterioration of the ozone layer, eutrophication or climate change). It should be easily degraded in the environment in harmless chemical substances. Safe chemicals should not be bioaccumulable both in plants and in lower trophic organisms such as algae and fish, which later form part of the human food network representing a potent risk of toxicity.

An example is the use of iodine X-ray contrast media during radiography. Because it requires high doses to obtain a good radiocontrast, the administration is intravenous. Iodine is a stable compound with extremely low toxicity, it is not surprising that due to compound stability it has a low biodegradability, so this it accumulates and is detectable in microgram levels in wastewater effluents and certain bodies of water, contributing to the loading of easily absorbed halogenated organic compounds (Steger-Hartmann et al., 1999; 2002). In an effort to reduce the risk of bioaccumulation, iodine compounds derived from sugars have been investigated as candidates for X-ray contrast media with low toxicity and are easily degradable (Crawford et al., 2017).

As in the previous case, several pharmaceutical products, whether anti-mycotic, antibiotics, anti-cancer, among others or even agricultural chemicals that are not easily biodegradable, can end up in surface water sources presenting a high risk to human health since water can be used for drinking. Due to this, measures have been presented to minimize persistence and bioaccumulation of chemicals on the environment, promoting the replacement of dangerous substances by more environmentally friendly molecules and to develop new compounds highly effective, efficient and easily biodegradable in the environment (Service, 2013).

VI. A safe chemical must offer environmental, human health and commercial advantages over other chemicals or existing chemicals with the same function. Comparing two production processes; the first for the synthesis of a new chemical designed with another of a commercialized chemical both with the same function, the first must present advantages such as; lower waste material production, solvents reduction or free solvents synthesis, just to mention some, that can lead to minimize environmental and economic impact. On the other hand if the chemical compound has less toxicity, it is easier to be biodegradable and has more potency than a commercial drug, and it will be better accepted.

The reason why it is not easy to design safer chemicals is due to the amount of requirements that must be met, for this reason it is necessary to understand the relationship between structural and energetic characteristics with chemical hazards. Even more so regarding human toxicity, that makes them carcinogenic, neurotoxic, why they have hepatotoxicity, or cardiotoxicity, why they have hematological or endocrine toxicity, which makes them immunotoxic, teratogenic or why they cause DNA mutation. In conclusion developing the fourth principle of green chemistry requires interdisciplinary knowledge since it is necessary to establish the relationship between the chemical structure and the toxic properties of a chemical compound, for this reason the advance of this principle in comparison with the other 11 has been slow (Anastas and Warner, 2005).

In the last decade, experimental chemical and toxicological data banks have been generated, they allow a partial overview of the relationship between the chemical molecule structures with their activity and toxicity, which have led to formulating some rules and principles that allow proposing molecular designs. On the other hand, green toxicology has occurred by means of which *in vitro*, *in vivo* and in silico mechanisms have been proposed allowing to analyze the toxicity of molecules more quickly (Rusyn and Greene, 2018).

In this chapter the design of safe chemicals are discussed from two points of view. First the substances chemical design in relation to their structure and some physicochemical properties that should be considered in the design of chemical compounds in order tofulfill their function but at the same time reduce the risk to human health and the environment. On the other hand, some methodologies used for toxicity determination of chemical compounds and to prevent risks to health and the environment will be described, as well as the development of new technologies using computational methods.

STRUCTURAL AND PHYSICOCHEMICAL CONSIDERATIONS IN THE DESIGN OF SAFE CHEMICALS

For many years, the design and synthesis of commercial chemicals or intermediates for the industry were focused only on the final product efficiency, considering mainly the operation of the chemical compound but not the impact over human health and the environment in the future. In 1928, Alice Hamilton observed that many chemicals used, presented a potential risk to human health and the environment (Marselos and Vainio, 1991). For this reason over the years some regulatory agencies have emerged such as Food and Drug Administration (FDA), European Medicines Agency (EMA) and the Evironmental Protection Agency (EPA) that have issued laws to control possible risks in new and existing drugs (FDA) and pesticides (EPA).

Through the "Federal Food Drug and Cosmetic Act" (FFDCA) and FDA, it was recommended to carry out an extensive analysis of: efficacy, bioavailability, distribution; metabolism; excretion and adverse toxic effects to identify and characterize chemical substances with possible pharmacological properties, but with low toxicity. All this information is available in data banks of free access. The chemical researcher can use this data to extract information that allows establishing a possible relationship between the structure-activity of the chemical compounds in order to design safe chemicals.

In 1997 Lipinski, carried out the analysis of approximately 900 commercial drugs available in the market (whose chemical and physicochemical properties had been described), he found that there was a relationship between the physicochemical properties and the absorption of the substances. With this information the "Five Rule" was formulated which states that chemical compounds have low bioavailability (absorption) when: a) there are more than five functional groups that donate hydrogen bonds (>5 H–Bonds) in the chemical structure, b) when there are more than 10 acceptors of hydrogen bonds (>10 H–bonds acceptors, where the sum of the nitrogen and oxygen atoms is more than 10), c) when the chemical compound has a molecular mass higher than 500 Da; d) when the lipophilicity of the compound is high (log Pow>5) (Lipinski et al., 2001; Hopkins and Groom, 2002). Although it has been a rule widely used in the pharmaceutical industry for drugs design, it does not establish a relationship between the chemical structure and the compound toxicity, which allows designing safe chemicals.

On the other hand, through studies where the chemical structure of certain chemical compounds has been related to toxicity, it was found that some substituents or molecular structure modifications, during the metabolism process are directly or indirectly toxic. Highly reactive chemical compounds, especially electrophilic substances, are highly carcinogenic since they can be reacted and covalently linked to nucleophilic molecules such as macromolecules (DNA). Some highly electrophilic species are shown in Fig. 5.1. All molecules with these chemical fractions within their structure, can lead to the tumor formations on the surface of exposure. Some highly electrophilic substances are: epoxies, amines, esters, oxaziridines, among others (Marselos and Vainio, 1991). These kind of chemicals are toxic because they are very reactive species, where the reaction constant (Kr) with macromolecules is greater than metabolic reaction constant (Km), through which enzymes carry out the biotransformation of chemical compound for its later excretion (Kr>Km). Therefore, it is best to avoid including these chemical species in the design of safe chemicals to reduce chemical compound toxicity.

Sometimes a chemical compound may have low or no toxicity, however, the degradation products through metabolism in humans or bacteria in the environment, can lead to the formation of highly toxic compounds. For example, for a long time it was thought that azo compounds used as dyes or pigments were carcinogenic substances. However, after several studies, it was determined that toxicity is not directly due to the compound, but to the molecules formed during metabolic degradation. The azo bonds of pigments can easily be broken down by chemical compounds or by enzymatic action, via reduction to form free aromatic amines, the aromatic amines can be easily absorbed and are responsible for the toxicity of these compounds (see below). In the textile industry, these types of dyes are widely used, therefore in some cases close by water sources can be contaminated, representing a source of exposure to both humans and animals and the environment (DeVito and Garrett, 1996).

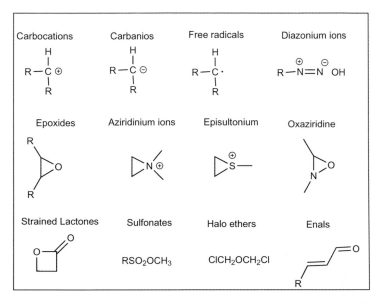

Figure 5.1 Chemical compounds electrophilic highly toxic that should be avoided in the design of safe chemicals, since they can be bonding covalently to DNA.

In order to reduce the toxicity of imines obtained during the azo compounds degradation, one strategy proposed is to include sulfonate groups into the aromatic ring, so azo compound degradation leads to the formation of highly hydrophilic amines. This prevents the imine penetration into cell membrane, and the small quantity that could be absorbed is quickly eliminated in urine (Fig. 5.2a). In some cases, the modification may alter the dye properties, so new strategies are required to reduce the amines toxicity which are formed during the degradation process without reducing the efficacy of the dye. Some structure modifications to azo dyes or pigments have been proposed in order to decrease the toxicity of the formed amines, for example, the inclusion of alkyl groups in ortho positions to the resulting amino groups in azo compounds derived from benzidines that help to reduce the toxicity (Fig. 5.2b) (Bae and Freeman, 2002; Hinks et al., 2000). Another modification is a substitution of a carbon into an aromatic ring by a nitrogen in order to make the aromatic ring deficient in electrons, which decreases the amines reactivity and therefore their toxicity (Fig. 5.2c) (Calogero et al., 1987).

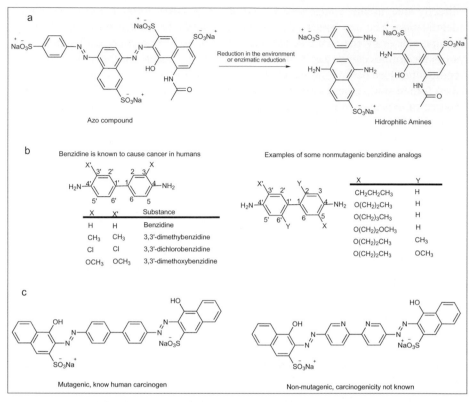

Figure 5.2 Modifications of the chemical structure in azo compounds to reduce the amines toxicity obtained by degradation; a) hydrophilicity increase of the amines resulting from metabolic degradation; b) chains substitution ortho to the amino group (introduction of steric hindrance to avoid reaction and deviate the molecules planarity); c) carbon substitution by a nitrogen (decreases the reactivity of the amines).

In addition to Lipinski's rules, the correlation between the chemical compounds structure and their physicochemical properties has been studied in order to identify a relationship with the biological activity and the effect on the toxicity of the chemical compound. In toxicology, the physicochemical properties are used to study specific biochemical interactions between a molecule and the enzyme, related to toxicity mechanisms (Hansch et al., 2002).

The binding affinity of a chemical compound with a macromolecule is a function of the Gibbs free energy (eq. 1), which can be interpreted as individual contributions of a series of attractive forces such as: Van der Waals, interaction of hydrogen bond and repulsive forces as hydrophobic effects that can lead the ligand out of the water and into the hydrophobic cavity of the protein (eq. 2) (Meanwell, 2016).

$$\Delta G_{binding} = \Delta H - T\Delta S = -RT \ln K_{eq} \qquad (1)$$

$$\Delta G_{binding} = \Delta G_{hydrogen\,bond} + \Delta G_{electrostatic} + \Delta G_{hydrophobic} + \Delta G_{vdw} \qquad (2)$$

The compounds-macromolecule interactions probability depends of the positive contributions of enthalpy and entropy. Maximizing the enthalpy contributions depends on the formation of hydrogen bond and the Van der Waals interactions, which are difficult to control since they are very sensitive to both; distance and the interaction angle, so very flexible molecules could avoid the interaction of a chemical compound and the target, decreasing the bioavailability of the compound. The molecule flexibility depends of the bonds number that can rotate freely, Veber found that chemical compounds that have less than 10 and more than two bonds with free rotation, have a high bioavailability (Voutchkova et al., 2010a).

TOXICOKINETICS CONSIDERATIONS

Derived from the analysis of experimental data where the toxicity of several chemical compounds with different structural characteristics has been studied, information has been obtained which are used to propose rules for designing chemicals with less toxicity. However, in order to establish a relationship between the chemical structure and its toxicity, the mechanisms of Adsorption, Distribution, Metabolism and Excretion (ADME) of a chemical must be understood.

For a living being to be exposed to chemical compound toxicity, the following must be accomplished: a) There should be exposure and absorption of the chemical which can be given through the gastrointestinal tract (GI), respiratory tract, and skin; b) the substance has to be bioavailable, namely, it must cross the biological interface and be available for distribution in the body; c) The chemical compound must be capable of causing an alteration of the normal cellular biochemistry and physiology, which is relayered with toxicodynamics.

In accordance with the above, blocking or reducing bioavailability selectively, helps to reduce the chemical toxicity. Due to absorption processes this can be a complicated process, however, medical chemistry has carried out several studies to understand and establish a relationship between the chemical structure and the bioavailability of the chemical compound (Hurst et al., 2007).

Absorption

Gastrointestinal absorption

The physicochemical properties, which determine the absorption of a chemical through the GI tract, can be listed as follows:

i)   The distribution coefficient ocatnol water (log P_{ow}), measures the lipophilicity and determines the absorption of the chemical compound. High values of log P_{ow} imply that the chemical compound is lipo-soluble and therefore it dissolves in the GI tract, on the other hand, low values of log P_{ow} increases the water solubility, decreasing the chemical absorption and favors the elimination.

ii) Physical state: liquids or solutions are absorbed better than solids

iii) Ionization and dissociation constants (pKa). The pH partition theory explains the influence of physiological pH in the absorption. Weak organic acids with pKa<7 are in its non-ionized form when the pH of the GI tract is lower than the pKa of the organic molecule, therefore, it is lipo-soluble and can be absorbed through the intestinal membrane into the torrent blood. On the other hand, weak organic bases with pKa>7 exist in a non-ionized form when the pH of the GI tract is higher than the pKa of the organic molecules, therefore, it is lipo-soluble and can be absorbed through the membrane of the intestine into the bloodstream.

iv) Molecular weight and the size of the molecule. Molecules with weights greater than 300 Da, are better absorbed than molecules between 300–400 Da, while molecules with a molecular weight greater than 500 Da are poorly absorbed (Johnson and Zheng, 2006).

Absorption by the respiratory tract

To reduce the bioavailability of a chemical compound through the respiratory tract, gases are the main chemical compounds that are absorbed, therefore the molecule lipophilicity is less important, while the blood-to-gas partition coefficient is the main factor that determines the absorption. Low values of the partition coefficient blood-to-gas reduce the adsorption speed of chemical, besides low vapor pressure values decrease the absorption of the chemical in the respiratory tract. It should also be considered that vapor pressure decreases with higher molecular weight and consequently the chemical absorption. Finally, if the particle size exceeds 5 μm, it is deposited in the respiratory tract and removed by the mucous layer.

Skin absorption

While there is only a small amount of chemicals that are applied directly on the skin, they are widely used in cosmetics and creams . These products contain a large amount of chemical compounds that can be absorbed representing a risk to health, so it is necessary to understand the mechanism that controls the skin absorption. Therefore to design chemicals with low absorption and availability, the following guidelines should be considered; a) physical state, since liquids are better absorbed than solids; b) polar molecules have low absortion, while lipophilic substances with high log P_{ow} values are well absorbed in the skin; c) molecules with high molecular weight (greater than 400 Da) are poorly absorbed (Schneider et al., 1999).

Distribution of a Chemical Compound

The distribution of chemical compounds refers to the movement of the compound from its entry site to other body parts through the bloodstream, where the distribution rate depends of blood flow and the diffusion capacity of the capillaries in the body tissues particular organ, therefore the distribution of the chemical can be completed in a short time.

On the other hand, the distribution selectivity is closely related to physicochemical and structural properties, the speed at which a chemical leaves the blood and the way in which it is distributed to the tissues depends of the physicochemical affinity for that tissue. A balance quantitative estimation between the concentration of the chemical into blood and in the tissue is given by the equation $V_D = TAXB/OCXB$, where TAXB is the total amount of xenobiotic (drug) in the body and OCXB is the Observed Concentration of Xenobiotic into Blood. Chemical compounds with high affinity for tissues are rapidly distributed, so low concentrations of chemicals remain in the blood and their V_D values are high. On the other hand, chemical compounds that remain in the bloodstream for long periods of time have high V_D values (Pohjanvirta et al., 1996; Huggins et al., 1963). Some models based on algorithms to calculate V_D indicate that some physicochemical properties that affect it are: lipophilicity, hydrogen bond, molecular size and acid-base properties (Sui et al., 2008; Obach et al., 2008).

Accumulation and Storage of Chemical Compounds

Due to the chemical and physicochemical characteristics some chemical compounds can accumulated and stored inside the body causing toxic effects, as they are constantly being released into the bloodstream. The four primary storage systems in the body are: Adipose tissue, Bones, Liver and Kidneys.

Chemical compounds that have high log P_{ow} value are more likely to be stored in the adipose tissue than to be metabolized and excreted. When the adipose tissue is consumed, the stored chemicals are released resulting in a redistribution of the stored chemical compounds. Figure 5.3 shows some molecules that have this phenomenon and are carcinogenic.

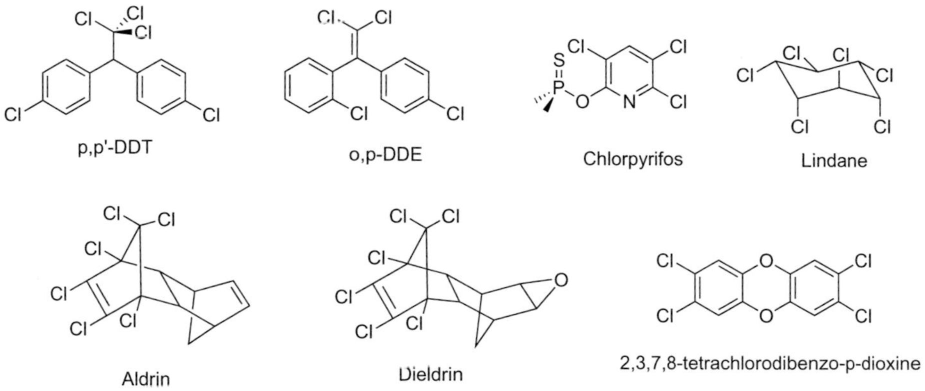

p,p'-DDT　　　　o,p-DDE　　　　Chlorpyrifos　　　　Lindane

Aldrin　　　　Dieldrin　　　　2,3,7,8-tetrachlorodibenzo-p-dioxine

Figure 5.3　Chemicals with high log P_{ow} that are stored in adipose tissue.

Compounds that contain fluorine atoms, as well as some heavy metals can be incorporated and stored in the bone matrix, by an equilibrium process these can be released over a long period. On the other hand, the liver and kidneys have a high capacity to store chemicals since they are the most exposed because they are responsible for metabolizing and excreting substances into the bloodstream. While molecules that have a large amount of halogen atoms tend to be stored (Smith, 2001; Longnecker, 2005).

Metabolism and Excretion

Biotransformation is the process to convert a lipophilic chemical compound into a hydrophilic one that is soluble in water and can be excreted in urine or bile, however, not all chemical compounds require biotransformation. As a general rule, chemical compounds having a log $D_{7.4}$ value greater than zero, require biotransformation to facilitate their excretion (where log $D_{7.4}$, refers to lipophilicity at pH 7.4). Most halogenated compounds are resistant to biotransformations, so these compounds tend to accumulate in the body until an alternative method to eliminate them enters.

The biotransformation of chemical compounds are catalyzed by several enzymes that are divided into four categories based on the catalytic reactions; Hydrolysis (e.g., carboxylesterease), reduction (e.g., carbonylreductase), oxidation (e.g., cytochrome P450) and conjugation (e.g., UDP-glucuronosyltransferase).

The degradation metabolism of chemical compounds occurs through reactions represented in two different phases; Phase I and Phase II, the reactions that are carried out through phase I are the most common, these are functionalization reactions such as; oxidation, reduction and hydrolysis reactions. The objective of functionalization is to introduce polar functional groups either by modifying the existing functional groups, introducing new groups (–OH–, –NH– or –COOH) or by exposing those already contained in the molecule (Fig. 5.4). The enzyme involved in phase I is the cytochrome P450 family (CYP) responsible for 90% of the metabolism in this phase (Evans and Relling, 1999). In Phase II there are several enzymes involved, but the most common reaction is conjugation, through this reaction glucuronidation, sulfonation, acetylation and glutathione conjugation can be carried out.

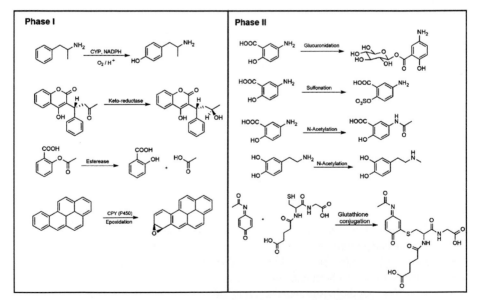

Figure 5.4 Biotransformation of chemical compounds before being excreted,
Phase I: oxidation, reduction and hydrolysis reactions.
Phase II: glucuronidation, sulfonation, acetylation and glutathione conjugation.

Toxicodynamics is responsible for the study and monitoring of the chemical compound biotransformation from entering or leaving the body. During the metabolic process of a chemical, bioactivation of the chemical can be carried out, this can cause an organic compound that is initially inert to become a toxic substance, which will depend on the chemical structure of the chemical compound and of the enzyme involved (Aschner et al., 2008). For example, in a compound with carbon-carbon double bond the metabolic route for its elimination leads to the double bond oxygen insertion in order to form an epoxide, which is a very reactive electrophilic chemical compound that increases the reaction probability with cellular macromolecules causing toxicity (Scheme 5.1).

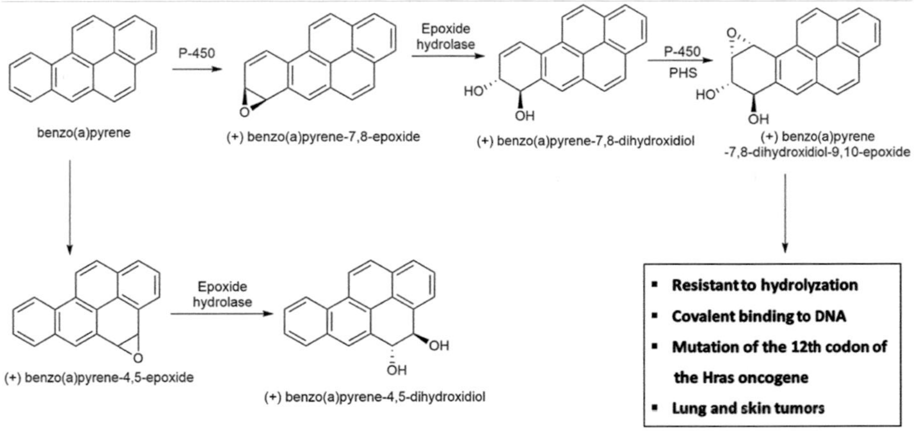

benzo(a)pyrene P-450 → (+) benzo(a)pyrene-7,8-epoxide Epoxide hydrolase → (+) benzo(a)pyrene-7,8-dihydroxidiol P-450 PHS → (+) benzo(a)pyrene -7,8-dihydroxidiol-9,10-epoxide

(+) benzo(a)pyrene-4,5-epoxide Epoxide hydrolase → (+) benzo(a)pyrene-4,5-dihydroxidiol

- **Resistant to hydrolyzation**
- **Covalent binding to DNA**
- **Mutation of the 12th codon of the Hras oncogene**
- **Lung and skin tumors**

Scheme 5.1 Biotransformation mechanism of benzopyrene by cytochrome P450 oxidation.

Because in many cases, bioactivation can lead to the formation of reactive species which provide toxicity to a chemical compound, it is necessary to establish rules to reduce biotransformation or prevent the formation of toxic substances. Williams et al. (Williams et al., 2003) described some functional groups or groups of molecules that can be biotransformed with toxic substances formation, as well as the enzyme used during metabolization. Subsequently Voutchkova et al. (Voutchkova et al., 2010b) supplemented this information by suggesting some structural modifications that allow decreasing the toxicity of the metabolite formed (Table 5.2).

As shown in Table 5.2, in many cases the chemical compounds do not cause the toxic effect, but after its biotransformation during metabolic degradation electrophilic molecules can be obtained, which can be bound covalently to DNA causing mutations. The molecular design can help in avoiding the biotransformation or ensuring that the products obtained from the metabolism are not toxic and can easily be excreted.

Table 5.2 Biotransformation of functional groups, toxicity and proposed structural modifications.

Chemical compound	Enzyme and mechanism	Structural modifications (to decrease toxicity)
Alkanes	The enzyme that catalyzes the metabolism is CYP, which is responsible for carrying out oxidation. For example, if we start with hexane, this can lead to the formation of 2,5-diones which can react with the nitrogens of lysine to form adducts, which are neurotoxic.	Alkyl branches that prevent oxidation to form ketones can be included.
Alkene Vinyl halides	The enzyme that catalyzes the biotransformation is CYP, which carries out the epoxidation of the carbon-carbon double bond. The epoxies formed are highly carcinogenic and mutagenic, since they react with cellular macromolecules and DNA.	Including halogenated substituents in the alkene decreases the electrophilicity of the resulting epoxide and therefore the toxicity. Including an alkyl group in the ring of arenes causes the product of the metabolism to be benzoic acid.
Arene	Nu = DNA and proteins	

Polyciclic aromatic hydrocarbons

The CYP enzyme metabolizes polycyclic aromatic hydrocarbons in order to form dihydrodiolepoxides, which are highly carcinogenic. Toxicity can be increased by increasing the number of fused rings and the substitution of methyl groups.

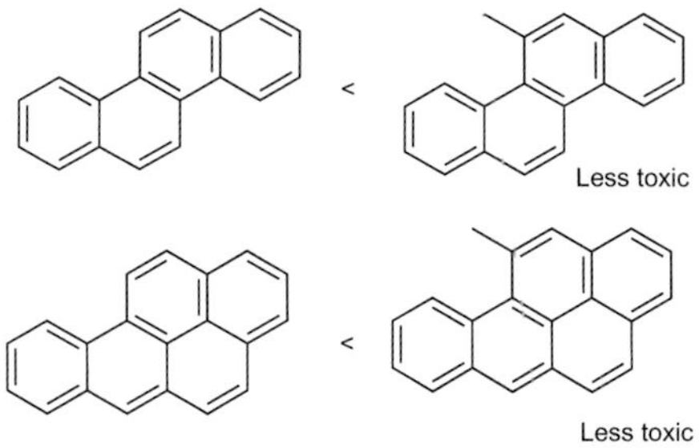

Less toxic

Less toxic

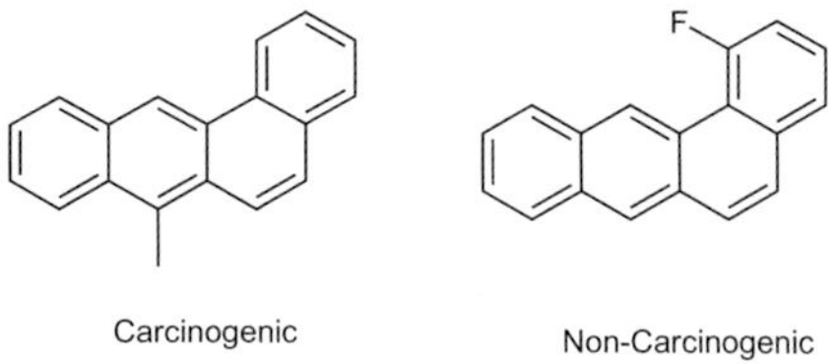

Furans

Furans are epoxidized by the CYP enzyme to obtain electrophilic species, which can cause cell death in the liver and lungs.

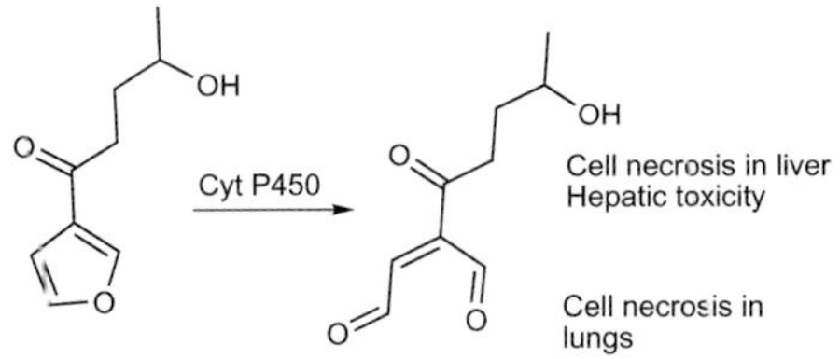

Cyt P450

Cell necrosis in liver
Hepatic toxicity

Cell necrosis in lungs

Replacing methyl groups by fluoride helps to reduce toxicity. For example, 7-methyl-benzo [a] anthracene is highly carcinogenic, but 1-fluoro7-benzo [a] anthracene is not. It is because fluoride prevents epoxidation at 1,2 positions.

Carcinogenic

Non-Carcinogenic

The furan methylation in positions 2 and 5 prevents epoxides formation. Although methylation does not prevent oxidation, the results products are less toxic.

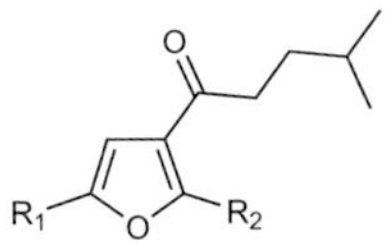

			LD_{50}
1	$R_1 = H$	$R_2 = H$	30 ± 1.8
2	$R_1 = Me$	$R_2 = Me$	2238 ± 1242
3	$R_1 = H$	$R_2 = Me$	193 ± 15
4	$R_1 = Me$	$R_2 = H$	8806 ± 3455

Benzoquinone

Benzoquinones are conjugated with GSH by the enzyme GST, forming mono, bis, tris or tetra GS conjugates. Of which the mono and tetra conjugates are not toxic, while bis and tris conjugated are very nephrotoxic as they are highly electrophilic.

GST/ GS Conjugation

Excretion, detoxification

The toxicity can be reduced by fusing aromatic rings to benzoquinone.

Less toxic

Aromatic amines

Several enzymes can catalyze the biotransformation of aromatic amines and therefore there are several mechanisms and metabolites that can be obtained. For example, the enzyme CPY catalyzes an N-hydroxylation or acetylation, the GST enzyme catalyzes the sulfonation or phosphorylation, and the ST enzyme catalyzes the glucouridation. The amines are highly carcinogenic, the planarity of the amines increases their toxicity, adding aromatic rings increases their toxicity, because of amine nucleophilicity increase.

2-Aminonaphthalene   CPY 1A2 N-hydroxylation   N-Hydroxy-2-naphthylamine

Activation of DNA-reactive metabolite that cause bladder or colon tumor

It is possible to obtain amines with low toxicities if:

- Using no-aromatic amines, these are less carcinogenic, since the nucleophilicity and planarity of the amines is diminished.
- An N-alkylation with bulky groups decreases the pKa and therefore the reactivity. In addition, the steric hindrance caused by the bulky group can impede the action of the enzyme over the nitrogen.
- Introducing electron withdrawing functional groups in the aromatic ring to decrease the nucleophilicity of the amine, as well as bulky groups that distort the molecule planarity.
- The introduction of hydrophilic groups allows the excretion of the molecule without having to be biotransformed and avoid storage.

For example, primary aromatic amines are known to be highly carcinogenic because they are nucleophilic and therefore reactive. It is possible to decrease the carcinogenic effect by changing the molecular geometry and physicochemical properties of imines, it should be considered that non-aromatic amines are less reactive than aromatic amines and could be considered less toxic, N-alkylation by using bulky groups causes steric hindrance and therefore cannot be catalyzed by enzymes or can not be intercalated with DNA. However, the metabolism will look for an alternate route to eliminate the chemical compound and avoid storage. Introducing functional groups in the aromatic ring as electro-acceptors help to decrease the reactivity of amines or introduce polar functional groups as sulfonates make the molecule more soluble, so biotransformation is not required to excrete the chemical.

Biodegradability of Chemicals

One objective of the design of safe chemicals is to make sure the compounds are easily degradable under environmental conditions either by enzymatic action, bacteria or by climatic conditions (effect of light, humidity, air, among others), and the product of degradation would not be toxic to human health or the environment. For this reason, designing and synthesizing chemical compounds that, in addition to fulfilling their function, are subject to easy biodegradation once they are in contact with the environment is a priority. Some generalizations have appeared regarding the chemical structure of a compound and its relationship to the ease of biodegrading. One of the "thumb rules" that predicts the effect of several functional groups and the number of times it repeats each in the structure as well as sub-structures in the molecule, in addition the position effects of the substituent in the biodegradability of the molecule Fig. 5.5.

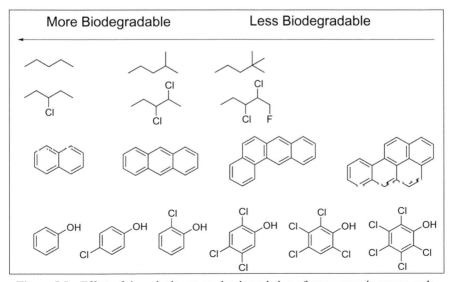

Figure 5.5   Effect of the substituent on the degradation of some organic compounds.

Among the structural characteristics related to an increase in resistance to aerobic biodegradation are: a) the halogens presence especially iodine, chlorine and fluorine, when there are more than three halogen atoms into a small molecule, the molecule is very stable; b) the presence of branched hydrocarbon alkyl chains, the quaternary carbons are difficult to biodegrade; c) tertiary amines, nitro groups, nitrozo, azo, and arylamino groups are difficult to biodegrade; d) polycyclic aromatic hydrocarbons, while more fused aromatic rings are more resistant to biodegradation; e) heterocyclic residues, for example imidazoles; aliphatic ester bonds (except in ethoxylates) (Boethling et al., 2007).

On the other hand, some functional groups are more likely to be attacked by enzymes and easily degraded especially phosphates esters and amides, those molecules, which contain oxygen atoms such as hydroxyls, aldehydes, ketones or carboxylic acids help in biodegradation. For example, cyclohexanol and cyclohexane degrade faster than cycloexhane. Another point that must be considered is that molecules with high molecular weights $MW > 1000$ Da are poorly biodegradable, even if they are organic compounds.

The use of heavy metals

Metals have been widely used, both in industry and in medicine, they are the key in the technological development and are present in biological processes. However, several of them represent a risk for both human health and the environment because they can form extremely toxic species or interrupt biological processes.

In ionic form, metals are very reactive and can interact with biological systems in a variety of ways. They can be easily coordinated with proteins by displacing essential metals as a preferred ligand, which can lead to a steric rearrangement or unfolding of the proteins impairing their function. Through mimicry, toxic metals can access and potentially disrupt a variety of important or even critical cellular functions through metals. For example, mimicry and zinc replacement is a toxic mechanism for cadmium, copper and nickel. Thallium can mimic potassium and manganese coordinating in place and interrupt its functions. Arsenate and vanadate mimic phosphate, allowing the cellular transport of these toxic elements, while selenate, molybdate and chromate mimic sulfate and can competing for sulfate carriers and in chemical sulfation reactions (Cousins et al., 2006; Kasprzak, 2002). In addition, many metals can act as catalysts of redox reactions with molecular oxygen or other endogenous oxidants, producing oxidative modification to macromolecules such as proteins or DNA, which is the key in the carcinogenicity of certain metals.

Some metals represent a high risk as they may have high toxicity such as, arsenic, beryllium, chromium, cadmium, lead and mercury, while those in high doses that can pose a health risk are nickel, titanium, zinc and iron.

Metals are mainly used in the synthesis of organometallic compounds for use as pesticides, insecticides or pesticides. Although the organic part is easily biodegradable, the metals do not, so plants and animals can absorb them. As they bioaccumulate, there is a high probability of human ingestion, increasing the risk to health. Avoiding organometallic compounds that include these metals either by replacing the toxic metal with one that is not or avoiding the use of metals in these compounds would be ideal.

TOXICOLOGICAL EVALUATION

Making the necessary efforts to design and synthesize safe chemical compounds, as they can lose their validity if toxicological analysis of the proposed molecules is not carried out. An efficient evaluation of the risks can be carried out through *in vitro* and *in vivo* analyzes, where toxicological studies are carried out to predict or identify compounds potentially harmful to human health. However, toxicological evaluation is complicated since there are more than 130 million compounds in the market, while about 500 to 1000 chemical compounds are produced each year, making toxicological evaluation of each chemical difficult by standard experimental methods (Cavaliere and Cozzini, 2018).

To establish a first relationship between the compounds chemical structure and the activity, High-Throughput Screening (HTS) technique is used. This method uses modern robotic techniques to evaluate thousands of samples per day for hundreds of endpoints including receptor ligands, enzymes inhibition, chemicals interaction with ion channels, RNA, DNA and many other assays (Anastas, 2016). This technique allows differentiate between chemical compounds that probably have little or no risk of causing adverse effects on health. Although HTS is a widely used technique, the selected compound is not always suitable for future medical examinations.

Once the *in vitro* assessment have been passed, *in vivo* toxicity of the compounds must be evaluated. This represents two challenges; the first is the use of animals for toxicological tests, which has caused a lot of controversy, because there are some organizations that are against such tests. On the other hand, the analysis of the chemical toxicity of a compound can take months or years, or in some cases the chemical toxicity of the compounds are not observed for several years later. In the last two decades an *in vivo* toxicological analysis technique has been developed that allows obtaining faster results using zebrafish as toxicological models. This method represents some advantages since it allows optimizing the safer chemicals design and detecting the toxicity of thousands of chemicals that are already in use. The use of zebrafish, allows to establish a relationship between the bioactivity of chemical structures and some physicochemical properties of molecules such as; functional groups, lipophilicity, molecular weight, among others (Zhang et al., 2013; Noyes et al., 2016). However, one important requirement is that the pure chemical compound should be available; however, the synthesis takes a long time and is expensive, although this toxicity assessment is mandatory.

On the other hand, the computer development as well as the implementation of new algorithms in computational chemistry such as; Molecular Mechanics (MM), molecular docking, virtual screening, among others, allows a quick alternative to perform the toxicology evaluation of chemical compounds, even before they are synthesized (Jorgensen, 2004). In silico toxicology, a theoretical methodology is used to evaluate, simulate or predict the chemical compounds toxicity by means of computational methods. The computational methods allow to complement the toxicity assessment *in vitro* and *in vivo*, minimizing the evaluation in animals, reducing the cost, improving the toxicity evaluations and allowing a safe evaluation (Cohen et al., 2018).

Depending on the information available, Virtual Screening (VS) can use several computational tools that include, database of chemical compounds, which may contain information such as toxicity and chemical properties; software to generate molecular descriptors; simulation tools for biological systems and molecular dynamics; methods of modeling and predicting toxicity; modeling tools such as statistical packages and software for the generation of prediction models; specialized systems to study interactions between ligand-targets that allow the prediction of toxicity; visualization tools (Raies and Bajic, 2016). The protocol for the determination of the toxicology of new chemical compounds *in silico* is summarized in Fig. 5.6 (Myatt et al., 2018).

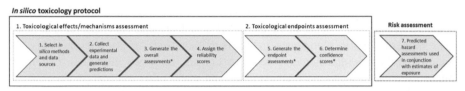

In silico **toxicology protocol**

| 1. Toxicological effects/mechanisms assessment | 2. Toxicological endpoints assessment | Risk assessment |

1. Select *in silico* methods and data sources → 2. Collect experimental data and generate predictions → 3. Generate the overall assessments* → 4. Assign the reliability scores

5. Generate the endpoint assessments* → 6. Determine confidence scores*

7. Predicted hazard assessments used in conjunction with estimates of exposure

* Based on rules/principles outlined in the IST protocols, including an expert review if warranted

Figure 5.6   Summary of the in silico protocol process. This image has been taken from Zhang et al., 2013 without modifications. doi.org/10.1016/j.yrtph.2018.04.014. Copyright Clearance Center, Inc.

In VS, the compounds are selected from a library for their ability to bind to macromolecules using computer programs that consider the physicochemical properties and general rules discussed earlier. The compounds under study do not necessarily exist so reagents, solvents, time, among others, are not used, decreasing risks to health and the environment. To be able to use VS requires some prerequisites such as knowledge of molecule spatial geometry and some energy criteria, which determines the molecule-receptor binding under study, this information, can be obtained from X-ray structures or NMR data (Klebe, 2006).

The VS process can be divided into three categories; virtual filtering, virtual profiling and virtual screening (Fig. 5.7). The first two focus mainly on criteria related to the fundamental issues concerning the pharmocological properties of the target. All those molecules specifically with respect to certain general requirements are eliminated, for example, rules of statistical exclusion based on the "five rule" which uses simple descriptors such as molecular mass, lipophilicity, hydrogen bonds acceptors/donors, among others.

Computational tools based on two-dimensional topological descriptors provide a rapid analysis of a huge amount of data using molecules known as pharmacophores, to generate potentially active new chemical compounds. Subsequently a higher computational level using three-dimensional pharmacophores is applied. However, this requires ligand tridimensional information as well as greater power and computational time since they are usually enzymes. However, there are databases where three-dimensional information can be consulted, for example, X-rays and NMR, which speeds up the calculations, otherwise Monte Carlo calculations must be performed in order to determine the most stable conformers (Bleicher et al., 2003).

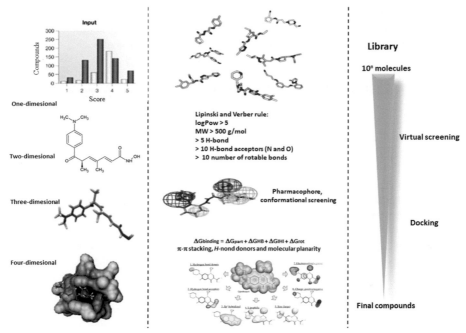

Figure 5.7   Design of new safer drugs using computational tools; visual screening; analysis of stable conformers by molecular mechanics; molecular docking).

The last step is to perform calculations in four dimensions, where the interaction between the chemical compound and the target is studied, this process is called molecular DOCKING, and predicts the more stable energetic orientation of the ligand when it is bound to the target through the interactions ligand-protein ($\Delta G_{binding}$, hydrogen bonds, lipophilic regions, charges, electro potential, among others). The stability of these interactions can determine the biological properties of the chemical compound. The interactional sequence of selection and refinement using one-dimensional descriptors, bidimensional ligands, three-dimensional pharmacophores, reduces the selection of chemicals to a small number of molecules to be used in the analysis of molecular docking that performs a filter based on biological activity (Bleicher et al., 2003; Sousa et al., 2006).

CONCLUSION

Although the design of safe chemicals is an area with little experimental development, it has increased in the last decade. There is a large amount of information not only on chemical structure but also chemical compounds toxicology, however, it needs to be analyzed in order to establish more specific rules to establish the relationship between structure-activity and toxicity that allow in developing specific guidelines to design and synthesize safe chemicals. Furthermore it has been proposed to include courses of toxicology for chemists to understand the mechanisms of Adsorption,

Distribution, Metabolism and Excretion (ADME), which will very useful to quickly develop the fourth principle of green chemistry, with the hope to see the benefits of this important principle, both in the environment and human health.

REFERENCES

Anastas, N.D., Warner, J.D. The incorporation of hazard reduction as a chemical design criterion in green chemistry. *Chemical Health and Safety*, 2005, 12, 9–13, doi:10.1016/j. chs.2004.10.001.

Anastas, N.D. Connecting toxicology and chemistry to ensure safer chemical design. *Green Chemistry*, 2016, 18, 4325–4331, doi:10.1039/C6GC00758A.

Anastas, P.T., Williamson, T.C., Hjeresen, D., Breen, J.J. Peer reviewed: Promoting green chemistry initiatives. *Environmental Science & Technology*, 1999, 33, 116A–119A, doi:10.1021/es992685c.

Anastas, P.T., Kirchhoff, M.M. Origins, current status, and future challenges of green chemistry. *Accounts Chemical Research*, 2002, 35, 686–694, doi:10.1021/ar010065m.

Aschner, M., Anand, S., Bloom, J.C., Brandt J.T. Cassarett and Doull's Toxicology: The Basic Science of Poisons, 2008. Vol. 12, ISBN 0071593519.

Bae, J.-S., Freeman, H.S. Synthesis and evaluation of non-genotoxic direct dyes. *Fibers and Polymers*, 2002, 3, 140–146, doi:10.1007/BF02912658.

Beach, E.S., Paul T. Green chemistry: The emergence of a transformative framework. *Green Chemistry Letters and Reviews*, 2007, 1, 9–24, doi:10.1080/17518250701882441.

Bleicher, K.H., Böhm, H.-J., Müller, K., Alanine, A.I. Hit and lead generation: Beyond high-throughput screening. *Nature Reviews Drug Discovery*, 2003, 2, 369.

Boethling, R.S., Sommer, E., DiFiore, D. Designing small molecules for biodegradability. *Chemical Reviews*, 2007, 107, 2207–2227, doi:10.1021/cr050952t.

Calogero, F., Freeman, H.S., Esancy, J.F., Whaley, W.M., Dabney, B.J. An approach to the design of non-mutagenic azo dyes: 2. Potential replacements for the benzidine moiety of some mutagenic azo dyestuffs. *Dyes and Pigments*, 1987, 8, 431–447, doi:https://doi.org/10.1016/0143-7208(87)85035-0.

Cavaliere, F., Cozzini, P. New in silico trends in food toxicology. *Chemical Research in Toxicology*, 2018, 31, 992–993, doi:10.1021/acs.chemrestox.8b00133.

Crawford, S.E., Hartung, T., Hollert, H., Mathes, B., van Ravenzwaay, B., Steger-Hartmann, T., Studer, C., Krug, H.F. Green toxicology: A strategy for sustainable chemical and material development. *Environmental Sciences Europe*, 2017, 29, 16. doi:10.1186/s12302-017-0115-z.

Cohen, J.M., Rice, J.W., Lewandowski, T.A. Expanding the toolbox: Hazard-screening methods and tools for identifying safer chemicals in green product design. *ACS Sustainable Chemistry & Engineering*, 2018, 6, 1941–1950, doi:10.1021/acssuschemeng.7b03368.

Cousins, R.J., Liuzzi, J.P., Lichten, L.A. Mammalian zinc transport, trafficking, and signals. *Journal of Biological Chemistry*, 2006, 281, 24085–24089, doi:10.1074/jbc.R600011200.

DeVito, S.C., Garrett, R.L. Designing Safer Chemicals: Green Chemistry for Pollution Prevention. ACS Symposium Series, vol. 640. American Chemical Society, 1996, ISBN 0-8412-3443-4.

DeVito, S.C. On the design of safer chemicals: a path forward. *Green Chemistry*, 2016, 18, 4332–4347, doi:10.1039/c6gc00526h.

Erythropel, H.C., Zimmerman, J.B., De Winter, T.M., Petitjean, L., Melnikov, F., Lam, C.H., Lounsbury, A.W., Mellor, K.E., Janković, N.Z., Tu, Q., Pincus, L.N., Falinski, M.M., Shi, W., Coish, P., Plata, D.L., Anastas, P.T. The Green chemisTREE: 20 years after taking root with the 12 principles. *Green Chemistry*, 2018, 20, 1929–1961, doi:10.1039/c8gc00482j.

Evans, W.E., Relling, M.V. Pharmacogenomics: translating functional genomics into rational therapeutics. *Science*, 1999, 286, 487–491, doi:10.1126/science.286.5439.487.

Garrett, R.L.R. Pollution prevention, green chemistry, and the design of safer chemicals. *ACS Symposium Series*, 1996, 640, 2–15, doi:10.1021/bk-1996-0640.ch001.

Hansch, C., Hoekman, D., Leo, A., Weininger, D., Selassie, C.D. Chem-bioinformatics: comparative QSAR at the interface between chemistry and biology. *Chemical Reviews*, 2002, 102, 783–812, doi:10.1021/cr0102009.

Hinks, D., Freeman, H.S., Nakpathom, M., Sokolowska, J. Synthesis and evaluation of organic pigments and intermediates. 1. Nonmutagenic benzidine analogs. *Dyes and Pigments*, 2000, 44(3), 199–207, doi:10.1016/S0143-7208(99)00078-9.

Hopkins, A.L., Groom, C.R. The druggable genome. *Nature Reviews Drug Discovery*, 2002, 1, 727.

Huggins, R.A., Smith, E.L., Deavers, S. Volume distribution of evans blue dye and iodinated albumin in the dog. *American Physiological Society*, 1963, 205, 351–356, doi:10.1152/ajplegacy.1963.205.2.351.

Hurst, S., Loi, C.-M., Brodfuehrer, J., El-Kattan, A. Impact of physiological, physicochemical and biopharmaceutical factors in absorption and metabolism mechanisms on the drug oral bioavailability of rats and humans. *Expert Opinion on Drug Metabolism & Toxicology*, 2007, 3, 469–489, doi:10.1517/17425225.3.4.469.

Johnson, S. R., Zheng, W. Recent progress in the computational prediction of aqueous solubility and absorption. *The American Association of Pharmaceutical Scientists Journal*, 2006, 8, E27–E40, doi:10.1208/aapsj080104.

Jorgensen, W.L. The many roles of computation in drug discovery. *Science*, 2004, 303, 1813–1818, doi:10.1126/science.1096361.

Kasprzak, K.S. Oxidative DNA and protein damage in metal-induced toxicity and carcinogenesis 1, 31. This article is part of a series of reviews on "Oxidative DNA Damage and Repair." The full list of papers may be found on the homepage of the journal.3Guest Editor: Miral Di. *Free Radical Biology and Medicine*, 2002, 32, 958–967, doi:https://doi.org/10.1016/S0891-5849(02)00809-2.

Klebe, G. Virtual ligand screening: Strategies, perspectives and limitations. *Drug Discovery Today*, 2006, 11, 580–594, doi:https://doi.org/10.1016/j.drudis.2006.05.012.

Lipinski, C.A., Lombardo, F., Dominy, B.W., Feeney, P.J. Experimental and computational approaches to estimate solubility and permeability in drug discovery and development settings. Advanced Drug Delivery Reviews, 2001, 46, 3–26, doi:10.1016/S0169-409X(00)00129-0.

Longnecker, M.P. Invited commentary: why DDT matters now. *American Journal of Epidemiology*, 2005, 162, 726–728, doi:10.1093/aje/kwi277.

Marselos, M., Vainio, H. Carcinogenic properties of pharmaceutical agents evaluated in the iarc monographs programme. *Carcinogenesis*, 1991, 12, 1751–1766, doi:10.1093/carcin/12.10.1751.

Matthews, S. J., McCoy, C. Thalidomide: A review of approved and investigational uses. *Clinical Therapeutics*, 2003, 25, 342–395, doi:10.1016/S0149-2918(03)80085-1.

Meanwell, N.A. Improving Drug design: An update on recent applications of efficiency metrics, strategies for replacing problematic elements, and compounds in nontraditional drug space. *Chemical Research in Toxicology*, 2016, 29, 564–616, doi:10.1021/acs. chemrestox.6b00043.

Melnikov, F., Mellor, K.E., Shen, L.Q., Coish, P., Zimmerman, J.B., Anastas, P.T., Steele, B., Brooks, B.W., Saari, G.N., Corrales, J., Kristofco, L.A., Mullins, M.L., Botta, D., Gallagher, E.P., Lasker, G.A., Mills, M., Simcox, N., Schmuck, S.C., Kavanagh, T. J., Voutchkova-Kostal, A., Kostal, J., Nesmith, S.M. The molecular design research network. *Toxicological Sciences*, 2017, 161, 241–248, doi:10.1093/ toxsci/kfx175.

Myatt, G.J., Ahlberg, E., Akahori, Y., Allen, D., Amberg, A., Anger, L.T., Aptula, A., Auerbach, S., Beilke, L., Bellion, P., Benigni, R., Bercu, J., Booth, E.D., Bower, D., Brigo, A., Burden, N., Cammerer, Z., Cronin, M.T.D., Cross, K.P., Custer, L., Dettwiler, M., Dobo, K., Ford, K.A., Fortin, M.C., Gad-McDonald, S.E., Gellatly, N., Gervais, V., Glover, K.P., Glowienke, S., Van Gompel, J., Gutsell, S., Hardy, B., Harvey, J.S., Hillegass, J., Honma, M., Hsieh, J.H., Hsu, C.W., Hughes, K., Johnson, C., Jolly, R., Jones, D., Kemper, R., Kenyon, M.O., Kim, M.T., Kruhlak, N.L., Kulkarni, S.A., Kümmerer, K., Leavitt, P., Majer, B., Masten, S., Miller, S., Moser, J., Mumtaz, M., Muster, W., Neilson, L., Oprea, T.I., Patlewicz, G., Paulino, A., Lo Piparo, E., Powley, M., Quigley, D.P., Reddy, M.V., Richarz, A.N., Ruiz, P., Schilter, B., Serafimova, R., Simpson, W., Stavitskaya, L., Stidl, R., Suarez-Rodriguez, D., Szabo, D.T., Teasdale, A., Trejo-Martin, A.,Valentin, J.P., Vuorinen, A., Wall, B.A., Watts, P., White, A.T., Wichard, J., Witt, K.L., Woolley, A., Woolley, D., Zwickl, C., Hasselgren, C. In silico toxicology protocols. *Regulatory Toxicology and Pharmacology*, 2018, 96, 1–17, doi:10.1016/j.yrtph.2018.04.014.

Noyes, P.D., Garcia, G.R., Tanguay, R.L. Zebrafish as an *in vivo* model for sustainable chemical design. *Green Chemistry*, 2016, 18, 6410–6430, doi:10.1039/C6GC02061E.

Obach, R.S., Lombardo, F., Waters, N.J. Trend analysis of a database of intravenous pharmacokinetic parameters in humans for 670 drug compounds. *Drug Metabolism and Disposition*, 2008, 36, 1385 LP-1405, doi:10.1124/dmd.108.020479.

Pohjanvirta, R., Laitinen, J.T., Vakkuri, O., Lindén, J., Kokkola, T., Unkila, M., Tuomisto, J. Mechanism by which 2,3,7,8-tetrachlorodibenzo-p-dioxin (TCDD) reduces circulating melatonin levels in the rat. *Toxicology*, 1996, 107, 85–97, doi:https://doi. org/10.1016/0300-483X(95)03241-7.

Raies, A.B., Bajic, V.B. In silico toxicology: Computational methods for the prediction of chemical toxicity. *WIREs Computational Molecular Science*, 2016, 6, 147–172, doi:10.1002/wcms.1240.

Rusyn, I., Greene, N. The impact of novel assessment methodologies in toxicology on green chemistry and chemical alternatives. *Toxicological Sciences*, 2018, 161, 276–284, doi:10.1093/toxsci/kfx196.

Schneider, T., Vermeulen, R., Brouwer, D.H., Cherrie, J.W., Kromhout, H., Fogh, C.L. Conceptual model for assessment of dermal exposure. *Occupational and Environmental Medicine*, 1999, 56, 765–773.

Service, B.I. Study on the environmental risks of medicinal products: Final report Prepared for Executive Agency for Health and Consumers. 2013; pp. 1–310.

Smith, A.G. DDT and its analogs. *In*: Krieger, R.I. (ed.), Handbook of Pesticide Toxicology, 2nd Ed. San Diego; Academic Press; 2001; pp. 1305–1355.

Sousa, S.F., Fernandes, P.A., Ramos, M.J. Protein–ligand docking: Current status and future challenges. *Proteins: Structure, Function, and Bioinformatics*, 2006, 65, 15–26, doi:10.1002/prot.21082.

Steger-Hartmann, T., Länge, R., Schweinfurth, H. Environmental risk assessment for the widely used iodinated x-ray contrast agent iopromide (ultravist). *Ecotoxicology and Environmental Safety*, 1999, 42, 274–281, doi:https://doi.org/10.1006/eesa.1998.1759.

Steger-Hartmann, T., Länge, R., Schweinfurth, H., Tschampel, M., Rehmann, I. Investigations into the environmental fate and effects of iopromide (ultravist), a widely used iodinated X-ray contrast medium. *Water Research*, 2002, 36, 266–274, doi:10.1016/S0043-1354(01)00241-X.

Sui, X., Sun, J., Wu, X., Li, H., Liu, J., He, Z. Predicting the volume of distribution of drugs in humans. *Current Drug Metabolism*, 2008, 9, 574–580.

Voutchkova, A.M., Ferris, L.A., Zimmerman, J.B., Anastas, P.T. Toward molecular design for hazard reduction—fundamental relationships between chemical properties and toxicity. *Tetrahedron*, 2010a, 66, 1031–1039, doi:https://doi.org/10.1016/j.tet.2009.11.002.

Voutchkova, A.M., Osimitz, T.G., Anastas, P.T. Toward a comprehensive molecular design framework for reduced hazard. *Chemical Reviews*, 2010b, 110, 5845–5882, doi:10.1021/cr9003105.

Williams, J.A., Hurst, S.I., Bauman, J., Jones, B.C., Hyland, R., Gibbs, J.P., Obach, R.S. Ball, S.E. Reaction phenotyping in drug discovery: moving forward with confidence? *Current Drug Metabolism*, 2003, 4, 527–534.

Zhang, J., Qian, J., Tong, J., Zhang, D., Hu, C. Toxic effects of cephalosporins with specific functional groups as indicated by zebrafish embryo toxicity testing. *Chemical Research in Toxicology*, 2013, 26, 1168–1181, doi:10.1021/tx400089y.

Chapter 6

Use of Green Chemistry for Extraction of Bioactive Compounds from Vegetal Sources

Adriana C. Flores-Gallegos, Ramsés Misael Reyes-Reyna,
Paloma Almanza-Tovanche,
Marisol Rodríguez-Duarte, Gerardo M. González,
Raúl Rodríguez-Herrera and J.A. Ascacio-Valdés*

Food Research Department, School of Chemistry,
Autonomous University of Coahuila, Saltillo, Coahuila, México.

INTRODUCTION

Vegetal Sources as Raw Material for the Extraction of Bioactive Compounds

Plants have two types of metabolism. The first is primary metabolism, which involves all the essential chemical processes for life. The next is secondary metabolism, whose function is producing and accumulating natural products, it also has diverse uses and applications such as medicines, insecticides, herbicides, perfumes, and colorings. The primary metabolitesare sugar, amino acids, fatty acids, nucleotides, polysaccharides, proteins, lipids, ARN and ADN, while in secondary metabolites, the natural pesticides, vitamins, pigments and bioactive compounds can be included (Ávalos and Pérez, 2009).

---

*For Correspondence: alberto_ascaciovaldes@uadec.edu.mx

A bioactive compound is an essential or non-essential compound that is found in nature and influences cellular activity, as well as in different physiological mechanisms (Biesalski et al., 2009). The National Institute of Cancer mentions that bioactive substances are able to provide and promote good health. These substances can be found in small quantities in plants and in some foods like fruits, vegetables, nuts, oils and grains (National Institute of Cancer). Although plants are a source of securing these types of molecules, their big diversity, the limited information that exists about their structure and the action mechanism of some phytochemicals, makes it difficult to take advantage of all their potential, since it is not possible to find all the bioactive compounds in all plants. Also, these compounds are synthesized in small quantities and not in a widespread form; that means, its production becomes restricted to certain parts of the plant or even to certain genera, families or species (Ávalos and Pérez, 2009).

The bioactive compounds have a very important value inside the cosmetic, pharmaceutical and food industries. In the past, a great number of these products were used for the treatment of different illnesses. For example, pomegranate leaves were used in Mexican traditional herbalist treatment for diarrhea or as an antimicrobial against different microorganisms. On the other hand, leaves of serrano pepper were used for the treatment of parts of the body affected by ulcers, abscesses or skin infections.

Terpenes are classified based on their number of units of isoprene: monoterpenes are terpenes with 10 C and contain two units of 5 C each one, sesquiterpenes have 15 C with three units, followed by those that have 20 C and four units 5 C, the diterpenes. On the other hand, triterpenes have 30 C, tetraterpenes have 40 C and finally, polyterpenes have above 8 units of isoprene (Ávalos and Pérez, 2009).

In the group of terpenes hormones, pigments, sterols, derivatives of sterols and essential oils are found. Many of these compounds have a high physiological and commercial value since they are used for their ability to generate aromas and fragrances in some foods and in the cosmetic industry; while other terpenes are used in medicine for their properties such as anti-carcinogenic, anti-ulcer, anti-microbial, etc. (Ávalos and Pérez, 2009). Limonene, for example, is the second most used terpene in the industry and can be obtained from lemon or another citrus husk.

On the other hand, a phenolic compound is an aromatic ring with a hydroxyl group from the synthesis of a phenol (Ávalos and Pérez, 2009). In this group polyphenols or phenylpropanoids are found. Polyphenols are molecules with antioxidant capacity that contain phenolic acids and flavonoids. In products of vegetal origin, a wide variety of phenolic acids such as caffeic, ferulic and chlorogenic acids present in fruits and seeds of coffee and soybeans, are found (Milner, 2004).

Flavonoids are polyphenolic compounds characterized by having a three ring structure with two aromatic centers and an oxygenated heterocycle in the central part. Within these x flavones, flavanones, catechins, and anthocyanins are found. In onion and lettuce quercetin (flavone) is found, while in citrus a flavonone called fisetin is found (Nijveldt et al., 2001).

According to its chemical structure, phenolic compounds are very varied, since there are simple molecules such as phenolic acids and very complex molecules such as tannins and lignin. Tannins are polymeric phenolic compounds that bind proteins, denaturing them. Its name derives from the practice of using vegetal extracts to convert skin into leather and these are divided into condensed tannins and hydrolyzable tannins (Ávalos and Pérez, 2009).

Condensed tannins are flavonoid polymers bound by C–C bonds, which cannot be hydrolyzed but oxidized by a strong acid to produce anthocyanidins. Hydrolyzable tannins are heterogeneous polymers, that contain phenolic acids and simple sugars; they are smaller in comparison with condensates and hydrolyze more easily (Ávalos and Pérez, 2009). Tannins can also be considered as toxins because of their ability to bind proteins. In plants, terpenes serve as protection against insects, herbivorous animals or high temperatures. As protection against animals, immature fruits contain tannins in the husk, so these compounds repel animal attacks since they generate bitter flavors. However, not all tannins have toxic effects, for example, tannins contained in red wine cause a beneficial effect on human health by blocking the formation of endothelin-1, a signal molecule that causes vasoconstriction (Ávalos and Pérez, 2009).

Glycosides owe their name to the glycosidic bond that is generated through condensation of a sugar molecule with a molecule that contains a hydroxyl group. Glycosides can be classified into three groups: saponins, cardiac glycosides and cyanogenic glycosides. However, it can also include a fourth group, the glucosinolates, because its structure is very similar to that of the glycosides (Ávalos and Pérez, 2009). Evaluated the presence of alkaloids, cyanogenic glycosides, and saponins in five of the most widely used medicinal plants in Ecuador: Artemisia absinthium, Cnidoscolus aconitifolius, Parthenium hysterophorus Linn, Piper carpunya Ruiz & Pav and Taraxacum officinale. These authors found that extracts from all plant leaves, were positive for saponins, and P. hysterophorus was the specie with the highest content, while the lowest content was found in P. carpunya. On the other hand, on quantification of cyanogenic glycosides, it was discovered that the plant that showed the highest concentration was C. aconitifolius and the one with the lowest concentration was A. absinthium.

Alkaloids are more than 15,000 secondary metabolites which share three characteristics:

1. They are soluble in water.
2. They contain at least one nitrogen atom in its molecule.
3. Exhibit biological activity.

Most of these compounds are in the form of heterocycles, although, some are non-cyclic, such as mescaline or colchicine. Alkaloids can be found in approximately 20% of vascular plants. Physiological and psychological responses are caused by alkaloids, because of its interaction with neurotransmitters (Ávalos and Pérez, 2009). If alkaloids are used in high doses, it may cause toxicity, on the contrary, if they are used in low or moderate doses may generate a therapeutic effect such as muscle relaxants, tranquilizers or analgesics. The classification of alkaloids is based on the rings that the molecule presents, some of these compounds are quinoline, isoquinoline, indole, tropane, quinolizidine, piperidine, purine and pyrrolizidine (Serrano et al., 2006).

Caffeine is an example of alkaloids and belongs to the xanthine group. More than 120,000 tons per year of this compound are consumed; within the food industry, caffeine is used during the formulation of carbonated drinks, bread and confectionery. Another use of caffeine is for the treatment of childhood apnea, acne and migraine. While in the pharmaceutical industry, caffeine is used in products such as analgesics and diuretics, to name just a few (Florez-Jaramillo and Barona-Cortés, 2016).

Bioactive Compounds Extraction

According to Ros et al. (2012), the vegetal material contains bioactive, aromatic and fatty compounds, vitamins, gelling materials, oils, flavors and aromas. With this background, it is interesting to transform by-products, residues and waste into products that generate economic and environmental benefits. There are several techniques for the extraction of chemical compounds from vegetal material. These techniques are commonly used to obtain substances or a mixture of them, so the best conditions for extractions can be found for every plant material (Parra et al., 2007). The objective of the different extraction methods is to separate the most important components from the remaining compounds. In general, these processes are based on four important issues for carrying out extraction and isolation of plant-derived assets, which are (Delgado, 2015):

- Nature of solvent.
- Amount of solvent.
- Temperature.
- Extraction time.

The combination of these factors will result in the best conditions for extracting the components (Azuola and Vargas-Aguilar, 2007), with the ultimate goal of extracting specific bioactive compounds from a material, increasing the concentrations of extracts, preparing them for the different analytical detection tests and producing a variable and reproducible method.

The authors mentioned that the process consists of a series of different stages or phases that must be carried out:

A. Pretreatment of biomass: which consists of processes such as drying, crushing or grinding or thermal treatments.

B. Processes of continued extraction with solvents (water, ethanol, methanol, etc.): this is where the methods and technologies are crossed to improve processes, combining them with different solvents to increase the extraction yields.

C. Purifications of extracts: by flocculation, centrifugation, treatment by resins, chromatography separation, etc.

D. Concentration and stabilization of products: the measures under which they will be stored.

E. Recovery of solvents: at this point, one of the green points in chemistry, the recovery and reintroduction of solvents as well as water to minimize consumption will be applied.

The most common classification for extraction processes of bioactive compounds from vegetal tissue is as follows (Sharapin, 2000):

A. Conventional extraction by solvents: based on selective transfers, which will depend on compounds solubility.

B. Steam drag extraction: the sample is heated in an aqueous medium and releases oils which, due to their minimum volatility, evaporates and is drawn to a condenser, freshening the mixture and finally, separating it.

C. Mechanical extraction: use pressing or screws in order to exert pressure on a material to extract oil.

D. Microwave-assisted extraction: is based on heating the raw material with magnetic waves for a short time, avoiding degradation of compounds.

E. Ultrasound-assisted extraction: mechanical vibrations are applied which create bubbles that explode producing high temperatures and pressures, favoring the release of compounds.

F. Extraction by enzymatic treatment: uses enzymes as membrane-breaking catalysts and releases the compounds of interest.

G. Solvents accelerated extraction: the material is brought to high pressures with solvents at high temperatures.

H. Supercritical fluids extraction: uses solvents with characteristics similar to a liquid and gaseous state, being able to penetrate cell pores.

Extraction Methods

Conventional extraction methods

There are two methods of extraction techniques, conventional (traditional) and unconventional (alternative) (Peredo-Luna et al., 2009). Among traditional techniques, steam distillation, maceration, extraction with solvents and Soxhlet, infusion or reflux can be mentioned (Azuola and Vargas Aguilar, 2007, Ballard et al., 2010). Conventional processes are useful for the isolation and identification of phenolic compounds. Successful extraction depends on type of solvent, nature and way in which extraction material are prepared, chemical structure of the compounds that the tissue contains, time and temperature of extraction, solid-liquid relation of the method used, as well as presence of substances that may interfere during the process (Bucić-Kojić et al., 2011). Extraction involves two physical phenomena: one is the diffusion through the cell wall and the other is the washing of cellular contents after the walls have been broken (Vinatoru, 2001).

Techniques such as maceration or infusion require high volumes of solvent, long extraction times and there is a risk of losing thermolabile compounds since these can be degraded during the process (Ballard et al., 2010; Wu et al., 2012). In the case of steam extraction, there are two immiscible phases, one organic and the other aqueous; therefore, each of them will have its own vapor pressure which corresponds to pure liquid at a specific temperature. This extraction technique is useful for obtaining essential oils; however, extraction times are long, the equipment is impractical and decantation is required to separate the water component, so there is a risk of losing part of the compounds (Peredo-Luna et al., 2009).

Extraction with solvents requires a dried and solid sample which is placed in contact with organic solvents such as alcohol, chloroform, among others (Peredo-Luna et al., 2009). With this technique, compounds can be separated according to their solubility (Azuola and Vargas Aguilar, 2007). In addition to solubilizing the essence of the compound, it is also possible to solubilize and extract substances such as fats and waxes, so that in the end an oleoresin or an impure extract can be obtained. Another disadvantage of this technique is that its use is not very feasible on an industrial scale since high volumes of solvents are required and some of these have restrictions on their use (Peredo-Luna et al., 2009).

Another conventional extraction technique is Soxhlet which is a type of glass material that is used to extract compounds of a lipid nature which are in a solid state, these compounds are extracted using a related solvent. The extraction equipment consists of an extractor, a bulb-type condenser, and a flask which functions acyclically, that is, in a first instance, solvent evaporates and rises to where the condenser is located, where it falls and then returns to the flask where the solvent is and thus separates the compounds (Azuola and Vargas Aguilar, 2007). The Soxhlet method allows to control the consumption of the solvent and generates high yields, at the laboratory level it is the most used method, in addition, extracts of good quality are obtained. However, its disadvantages include long extraction time (it can take more than 20 hours) and is not recommended for thermolabile compounds (Azuola and Vargas Aguilar, 2007; Peredo-Luna, et al., 2009; Ribeiro et al., 2009).

Non-conventional extraction methods

In recent years, alternative technologies have been used for the extraction of bioactive compounds from plant material in order to minimize damage to the environment and improve reaction yields. Alternative technologies, also known as "green techniques", seek to improve the efficiency of a traditional method of extraction using physical action on the environment, in addition to decreasing energy depletion and using unconventional solvents to guarantee the quality of the extracts (Ballard et al., 2010; Saini et al., 2014).

Some of these technologies are ultrasound-assisted, microwave extraction and supercritical fluid extraction. Ultrasound-Assisted Extraction, also known as UAE, is a rapid extraction method compared to many traditional methods; This is due to the fact that there is a large contact surface between solid and liquid phases, and that a reduction in particle size is generated. The ultrasound equipment is in the frequency region of 18 kHz to 100 MHz and can be divided into high-intensity ultrasound (20–100 kHz) and diagnostic ultrasound (1–10 MHz) (Peredo-Luna et al., 2009). High-intensity ultrasound causes acoustic cavitation which produces the rupture of cell walls, generates greater mass transfer and improves the efficiency of solvent penetration into plant cells and capillary tissue so that the release of active ingredients becomes easier.

Ultrasound-Assisted Extraction (UAE) offers many advantages, some of these are selectivity, high efficiency, productivity, good quality, low energy consumption, reduction in extraction time and solvents consumption, as well as, reduction of both chemical and physical risks. On the other hand, ultrasound-assisted extraction is friendly to the environment, has a low cost and a high level of automation, compared to conventional extraction techniques.

Another green technique is Microwave-Assisted Extraction (MAE), also called microwave extraction, which is a technique that combines extraction using microwave equipment and traditional solvents. The MAE has been used in a wide variety of vegetal materials as it offers rapid delivery of energy between the solvent and the vegetal matrix. Both the solvent and the vegetal matrix are heated homogeneously and migration of ions and a rotation of dipoles of the solvent are generated; since there is water within the plant matrix, it absorbs the energy generated by the microwave and causes cellular rupture due to internal overheating, which facilitates the absorption of the matrix's chemical products, and improves recovery of nutraceuticals (Azuola and Vargas Aguilar, 2007).

Microwave-assisted extraction (MAE) has many advantages, for example, reduction in extraction time and solvent volume, higher extraction rate and better products. Microwave equipment is simpler and cheaper than a supercritical fluid extraction equipment, and can be used for different materials because it does not limit polarity of the extract, making it an interesting alternative for compound extraction from vegetal materials (Ballard et al., 2010).

For its part, Supercritical Fluid Extraction (SFE) is a unitary operation in which the dissolving power of supercritical fluids is exploited under conditions of high temperature and high pressure (Peredo-Luna et al., 2009). A supercritical fluid is any substance that is at a temperature and pressure higher than its critical point, in addition, its density is close or superior to that of its critical density.

Supercritical fluids have the ability to diffuse through solids, gas and to dissolve liquid materials, thus generating a low viscosity solvent and increasing diffusion without increasing surface tension, which is why by using supercritical fluids it is possible to obtain solvent-free extracts (Azuola and Vargas Aguilar, 2007, Peredo-Luna et al., 2009). Other advantages granted by the SEF are: shorter extraction time, less solvent, higher yields and lower energy consumption; however, these advantages are overshadowed by the high investment required to assemble the equipment as well as the unwanted extraction of high molecular weight compounds (Peredo-Luna, et al., 2009). Carbon Dioxide (CO_2) is one of the most frequently used fluids mainly in the extraction of essential oils, pigments, carotenoids, antioxidants, antimicrobials, substances used as ingredients for food, medicines and perfumery products which can be obtained through spices, herbs and other biological materials.

Currently, the possibility of combining ultrasound technology with microwave technology is being explored, since the combination of these energies in a certain sense is complementary, if the material is treated with the two technologies it is more likely that greater number of compounds are extracted because when used simultaneously, a large number of synthetic reactions can be improved, as well as the extraction of natural products and preparation of samples for a chemical analysis.

Green Chemistry in the Extraction Processes

The fundamental purpose of green chemistry is to reduce and eliminate substances hazardous to the environment and health. According to the US Environmental Protection Agency (EPA), green chemistry is the use of chemistry for the prevention of contamination, in this context, the objective is the design of products or processes for the reduction and elimination of use or production of hazardous substances.

Green Chemistry also propounds 12 principles, which were originally proposed by Paul Anastas and John Warner in their book Green Chemistry, theory and practice in 1998, that expresses the application of tactics for implementation of procedures that will contribute to environmental sustainability (Pájaro and Olivero, 2011).

Currently, the intention is to design processes that prevent pollution and are safe for humans and the environment, which reduces the generation of toxic waste from various industrial activities. Therefore, one of the tools used by the chemical industry is the use of alternative techniques, incorporating natural processes such as photochemical synthesis, use of renewable materials, alternative reaction conditions, use of solvents that are harmless to the environment and use of catalysts which are easily recovered and reused (Clark, 2001, Li and Trost, 2008).

To obtain the components of interest in plant extracts, alternative techniques have been developed, also known as "green techniques" because they allow a greater amount of bioactive compounds to be recovered, using shorter extraction periods, increasing quality of extracts with lower costs of processing. Therefore, some of the techniques used are ultrasound-assisted and microwave-assisted extraction.

In addition to the previously mentioned benefits, ultrasound-assisted extraction offers high reproducibility in short periods, reducing solvent consumption, increasing yields and using a small amount of energy. It has been reported that this technique is more efficient than traditional extraction methods (Azuola and Vargas, 2007).

Another technique considered "green" is microwave assisted extraction; this technique is quick and high-quality compounds can be obtained since heat is transferred to the solvent, which allows penetration directly into the sample and instantly heating the solvent trapped between the sample pores. This technique reduces energy consumption, which reduces environmental pollution with shorter extraction times.

Rojas et al., 2014 evaluated the effect of water, acetone, ethanol and methanol at different concentrations using an ultrasonic bath during 2 minutes of sonication, for extraction of the total content of phenolic compounds and antioxidant capacity of Rubus glaucus Benth. Valadez et al., 2017 studied the polyphenolic compounds and antioxidant capacity of chipilin extracts (Crotalaria maypurensis HBK) using mixtures of 100% methanol, methanol-water 75, 50, 25%, and 0% water, obtaining extracts by microwave assisted and ultrasound in extraction times of 30, 60 and 120 minutes.

IMPORTANCE OF THE SOLVENT

In addition to alternative technologies, there is also the concern to implement Green Chemistry in terms of solvents, which involves avoiding solvents such as chloroform and methanol, among others. It is also known that the choice of extraction solvent affects yield and the number of metabolites detectable in the metabolomic analysis (Ribeiro et al., 2009).

Due to its polarity, solvents such as chloroform/methanol, are able to dissolve chlorophyll pigments, as well as lipids. Chloroform and methanol are toxic and flammable, which affects the environment (Kumar et al., 2017).

On the other hand, there are pressurized liquids, mainly used for the extraction of phenolic compounds, lignans and carotenoids. This type of extraction uses liquid solvents at high temperatures, above its boiling point, and under controlled pressure, improving solubility and mass transfer (Rodríguez et al., 2016). Sometimes ethanol, water or CO2 are used for the extraction of plant nutraceuticals (Saini et al., 2014).

In the field of sustainable alternatives, lipid extractions can be done with bio-solvents, which are efficient and eco-friendly. Examples of this are terpenes, extracted from citrus plants, which can be used as solvents to extract algal lipids and their quality is similar to that extracted with hexane. There are other types of green solvents called bio-derived solvents, which are present in agriculture such as ethyl lactate and methyl soyate, from corn, citrus and soybeans (Kumar et al., 2017).

To extract alkaloids, low polarity solvents such as toluene, dichloromethane, chloroform or mixtures of them can be used; in the earlier years, benzene (a carcinogen) and trichloroethylene (production interrupted after 1995) were used but both were discontinued. Solvents of low polarity dissolve alkaloids in the form of free bases, but not in the form of salts. These compounds are found in plants together with organic acids, so prior to dissolution of alkaloids, a previous treatment with alkaline solutions is necessary to release alkaloids from their salts, and includes a method where vegetal material is moistened with sodium carbonate solution, that involves extraction with organic solvent, extraction with sulfuric acid, alkalization with ammonium hydroxide until pH 9, extraction with methane or chloroform dichloride and evaporation in vacuum (Sharapin et al., 2000).

Another alternative is supercritical fluids (substances with intermediate physicochemical properties between gases and liquids at temperature and pressure above their critical points) (Rodríguez et al., 2016). However, the equipment necessary for this task requires a high investment (Périno et al., 2016).

For compounds of medium polarity, it is convenient to use co-solvents such as ethanol in a concentration of 1–10%, which is widely used for its availability in high purity, low cost, complete biodegradability and generally recognized as a safe solvent in the food industry. This solvent, above all, favors the extraction of polyphenols and flavonols (Rodríguez et al., 2016).

Currently, there is a need to consider non-toxic, sustainable and highly efficient solvents to carry out extractions, so that, it does not interfere with the compounds of interest. When using organic solvents such as methanol, chloroform, hexane, among others, there is a risk that residuals remain present in the extract.

The Food and Drug Administration (FDA) lists in a table called Q3C, three classes of solvents and their respective recommended concentrations for use in food, where class 1 includes solvents such as benzene and carbon tetrachloride, which are unacceptably toxic and harmful to the environment, so its use is exclusive for drugs with a significant therapeutic use. Class 2 includes solvents such as acetonitrile, chloroform, cyclohexane, dichloromethane, methanol, hexane, toluene, among others, which recommends that its use is limited in pharmaceutical products due to its inherent toxicity. Class 3 considers solvents that are less toxic or of low risk to human health where the amounts of these residual solvents are considered to be 50 mg per day or less, although larger quantities would be acceptable, among these are acetic acid, acetone, ethanol, formic acid, among others (CDER, 2012).

REFERENCES

Anastas, P., Warner, J.C. Green Chemistry: Theory and Practice. Oxford University Press. 1998, p. 135.

Ávalos, A., Pérez, E. Metabolismo secundario de plantas. *Reduca Biología Serie Fisiología Vegetal*, 2009, 2(3), 119–145.

Azuola, R., Vargas-Aguilar, P. Extracción de sustancias asistida por ultrasonido (EUA). *Revista Tecnología en Marcha*, 2007, 20(4).

Ballard, T.S., Mallikarjunan, P., Zhou, K., O'Keefe, S. Microwave-assisted extraction of phenolic antioxidant compounds from peanut skins. *Food Chemistry*, 2010, 120(4), 1185–1192.

Biesalski, H.-K., Dragsted, L.O., Elmadfa, I., Grossklaus, R., Müller, M., Schrenk, D., Walter, P., Weber, P. Bioactive compounds: Definition and assessment of activity. *Nutrition*, 2009, 25(11–12), 1202–1205.

Bucić-Kojić, A., Planinić, M., Tomas, S., Jokić, S., Mujić, I., Bilić, M, Velić, D. Effect of extraction conditions on the extractability of phenolic compounds from lyophilised fig fruits (*Ficus Carica* L.). *Polish Journal of Food and Nutrition Sciences. Versita*, 2011, 61(3), 195–199.

CDER. U.S. Department of Health and Human Services Food and Drug Administration Center for Drug Evaluation and Research Center for Biologics Evaluation and Research (CBER). Guide of Industry. Q3C tables and list, 2012. [https://www.fda.gov/downloads/drugs/guidances/ucm073395.pdf].

Clark, J. Catalysis for green chemistry. *Pure and Applied Chemistry*, 2001, 73, 103–111.

Delgado, J.O. Aplicación del ultrasonido en la industria de los alimentos. *Publicaciones e Investigación*, 2015, 6, 141–152.

Flórez Jaramillo, L.Á., Barona Cortés, E. Diversity of reptiles associated withthree contrasting areas in a tropical dry forest (La Dorada and Victoria, Caldas). *Revista de Ciencias*, 2016, 20, 109–123. Available at: http://www.ub.edu.ar/revistas_digitales/Cienpara ser utilizados en las indus- trias de alimentos tipo botana (snacks), pastelería y confiteríacias/A2Num5/articulos.htm

Kumar, J., Kumar, V., Dash, A., Scholz, P., Banerjee, R. Sustainable green solvents and techniques for lipid extraction from microalgae: A review. *Algal Research*, 2017, 21, 138–147.

Li, C., Trost, B. Green chemistryforchemicalsynthesis. *Proceedings of the National Academy of Sciences*, 2008, 105, 13179–13202.

Milner, J.A., Molecular targets for bioactive food components. *The Journal of Nutrition*, 2004, 134(3), 2492S–2498S.

Nijveldt, R.J., van Nood, E., van Hoorn, D.E., Boelens, P.G., van Norren, K., van Leeuwen, P.A. Flavonoids: A review of probable mechanisms of action and potential applications. *The American Journal of Clinical Nutrition*, 2001, 74(4), 418–25. Available at: http://www.ncbi.nlm.nih.gov/pubmed/11566638 (Accessed: 9 November 2017).

Pájaro Castro, N.P., Olivero Verbel, J.T. Química verde: Un nuevo reto. *Ciencia e Ingeniería Neogranadina*, 2011, 21(2), 169–182.

Parra Henao, G., Garcia Pajón, C.M., Cotes Torres, J.M. Actividad insecticida de extractos vegetales sobre Rhodnius prolixus y Rhodnius pallescens (Hemiptera: Reduviidae). *Boletín de Malariología y Salud Ambiental*, 2007, 47(1), 125–137.

Peredo-Luna, H.A., Palou-García, E., López-Malo, A. Aceites esenciales: Métodos de extracción, *Temas Selectos de Ingeniería de Alimentos*, 2009, 3, 24–32.

Périno, S., Pierson, J.T., Ruiz, K., Cravotto, G., Chemat, F. Laboratory to pilot scale: Microwave extraction for polyphenols lettuce. *Food Chemistry*, 2016, 204, 108–114.

Ribeiro, E.B., Reis, R., Alfonso, S., Scarminio, I.S. Enhanced extraction yields and mobile phase separations by solvent mixtures for the analysis of metabolites in Annona muricata L. leaves. *Journal of Separation Science*, 2009, 32, 4176–4185.

Rodríguez, C., Mendiola, J., Quirantes, R., Ibáñez, E., Segura, A. Green downstream processing using supercritical carbon dioxide, CO2-expanded ethanol and pressurized hot water extractions for recovering bioactive compounds from Moringa oleifera leaves. *The Journal of Supercritical Fluids*, 2016, 116, 90–100.

Rojas, P., Martínez, J., Stashenko, E. Contenido de compuestos fenólicos y capacidad antioxidante de extractos de mora (Rubus glaucus Benth) obtenido bajo diferentes condiciones. *VITAE, Revista de la Facultad de Química Farmacéutica*, 2014, 21(3), 218–227.

Ros, M., Pascual, J.A., Ayuso, M., Morales, A.B., Miralles, J.R., Solera, C. Salidas valorizables de los residuos y subproductos orgánicos de la industria de los transformados de frutas y hortalizas: proyecto Life Agrowaste. CEBAS-CSIC, CTC y AGRUPAL. España. 2012.

Saini, R.K., Shetty, N.P., Giridhar, P. GC-FID/MS analysis of fatty acids in Indian cultivars of Moringa oleifera: Potential sources of PUFA. *Journal of the American Oil Chemists' Society*, 2014, 91, 1029–1034.

Serrano, M.E.D., López, M.L., Espuñes, T.D.R.S. Componentes bioactivos de alimentos funcionales de origen vegetal. *Revista Mexicana de Ciencias Farmaceuticas*, 2006, 37(4), 58–68.

Sharapin, N., Machado, L., Pinzón, R. (ed.). Fundamentos de tecnología de productos fitoterapéuticos. Primera edición. Programa Iberoamericano de ciencia y tecnología para el desarrollo, 2000.

Valadez, A., López, E., García R., Ruiz, F.L. Comparación de dos técnicas de extracción de fenólicos totales y capacidad antioxidante a partir de Chipilín (Crotalariamaypurensis H.B.K.). *Investigación y Desarrollo en Ciencia y Tecnología de Alimentos*, 2017, 2, 481–487.

Vinatoru, M. An overview of the ultrasonically assisted extraction of bioactive principles from herbs. *Ultrasonics Sonochemistry*, 2001, 8(3), 303–313.

Wu, T., Yan, J., Liu, R., Marcone, M.F., Aisa, H.A., Tsao, R. Optimization of microwave-assisted extraction of phenolics from potato and its downstream waste using orthogonal array design. *Food Chemistry*, 2012, 133(4), 1292–1298.

7

Design for Energy Efficiency

Aidé Sáenz-Galindo*, José Juan Cedillo-Portilo, Karina G., Espinoza-Cavazos, Patricia Adriana de Léon-Martínez and Adali Oliva Castañeda-Facio

Facultad de Ciencias Químicas.
Universidad Autónoma de Coahuila. Saltillo, Coahuila, Mexico. CP. 25290.

INTRODUCTION

In recent years the use of energy has increased to meet the needs of the high consumption required to develop activities of today's life which has led to excessive use of natural resources in addition to the excessive generation of toxic waste, with a consequence of environmental deterioration that has led to great climatic changes. Therefore it was decided to look for alternatives in the industry that are friendly to the environment (Ivankovie et al., 2017).

Currently green chemistry is a viable alternative applicable in countless chemical processes, which have important applications in the design, synthesis and application of new and improved chemical compounds, at the laboratory level and pilot plant up to the industrial level. For this reason, it is very important to try to implement the 12 principles of green chemistry, in the different processes, Principle 6, refers to the design of efficient energies, such as the use of microwave, ultrasound, infrared or its combination, in order to reduce environmental impact, by minimizing the waste of natural resources.

Sonochemistry is an area of great interest in green chemistry, which is based on sound waves that occur in chemical reactions, this area is considered an excellent

*For Correspondence: aidesaenz@uadec.edu.mx

option for industrial applications in various fields, it conserves energy minimizing chemical waste (Mason et al., 2002). Thus leading to a great interest in developing new equipment that generate this type of energy. This area includes methods such as ultrasound, which is a not destructive technique based on the propagation of sound waves through a medium with a controlled speed which have high frequencies and short conda lengths (Santos et al., 2005), on the other hand a microwave and infrared that improves the performance of chemical reactions by being friendly to the environment. It is also considered a very economical and easy technique to use, with varied frequencies and powers.

Microwave is an interesting method, in this energy is electromagnetic waves with wavelengths from 1mm to 1m and frequencies approximately between 0.3GHz and 300 GHz, respectively. It is a method with numerous applications in different areas, besides being considered green, friendly to the environment, effective, economical, fast, etc., as its main characteristic includes reaching high temperatures in the shortest time of the process.

The objective of this chapter is to know the techniques that are based on principle 6 of green chemistry, while being sonochemistry and microwave energy which is the main area of interest for its study and development.

ULTRASONIC

Currently the use of the ultrasonic radiation has been shown to be an excellent method for dispersion and mixing of particles of different sizes as well, recently it has also been used for the preparation of countless compounds.

The phenomenon of cavitation is the most important feature in sonochemistry, it is defined as the generation, growth and collapse of microbubbles or cavities under ultrasound irradiation that generates a large amount of energy locally, which covers the frequency range of 20 kHz and 5 MHz. These effect the increase of temperature and pressure (Asgharzadehahmadi et al., 2016), It is important to emphasize that within the philosophy of green chemistry, sonochemistry is considered as an alternative energy (Figure 7.1).

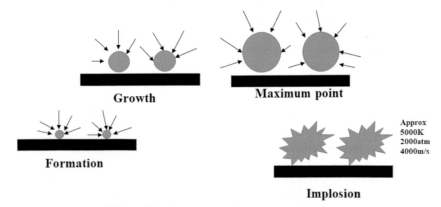

Figure 7.1   Process Cavitation. Sonochemistry.

The use of ultrasound has focused on developing processes in different and numerous applications of physical and chemical processes based on the effects of cavitation, the application of this type of energy can improve reactions when carried out in shorter times, obtaining a better performance as well as the reduction of large quantities of toxic or expensive reagents (Sancheti et al., 2017). It is considered a green method friendly to the environment, reducing the time of chemical processes, increasing yield and resulting in low costs of the processes.

The extraction of natural compounds "Phytochemical" is an area that is currently attractive, using renewable raw materials, the use of ultrasound is an excellent alternative for obtaining compounds of this type, decreasing extraction times and increasing extraction yield of natural compounds, such as phytochemicals: polyphenols, organic acids, tannins, terpenes, etc.

In the materials area, the use of ultrasonic energy presents numerous applications, in the context of dispersion and homogenization of particles and nanoparticles in different matrices for obtaining nanocomposites.

The implementation of ultrasonics in the organic synthesis has been potentialized because this type of alternative energy can be considered as a green method for reactions more friendly to the environment, which is due to reaction times that are shorter or higher yields of the desired product as well as less severe reaction conditions (Sancheti et al., 2017).

Acar Bozkurt in 2017 reported the study of synthesis of the nanocomposites Ag/graphene, using sodium citrate as a reducer assisted the synthesis by the ultrasonic method, found that graphene was oxidized and silver ions to silver nanoparticles white sodium citrate which contributed green attributes, also reported the size of Ag nanoparticles with an averages size of 20 nm and a spherical morphology using the TEM and SEM images also indicated that the green sonochemical synthesis could become a promising method for the preparation of graphene-based nanocomposites (Acar Bozkurt, 2017).

Price in 2018, reported surface modification of multi-walled carbon nanotubes with the ultrasound method in dilute acids H_2SO_4 and HNO_3, these nanomaterials were incorporated into a natural polymer matrix, chitosan, observed a 10-fold increase in tensile strength (Price et al., 2018).

Bibi et al., reported in 2018, that in the study the preparation assisted by ultrasound of nancomposite of chitosan whit Ag nanoparticle and carbon nanotubes, the ultrasound is used for dispersion and crosslinking of the nancomposites finding that the Ag nanoparticles as well the carbon nanotubes increased the mechanical strength and the resistence to rupture, and also reported that the use of ultrasound also helped inobtaining a good dispersion of the nanonstruture which was demonstrated by microscopy TEM, maintaining the properties of Ag nanoparticle and carbon nanotubes (Bibi et al., 2018).

Abdolmaleki in 2018 reported the preparation of nanocomposites of multi-walled carbon nanotubes functiolalized with ribolavine and its integration in polyvinyl chloride (PVC), used green methodologies such as microwaves and ultrasound to obtain nanocomposites, reporting some improvements in dispersion and mechanical and chemical properties (Abdolmaleki et al., 2018).

Mallakpour reported in 2018, that the surface modificaction of carbon nanotubes with a big organic molecule such as gluocose, and integration in a natural polymer of starch to obtain films, this process of integration was assisted by ultrasonics to obtain glucosated nanocomposites, finding a good modification and excellent integration of the carbon nanotubes reduced the natural polymeric in the nancomposites through the C–O bond, which was demostrated with different characterization techniques, highlighting the solubility test in polar solvents, in addition the authors reported that the sonochemcial method is economical, fast, effective and the ecological technique for obtaining nanocomposites polymeric, (Mallakpour et al., 2018).

Nanotechnology is an area of great interest these days , in here diverse carbon-based nanoparticles, such as carbon nanotubes: single wall carbon nanotubes (SWCNT) and multiple wall carbon nanotubes or graphene, metal nanoparticle are studied, as different metal oxides. Recently Andrade-Guel et al., reported the study of surface modification of graphene nanoplatelets by organic acids and ultrasonic radiation, finding that this type of nanomaterials were tested as adsorbent material for uremic toxins using the equilibrium isotherms where the adsorption isotherm of urea was adjusted for the Langmuir model. From the solution, 75% of uremic toxins were removed and absorbed by the modified nanoplatelets (Andrade-Guel et al., 2019).

In 2018, Zhao et al. studied the rapid impregnation assisted by ultrasonic energy of thermoset and thermoplastic polymers, which include epoxy resins without solvents, polydimethylsiloxane resins and polypropylene in multiple wall carbon nanotubes (MWCNT), this type of impregnation process is complicated by conventional methods, for this reason it was studied with alterative methods, with the objective of obtaining MWCNT impregnated with thermoplastic and thermoset polymers, is to obtain macroscopic sheets similar to paper (also called "nanopapers") through a dispersion filtration method and then impregnate the mesoporous nanopapers with polymeric resins using an ultrasonic infiltration method to form a prepreg or nanocomposite, the nanocomposites with MWCNT presented excellent results of mechanical strength, sand erosion and scratch resistance, electrical and thermal properties and good processing flexibility (Zhao et al., 2018).

MICROWAVE

Microwave radiation is a part of the electromagnetic spectrum in a wavelength range from 1m to 1mm (between infrared and radio waves), which corresponds to a frequency in the range of 0.3 to 300 GHz respectively. Microwave heating occurs when a material has the ability to absorb high frequency electromagnetic energy (microwaves) and convert it to heat (Jacob et al., 1995), (Prado-Gonjal and Moran, 2011). This phenomenon can occur in two ways: i) Polarization of dipoles; which occurs when the dipoles of the material are aligned or oriented with the electric field of the microwaves; this movement causes friction (kinetic energy) and is transferred to the medium as thermal energy. ii) Ionic conduction; the ions oscillate under the influence of the electric force of microwaves so that the movement of the ions causes the collision between atoms and molecules, which generates an increase in the temperature of the material (Ebner et al., 2011), (Chandrasekaran et al., 2012).

Microwave irradiation is a very useful method in many homes and in industrial applications with a series of advantages such as i) non-contact heating, ii) rapid heating, iii) and specific heating (Wiesbrock et al., 2004). In addition, it is considered a green technique, since it is possible to work with low boiling point solvents (decreasing toxic residues), likewise with the rapid and homogeneous heating of the molecules the reaction rate is reduced, decreasing the secondary reactions. High yields and cleaner products are obtained, as well as reproducible results (Hoogenboom and Schubert, 2007) (Figure 7.2).

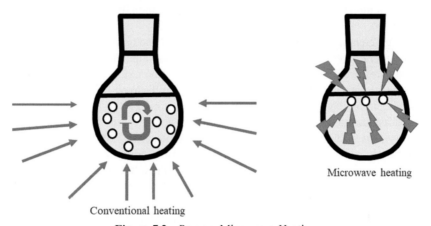

Conventional heating

Microwave heating

Figure 7.2   Process Microwave Heating.

Due of its advantages, it is an interesting topic for the scientific community in the field of organic synthesis, nanocomposites and nanoparticle surface modification. In organic synthesis, several works have been carried out in reactions of free radicals, controlled radical, living and by ring opening polymerizations (Zhang and Schubert, 2004; Pang et al., 2006).

Zhang et al. studied the polymerization of methyl methacrylate (MMA) by the microwave-assisted atom transfer polymerization (ATRP) technique. They found that the main advantage of microwave-assisted reactions is the significant increase in polymerization rates, even in ATRP systems characterized by the need for long reaction times to obtain high molecular weights. In addition to controlled reactions, a linear increase in Mn (molecular weight) and Low Polydispersity Indexes (PDI) are obtained (Zhang and Schubert, 2004).

On the other hand, Liao et al. studied the ε-caprolactone microwave-assisted ring-opening polymerization, comparing the heating methods (conventional and microwave), finding an increase in the reaction rate when using microwave heating, and therefore the reaction time decreased (Liao et al., 2002).

Frederick et al. studied the microwave-assisted synthesis of PET, finding an increase in the polymerization rate, further demonstrating that microwave energy could induce an increase in the molecular weight of PET; due to the diffusion speeds that occur in the system when using the microwave irradiation (Frederick et al., 2005).

In polymers based on nanocomposites, the nanoparticles (NPs) are generated in a medium (solvent) or *in situ* in the polymer or its monomers (Akamatsu et al., 2003), (Zhang et al., 2003).

Wade and Burdick developed a method for the preparation of nanocomposites *in situ*, under microwave radiation, in which they synthesized silver NPs in the presence of acrylate monomers, subsequently the photo-polymerization of the excess monomer was carried out. Finding good dispersion and distribution of NPs in the polymer matrix, in addition to a new method for obtaining nanocomposites (Wade and Burdick, 2014).

Surface modification of NPs is possible by adding a solvent with functional groups or a monomer in order to coat the surface of the NPs (Moongraksathun and Chen, 2016).

The given modified NPs can be easily dispersed and homogeneously distributed in a polymeric matrix to produce a functional nanocomposite (Ebner et al., 2011).

With this premise, Rodríguez et al. modified zinc oxide nanoparticles (nZnO) superficially with Polyacid Lactic Acid (PLA), synthesized by microwaves. From this study, a possible grafting mechanism, between nZnO and PLA were proposed, establishing the existence of different reactions that explain the obtaining of the hybrid system (nZnO-graft-PLA) (Rodríguez-Tobías et al., 2015).

Similarly, Zhu et al. obtained nanocomposites from the polymerization of acrylamide and the simultaneous reduction of metal salts (Ag, Pt, and Cu), using ethylene glycol (EG) as a solvent and reducing agent, through an *insitu* synthesis assisted by microwaves (Zhu and Zhu, 2006). Finding a homogeneous dispersion of the metallic NPs in the poly (acrylamide) matrix (PAM). In this same sense, Li et al. synthesized nanocomposites of cellulose and nAg using microwave radiation, reducing silver nitrate in Ethylene Glycol (EG). Finding that the nAg are homogeneously dispersed in the cellulose matrix and having a good antimicrobial activity (Li et al., 2011).

In the area of materials, the use of microwave energy is an area of great interest for obtaining polymeric nanocomposite, in 2018, Gonzalez et al. reported the study of obtaining nanocomposite Nylon-6-graphene, by microwave-assisted polymerization, using a heterogeneous multicomponent system, they elucidated four distinct steps that take place during the process *in situ* MAP of Nylon 6 from ε-caprolactam and 6-aminocaproic in the presence of graphite oxide and was possible to increase the molcecular weights of the polymeric nanocomposite Nylon-6-graphene, (Gonzalez et al., 2018).

Conclusion

Energetic consumption is a topic of great interest, at both the academic and industrial level. The philosophy of green chemistry is to design energy efficiencies, to minimize the energy requirements for chemical processes, which will be evaluated for their environmental and economical impact, as well as to reduce them to the maximun, trying to carry out the synthesis methods at temperature and ambient pressure. Thus for Earth the use of alternative energy sources, such as the use of microwaves and ultrasounds, are the main advantages which reduce the time of the different chemical processes in diverse areas.

Acknowledgments

This project was financially supported by Ciencia Basica A1-S-44977.

REFERENCES

Andrade-Guel. M., Cabello-Alvarado, C., Cruz-Delgado, V.J., Bartolo-Perez, P., De León-Martínez, P.A., Sáenz-Galindo, A., Cadenas-Pliego, G., Ávila-Orta, C.A. Surface Modification of Graphene Nanoplatelets by Organic Acids and Ultrasonic Radiation for Enhance Uremic Toxins Adsorption. *Materials,* 2019, 12, 715–730.

Abdolmaleki, A., Mallakpour, S., Azimi, F. Microwave and ultrasound-assisted synthesis of poly(vinyl chloride)/riboflavin modified MWCNTs: Examination of thermal, mechanical and morphology properties. *Ultrasonics Sonochemistry*, 2018, 41, 27–36.

Acar Bozkurt, P. Sonochemical green synthesis of Ag/graphene nanocomposite. *Ultrasonics Sonochemistry*, 2017, 35, 397–404.

Akamatsu, K., Ikeda, S., Nawafune, H. Site-selective direct silver metallization on surface-modified polyimide layers. *Langmuir*, 2003, 19(24), 10366–10371.

Asgharzadehahmadi, S., Abdul Raman, A.A., Parthasarathy, R., Sajjadi, B. Sonochemical reactors: Review on features, advantages and limitations. *Renewable and Sustainable Energy Reviews*, 2016, 63, 302–314.

Bibi, S., Jamil, A., Yasin, T., Rafiq, M.A., Nawaz, M., Price, G.J. Ultrasound promoted synthesis and properties of chitosan nanocomposites containing carbon nanotubes and silver nanoparticles. *European Polymer Journal*, 2018, 105, 297–303.

Chandrasekaran S., Ramanathan, S., Basak, T. Microwave material processing—A review. *American Institute of Chemical Engineers*, 2012, 58(2), 405–410.

Ebner, C., Bodner, T., Stelzer, F., Wiesbrock, F. One decade of microwave-assisted polymerizations: Quo vadis? *Macromolecular Rapid Communications*, 2011, 32(3), 254–288.

Frederick, R., Eirich, J.E., Mark, B.E. Science and Technology of Rubber, 3rd Ed. San diego, California; Elsevier; 2005.

Gonzalez, M., Hernandez, H.E., Fernandez, T.S., Ledezama, R.R., Saenz-Galindo, A., Cadenas, P.G., Avila-Orta, C.A., Ziolo, F.R. Exfoliation, reduction, hybridization and polymerization mechanisms in one-step microwave-assist synthesis of nanocomposite nylon-6/Graphene. *Polymer*, 2018, 146, 73–81.

Hoogenboom, R., Schubert, U.Ş, Microwave-assisted polymer synthesis: Recent developments In a rapidly expanding field of research. *Macromolecular Rapid Communications*, 2007, 28(4), 368–386.

Ivankovie, A., Dronjié, A., Bevanda, A., Talié, S. Review of 12 principles of green chemistry in practice. *Science Publishing Group*, 2017, 6(3), 39–48.

Jacob, J., Chia, L.H.L., Boey, F.Y.C. Thermal and non-thermal interaction of microwave radiation with materials. *Journal of Materials Science*, 1995, 30(21), 5321–5327.

Li, S-M, Jia, N., Ma, M-G., Zhang, Z., Liu, Q-H., Sun, R-C. Cellulose-Silver nanocomposites: Microwave-assisted synthesis, characterization, their thermal stability, and antimicrobial property. *Carbohydrate Polymer*, 2011, 86, 441–447.

Liao, L.Q., Liu, L.J., Zhang, C., He, F., Zhuo, R.X., Wan, K. Microwave-assisted ring-opening polymerization of ε-caprolactone. *Journal of Polymer Science. Part A. Polymer Chemistry*, 2002, 40(11), 1749–1755.

Mallakpour, S., Khodadadzadeh, L. Ultrasonic-assisted fabrication of starch/MWCNT-glucose nanocomposites for drug delivery. *Ultrasonics Sonochemistry*, 2018, 40, 402–409.

Mason, T.J., Cintas, P. Sonochemistry. *In*: Clark, J., MacQuarrie, D. (eds), Handbook of Green Chemisty and Technology. Oxford: Blackwell Science; 2002; pp. 372–396.

Moongraksathum, B., Chen, P.H.Y. Photocatalytic activity of ascorbic acid-modified TiO_2 sol prepared by the peroxo sol–gel method. *Journal of Sol-Gel Science and Technology*, 2016, 78(3), 647–659.

Pang, K., Kotek, R., Tonelli, A. Review of conventional and novel polymerization processes for polyesters. *Progress in Polymer Science*, 2006, 31(11), 1009–1037.

Prado-Gonjal, J., Morán, E. Síntesis asistida por microondas de sólidos inorgánicos. *Anales de Química*, 2011, 107, 129–136.

Price, G.J., Nawaz, M., Yasin, T., Bibi, S. Sonochemical modification of carbon nanotubes for enhanced nanocomposite performance. *Ultrasonics Sonochemistry*, 2018, 40, 123–130.

Rodríguez-Tobías, H., Morales, G., Olivas, A., Grande, D. One-Pot Formation of ZnO-*graft*-Poly(D,L-Lactide) hybrid systems via microwave-assisted polymerization of D,L-Lactide in the presence of ZnO nanoparticles. *Macromolecular Chemistry and Physics*, 2015, 216, 1629–1637.

Sancheti, S.V., Gogate, P.R. A review of engineering aspects of intensification of chemical synthesis using ultrasound. *Ultrasonics Sonochemistry*, 2017, 36, 527–543.

Santos, E., Cancino, V., Yenque, D., Ramirez, D., Palomino, P. El ultrasonido y su aplicación. *Revista de la facultad de ingeniería industrial*, 2005, 1(8), 25–28.

Wade, R.J., Burdick, J.A. Advances in nanofibrous scaffolds for biomedical applications: From electrospinning to self-assembly. *Nano Today*, 2014, 9(6), 722–742.

Wiesbrock, F., Hoogenboom, R., Schubert, U.S. Microwave-assisted polymer synthesis: State-of-the-art and future perspectives. *Macromolecular Rapid Communications*, 2004, 25(20), 1739–1764.

Zhang, H., Cui, Z., Wang, Y., Zhang, K., Ji, X., Lü, C., Yang, B., Gao, M. From water-soluble CdTe nanocrystals to fluorescent nanocrystal-polymer transparent composites using polymerizable surfactants. *Advance Materials*, 2003, 15 (10), 777–780.

Zhang, H., Schubert, U.S. Monomode microwave-assisted atom transfer radical polymerization. *Macromolecular Rapid Communications*, 2004, 25(13), 1225–1230.

Zhao,Y., Duarte, C.E., Zhang, D., Sun, J., Kuang, T., Yang, W., Lertola, M.J., Benatar, A., Castro J.M., Lee, L.J. Ultrasonic processing of MWCNT nanopaper reinforced polymeric nanocomposites. *Polymer*, 2018, 156, 85–94.

Zhu, J., Zhu, Y. (2006). Microwave-assisted one-step synthesis of polyacrylamide–Metal (M = Ag, Pt, Cu) nanocomposites in ethylene glycol. *Journal of Physical Chemistry B*, 2006, 110, 8593–8597.

Chapter 8

Section A:
Use of Renewable Feedstocks
The recovery of high value molecules from waste of renewable feedstocks: Soybean Hull

**Paola Camiscia, Nadia Woitovich Valetti
and Guillermo Picó***

Instituto de Procesos Biotecnológicos y Químicos (CONICET - Rosario)
Universidad Nacional de Rosario, Suipacha 570 (S2002RLK) Rosario, Argentina.

INTRODUCTION

Biomass is a word which refers to any material derived from living organisms, animals, plants, microorganisms, algae, etc. Biomass production in Latin America has led to great amounts of crops such as corn, wheat, barley rice, coffee, etc. The processing of renewable biomass from agriculture produces millions of tons of waste every year and they are discarded into the environment, producing a negative impact on it. At the same time, these wastes are discarded near urban areas creating health problems for the population. Due to the great biodiversity of crops distributed in Latin America, the types of waste found are very different, with different compositions, producing a variety of environmental problems, which implicates different kinds of treatments to dispose of these wastes.

Large crops produce the greatest amounts of waste, especially those that need to process the entire plant to recover only a small part of it, which will be used as food or raw material to produce food. Among the most abundant crops are:

*For Correspondence: pico@iprobyq-conicet.gob.ar

corn (Mexico, Argentine, Brazil), wheat (Argentine), soybean (Brazil, Argentine), sugarcane (Brazil, Mexico, Argentine, Colombia) and coffee (Colombia, Brazil). These biomasses undergo an earlier processing before obtaining the final product. This process consists on eliminating a membrane, husk or mesocarp that covers the seed, it acquires different names according to the cereal it comes from, but the term husk or hull is generally preferred. The waste produced by the processing of massive crops represents a variable percentage of the total mass of the biomass, which depends on the type of crop. In the case of corn, the waste produced represents between 40 to 60% of its total mass but for most of the grains it is only 2 to 5%, while for soybean it represents less than 1–2% of its total mass (Ferrer et al., 2016). Intensive crop production generates thousands or millions of plant biomass residues annually, which are generally not harvested and discarded into the environment. Table 8.1 shows the major crops waste estimated per year in America.

Table 8.1   Major crops waste in America

Crop	**Millon tons/year***	**% *Waste***
Soybean	293	1–2
Corn	547	50–60
Wheat	126	≈15
Coffee	5	≈65
Rice	36	20–25
Sugar cane	104	11–16

Source: FAO (2018)

The use of biomass for projects other than feedstock, such as the production of high-performance materials, is a potential commercial application that would unlock a generation of value-added products from agro-industrial commodities. Since soy bean (*Glycine max*) is the second most produced crop in America (USA, Brazil and Argentina are the main producers with around 230 million ton in 2018), it is interesting to analyze the waste produced by the industrialization of this crop and its possible uses. A review is summarized here with knowledge on the composition, application and possible conversions of the soybean hulls since it presents a great number of high value molecules with a wide application in different biotechnological and medical fields (Benkő et al., 2007; Maran et al., 2013; Phinichka and Kaenthong, 2018; Preece et al., 2017). Figure 8.1 summarizes the various applications of this waste.

The first step in processing soybean, for oil and meal production, is the removal of its hulls. Dehulling is accomplished on an industrial scale by the following sequence: warming/drying to cause some shrinkage of the seed away from the hulls, gentle rolling to crack the hulls without producing small particles of it and air separation of the bean from the hull fragments by counter-current flow. Soybean hulls are light, flaky and bulky, and it requires special considerations when handling it. For example, cattle feeders are at a disadvantage when soybean hulls are used as nourishment since they are small in size and not very dense, so the wind tends to blow away the hulls. Therefore its transportation requires closed and covered trailers. In many cases, the properties of soybean husk are considered negative, so no attention is paid in developing methods for its use and value-added allocation.

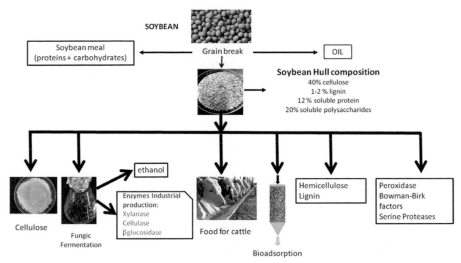

Figure 8.1   Applications of Soybean hulls.

Soybean hulls (SBH) are the pericarpic of the bean, it contains more than 70% of fiber formed by different polysaccharides such as: 30–50% cellulose, 15–20% hemicellulose (xylan) and 12% pectin, also 1–5% lignin and 7–14% proteins (Al Loman and Ju, 2016), these percentages depend on the genetic type of the bean. SBH have not received much attention beyond the use as animal feed. They are typically sold as is or as compressed pellets and fed to cattle and pigs.

Because of its high fiber content, SBH has many advantages over other hulls such as rice, wheat, barley, etc., in the sense that it is possible to reduce its volume 5 to 7 times by grinding it achieving a more compact and manageable product. However, its lightweight structure allows to recover a series of valuable molecules which it contains using less aggressive and environmentally friendly methods (Al Loman and Ju, 2016).

THE PRIMARY APPROACH OF SOYBEAN HULL: ANIMAL FEED

In some countries, such as Argentina and Brazil, that have a high cattle production, SBH has been used as a dietary supplement for these animals. Studies show that more than 15% of SBH cannot be added to an animal's diet, because its digestion rate is slow which can limit its capacity of energy consumption (Kornegay, 1981). Thus, the digestion can be incomplete in animals that have a rapid gastrointestinal transit in the rumen and are fed with SBH. On the other hand, adult ruminants tolerate this type of food very well, allowing its use in high quantities without causing harm to the animal. It must be taken into account to limit this type of food due to its high fiber content when used to feed young animals (up to 170 kg), because their rumen is not fully developed, and it would not obtain the same amount of necessary nutrients. For example this conditions its use in calves, where it is better to offer another type of supplement so as not to affect their development.

ENERGY AND ETHANOL PRODUCTION

The first given application to a biomass was the generation of energy: through its direct combustion when burnt in ovens (Panwar et al., 2011). This produced a major environmental contamination due to its low calorific content. Also, the fact that their mass is very voluminous, its handling becomes complicated, raising its cost higher than the energy yield obtained.

Ethanol is produced from sugarcane and corn in Brazil and USA, respectively (Mielenz, 2001). In Argentine, the sugarcane's waste was recently used in a cellulosic ethanol production also called "second-generation ethanol", where the main source is cellulose which needs to be decomposed into its monomers in order for it to release fermentable sugars (Al Loman and Ju, 2016; Qing et al., 2017).

SBH is basically lignocellulosic material composed of fermentable sugars, cellulose and hemicellulose (xylose and arabinose), and a very small proportion of lignin, compared to other residues such as rice hulls and straw, sugarcane bagasse and wheat straw (Camiscia et al., 2018). Therefore, its hydrolysis to liberate fermentable sugars is potentially appealing for bioconversion into bioethanol and other fine chemicals such as xylitol (Mielenz et al., 2009). There are still unsolved questions related to the economic viability of the hydrolysis processes of lignocellulosic substrates.

A diluted acid hydrolysis is performed to obtain its polysaccharides. The cellulose extracted by this process is hardly affected, and the polysaccharides obtained require other physicochemical hydrolyses at higher temperatures to obtain its constituent monosaccharides for the ethanolic fermentation (Qing et al., 2017). The traditional hydrolysis method for the conversion of polysaccharides into monosaccharides uses a strong acid or a basic medium at high temperatures, which has its disadvantages because it consumes high energy and produces toxic wastes that are discarded into the environment.

The use of diluted sulfuric acid, which has been extensively investigated using experimental and theoretical approaches as a pretreatment for SBH can effectively degrade hemicellulose present in the cell wall network by the catalytic effect of the proton. It was reported that over 90% of hemicellulose could be successfully removed during this acid pretreatment; however, a highly digestible residue remains which is mainly composed of cellulose and a small portion of lignin (Qing et al., 2017). The strong protonation effect decomposes the monosaccharides obtained into their degradation forms, such as furfural, 5-hydroxymethyl furfural, levulinic, etc.

On the other hand, alkaline pretreatment is considered as the most effective pretreatment with less sugar degradation, lower energy consumption, more lignin removal, due to its high capacity in breaking the linkages between lignin and carbohydrates and disrupting the lignin structure (Qing et al., 2017). Alkaline pretreatment could also swell the material, thus increasing the internal surface and favoring the swelling of the end product.

Another way to transform the cellulose and other polysaccharides present in the SBH into glucose is by using an enzymatic treatment with hydrolases, which includes three classes of enzymes: endo-1,4-β-glucanases that randomly cleaves internal bonds in the cellulose chain; exo-1,4-β-glucanases which attacks the reducing or non-reducing end of the cellulose polymer and β-glucosidases that converts cellobiose, the major product of the endo- and exo-glucanase mixture, into glucose (Deshpande et al., 1984; Gao et al., 2008). These enzymes act synergistically because the endo-acting enzymes generate new reducing and non-reducing chain ends for the exo-acting enzymes, which releases cellobiose that is then converted to glucose by β-glucosidases. It is important to note that natural cellulolytic enzyme systems often contain several exo- and endo- acting enzymes which may have varying preferences for different forms of cellulose. All these enzymes cleave glycosidic bonds by adding a water molecule (Rabinovich et al., 2002). However, pretreatment of lignocellulosic biomass is often required to facilitate the enzymatic hydrolysis (Al Loman and Ju, 2017).

The major challenge for the production of second-generation ethanol by enzymatic methods is the availability of efficient and economical enzymatic preparations (Mielenz, 2001). Cellulase complexes are commonly produced by fungi, especially of the genera *Trichoderma, Penicillum* and *Aspergillus* (Adsul et al., 2007; da Silva Menezes et al., 2017; Gao et al., 2008). Although these enzymatic complexes are usually used as raw preparations, their production costs are still very high and these commercial cellulase mixtures produce an 85% yield of reducing sugars from pretreated biomass (Brodeur et al., 2011).

From a biotechnological point of view, the conversion of biomass into monosaccharides cannot be done by only one enzyme, so there are different ways to carry out this transformation by using a complex mixture formed by different hydrolases (Brijwani et al., 2010). There are commercial enzymes mixtures available for the treatment of lignocellulosic biomass to obtain monosaccharides. These enzyme cocktails are often used in the production of glucose, for example: Novozyme 188™ (β-glucosidase) and a cell-wall degrading enzyme complex, Viscozyme L™ is used for enzymatic saccharification of pretreated SBH (Corredor et al., 2008). Viscozyme L™ contains a wide range of carbohydrases, including arabinase, cellulase, β-glucanase, hemicellulose and xylanase, which acts on branched pectin-liked substances found in plant cell walls (Guan and Yao, 2008). Optimash VR™ has a wide enzymatic activity for endoglucanase, xylanase, β-glucosidase, exoglycosidase, being effective in transforming cellulose into monosaccharides (Zieminski et al., 2012).

The production of hydrolases such as cellulolytic and xylanolytic enzymes using submerged fermentation (in a liquid phase) may be uneconomical due to the high maintenance costs of the bioreactor, the need of components that induce microbial growth such as a high concentration of proteins, etc. A possible solution is the use of a solid-state fermentation process which presents many advantages including high volumetric productivity and a relatively high concentration of the enzymes produced (Hölker et al., 2004). Also, it will involve a lower capital investment and lower operating cost. Another important feature of a solid-state fermentation is that it utilizes heterogeneous agricultural residues from agro based industries (Julia et al., 2016; Orzua et al., 2009; Taddia et al., 2018). In a solid-state fermentation for the production of cellulase, the cellulosic substrate acts as both a carbon source and as an inducer for the production of cellulase (Julia et al., 2016). Both, bacteria and fungi can use cellulose as a primary carbon source. However, most bacteria are incapable of degrading crystalline cellulose since their cellulase systems are incomplete, so enzymes produced by fungi are generally preferred because they produce all types of enzymes (Brijwani et al., 2010). Therefore, fungi hydrolases are very useful in the saccharification of pretreated lignocellulosic material and it presents itself as an emerging alternative for the reduction of the production costs of "second-generation ethanol".

OBTAINING CELLULOSE FROM SBH

Cellulose is the most abundant organic compound in nature and it is the main constituent of plant fibers, providing rigidity. It is a bio-polymer present in all plants in different amounts, for example cotton contains the largest amount of cellulose in its composition reaching 90% of its total mass (de Morais Teixeira et al., 2010). SBH has around 40–50% of cellulose, so it is an attractive and a inexpensive cellulose source (Al Loman and Ju, 2016).

As a first step in cellulose extraction from SBH and other plant sources, an alkaline pre-treatment has been the most successful delignification method due to its ability to disrupt the ester bonds that crosslink lignin and xylan, providing fractions enriched with cellulose and hemicelluloses (McIntosh and Vancov, 2011). The important modification done by this alkaline pretreatment is the disruption of –OH bonding in the fiber structure producing ionization and leading to an alkoxide:

$$\text{Fiber–OH+NaOH} \rightarrow \text{Fiber–O–Na+ (alkoxide)+H}_2\text{O}$$

This produces the separation of interfibrillar regions of cellulose fibers. An alkaline treatment of SBH with a low alkali concentration solution (NaOH) removes alkali-soluble substances such as lignin and pectin and other soluble impurities covering the external surface of the fiber cell wall (Ng et al., 2015). The alkaline pre-treatment changes the chemical structure, the physical structure, dimensions, morphology and mechanical properties of the biomass (Corapcioglu and Huang, 1987; Crini, 2005).

Many authors (Ferrer et al., 2016) used an alkaline medium (NaOH or KOH) with a concentration between 5% (w/v) to 20% (w/v) and temperature that ranges from 20°C to 120°C, or instead used sulfuric acid 65% (v/v) at 80°C–120°C and were able to solubilize the lignin, xylan and pectin from wood and other biomass (Ferrer et al., 2016; Johar et al., 2012; Neto et al., 2013; Sun et al., 2004). Due to the fact that during the alkaline pre-treatment not all the lignin is lost, a second bleaching process was applied by using chlorinated agents such as $NaClO_2$ or oxidant agents such as H_2O_2 in an alkaline medium, and also in a temperature range between 20°C to 120°C (Johar et al., 2012; Neto et al., 2013). The final product obtained is a mixture of amorphous, fibrillary and crystalline cellulose. Another method treated SBH with water at temperatures between 130°C–180°C to induce the release of saccharides (Liu et al., 2016c). Other treatments used four different chemical pre-treatments to obtain cellulose from SBH: NaOH solution 5% (w/v) or 17.5% (w/v) under two conditions: 90°C for 1 hour, which was repeated twice, followed by 15 hours at 30°C (Martelli-Tosi et al., 2016). The fibers were then bleached by using two types of solutions: a solution containing 0.7% (v/v) acetic acid and 3.3% (w/v) $NaClO_2$, agitated at 75°C for 4 hours; and another solution containing H_2O_2, NaOH 2%, and $MgSO_4·7H_2O$ 0.3% (as a stabilizer) at 90°C for 3 hours. The fibers were then filtered and washed with distilled water until reaching neutral pH, and cleaned with ethanol and acetone. These traditional methods were created by various steps which used corrosive reagents that cannot be discarded easily, which implies a significant increase in the total cost of the process.

Camiscia et al. (Camiscia et al., 2018) reported a simple extraction method using an alkaline pretreatment with NaOH at 5% (w/v) during 24 hours at 50°C as a delignification step of the SBH. The final product obtained has negligible lignin content as it was demonstrated by FTIR spectroscopy. These authors found that this pre-treatment had 43% performance and that the structure of the cellulose obtained is 50% amorphous. Taking into account these results, the total cost of cellulose production would be about US$ 150 per ton, compared to the cost of US$ 800 to produce cellulose paste by using wood as the main source.

SBH AS A BIOREMEDIATION TOOL: THE BIOADSORPTION PROCESS

Traditionally, heavy metals, hydrophobic molecules and other toxic molecules present in industrial residual waters have been removed by physical and chemical techniques, such as ion exchange chromatography, precipitation, reverse osmosis, electrodialysis, flocculation, coagulation and then sedimentation by aggregation of polymers, and adsorption using activated carbon, among others (Corapcioglu and Huang, 1987; Crini, 2005; Charerntanyarak, 1999). Activated carbon is perhaps the most widely used sorbent for the removal of dyes and heavy cations from industrial wastewater. However, the regenerative process of the activated carbon is a bit difficult, and together with the high operating costs hampers the application of this compound on a large-scale, sometimes making the use of this adsorbent uneconomical (Purkait et al., 2007). Therefore, low cost-effective methods are needed for a simple and environmentally friendly technology.

An emerging technique known as biosorption has been proposed to be both inexpensive and effective in comparison to chemical and physical technologies. The term biosorption refers to the use of any biomass as the binding material of metal ions and other molecules into the complex structure of biological materials. One possible source for biosorption systems would be agricultural wastes. Some reviews provide results from the use of different types of agricultural wastes in pollution removal (Demirbas, 2008; Rafatullah et al., 2010; Sud et al., 2008). However, not all wastes can be applied in a biosorption process. The general rule is that the plant waster should have a small size, with a high porosity degree and an open structure to allow the penetration of the molecules favoring the adsorption process (Neris et al., 2018). In the list of agricultural wastes, the so called hulls are the best candidates to be used in the biosorption process because they have a high surface/mass ratio.

The industrialization of agriculture produces many types of hulls. The most important being those derived from the industrialization of soy, rice and wheat. Soybean hulls and rice hulls are two economical and readily available wastes for its use. Several studies have demonstrated that soybean and rice hulls can be used as an ion exchange resin for the adsorption of chromium and copper ions, as well as arsenate and seleniate anions, heavy metal such as mercury, cadmium, copper, zinc, cobalt, manganese, chromium and nickel from contaminated waters, hydrophobic molecules such as an aromatic contaminant present in industrial waters removing these contaminates with high efficiency (Jin-fa, 2013; Marshall et al., 1999; Neris et al., 2018; Roy et al., 1993; Wang and Qin, 2005).

SBH has the advantage that they are very voluminous, with very low density, with a very open polysaccharide structure which allows easy water adsorption, increasing its volume and therefore its adsorptive properties. So, SBH are preferred because it is easy to decrease its size by simply grinding it and selecting the appropriate fraction through sieving. The biosorption potential of SBH is an open subject due to its physical properties which makes it a special material.

SBH has been used with excellent results for the adsorption of dyes that are released in the effluents of the textile and food industries (Arami et al., 2006). The presence of dye in effluents from the textile industry is produced because the dye cannot bind completely to the cotton, because it has a maximum binding of 80% and a minimum of 25%. Dyes are effluents that are toxic by nature and affect the lives of animals and humans; they are easy to spot in bodies of water since its color gives it away. Therefore, the presence of dye effluents in water sources is unacceptable (Garg et al., 2004).

The quaternization of lignocellulosic materials for the production of an anion exchange resin has been used to obtain an inexpensive and effective method for the removal of anions from wastewater. The introduction of amine groups is produced by a reaction with a number of quaternary ammonium compounds. These compounds primarily combine with alcoholic −OH groups present in the polyssacharide chains of the biomass with various cross-linking agents, like the quaternizing agent, N-(3-chloro-2-hydroxypropyl) trimethylammonium chloride, in the presence of a strongly alkaline environment. This modification increases the amount of positive charge on the hulls surface as evidenced by an increase in the nitrogen content and an increased uptake of anions compared with the unmodified hulls (Hashem et al., 2003; Prado and Matulewicz, 2014).

In the same way, carboxylic groups can be introduced into the biomass by reaction with tricarboxylic acid such as citric and tartaric acid (Marshall et al., 1999). When heated, tricarboxylic acids will dehydrate producing a reactive anhydride which can react with the hydroxyl groups present in the biomass. An ester linkage is formed with the −OH of the polysaccharides chains, losing a molecule of water by condensation. Citric acid and tartaric acid are low cost materials, used extensively in the food industry. The grafted carboxyl groups onto the SBH increases the negative charge on the fiber, thereby increasing its binding potential for cationic contaminants. This modified SBH shows high capacity for the adsorption of Cu^{++} (Marshall et al., 1999).

SOLUBLE PROTEINS FRACTION OF THE SBH: PROPERTIES AND APPLICATIONS

The SBH contains a fraction of soluble proteins, 7 to 12% of the total mass, which can be extracted by a simple lixiviation using an aqueous phase. There are some works that are related to this process of extraction of the soluble proteins fraction contained in the SBH (Bracco et al., 2017).

The commonly used procedure for protein extraction from SBH is by using diluted sodium phosphate and ethanol, but the yield of extracted proteins was not desired (Liu et al., 2016b). Most of the papers report the extraction procedure using water or buffer as the extraction phase (Bracco et al., 2017; Liu et al., 2016a; Steevensz et al., 2013). No study was done on how certain variables influence the lixiviation process such as, pH, contact time between the solid phase and the aqueous phase, biomass weight/liquid phase volume ratio, particle size, etc. Another variable to evaluate is the scaling of the process from the laboratory to at least a pilot plant, with the goal of increasing the efficiency of the process and the handling of appropriate biomass weight/liquid phase volume ratio to decrease the cost of bioseparation of the protein with biological activity, making the extraction process highly competitive.

Soy hull proteins have been investigated for almost two decades, but only a few numbers of proteins have been identified and their biological activities and functions studied:

a) **Soybean peroxidase (SBP)**: with a pI of 4.1 and molecular mass of 37 kDa, is a glycoprotein from soybean hulls (Sessa, 2004). Peroxidases have been used in the bread-baking industry as a cross-linking agent for proteins and as an oxidative coupling agent for the production of phenolic resins (Hilhorst et al., 1999; Matheis and Whitaker, 1984). Peroxidase can replace the potassium bromate used in the baking industry, the formaldehyde used as a cross-linking agent in the chemical industry, and the horseradish peroxidase used in the medical diagnostics industry (Steevensz et al., 2013).

b) **Protease inhibitor**: The soybean's trypsin inhibitor can be divided into two types: Kunitz Trypsin Inhibitor (KTI) and Bowman-Birk Inhibitor (BBI) , and their molecular weights are 20 kDa and 8 kDa, respectively (Sessa and Wolf, 2001). The KTI has two disulfide bridges and is very stable between pH 3 to 10 and has a relatively high thermal stability in low moisture conditions (Kunitz, 1947). BBI, from soybean cotyledon, is known as a cancer-preventive and anti-carcinogenic agent. BBIs could be a potential replacement for the synthetic serine protease inhibitors now used as anti-carcinogenic agents (Hsieh et al., 2010).

c) **Pathogenesis-related protein family**: This protein family is known to function in higher plants as a protein-based defensive system against abiotic and biotic stress, especially against pathogen infection. One of the PR proteins, pathogenesis-related 5 (PR-5), is a group of antimicrobial proteins with diverse functions in plants. Members of this family show similarity to thaumatin, a low-calorie sweetener. The antimicrobial activity of PR-5 proteins has been shown by *in vitro* bioassays or by the overexpression of these proteins in transgenic plants; therefore, these proteins should be a good resource material for the molecular breeding of pathogen-resistant plants. However, the underlying mechanism of this antimicrobial activity is still not fully understood. (Liu et al., 2016a)

d) **Class I Chitinase**: Chitinases are enzymes that catalyze the hydrolysis of β-1,4-N-acetylglucosamine linkages present in chitin (Tharanathan and Kittur, 2003). As chitin is a major component of fungal cell walls, and is absent in plants, chitinases plays a role in plant defense against pathogens. In addition to their role in plant defense, class I chitinases are emerging as a distinct group of panallergens causing cross-sensitization to different foods and materials in susceptible persons (Gijzen et al., 2001). These types of enzymes have been identified in SBH, demonstrating the richness of this material in proteins with high importance for the plants defense (Gijzen et al., 2001).

OVERVIEW

In this chapter we have tried to put into perspective the importance given to added-value of the agricultural wastes produced on our continent. Special attention has been paid to soybean hulls. Soybean is one of the main crops on the American continent, and its processing generates a large amount of residues, among them the so-called "hull". This soybean hull is produced in large amounts and it is mainly discarded into the environment or burned. However, it also discards a great number of valuable molecules with very important applications, such as proteins, polysaccharides and polyphenols.

So the question arises: *Why isn't this waste being taken advantage when presenting valuable molecules instead of being discarded?* The possible answer to this question might be due to the fact that this waste is only present in the major soybean producing countries (USA, Brazil and Argentina). The second reason is the difficulty of using traditional methods in the recovery of the molecules of interest contained in the hull. This is why; various investigation groups around the world are focusing their efforts in the study of this material and its possible applications.

The simplest and oldest application for this waste is its use, in low percentages, as an additive in cattle feed. However, by studying the composition of this residue it shows that it is composed of 12% soluble proteins, 9–11% galactomannans, 10–12% pectin, 9–10% hemicellulose (xylan), approximately 40% cellulose and 1–3% lignin. All these compounds can be extracted and/or exploited by valorizing, in this way, the SBH and at the same time it could reduce the amount of waste produced during the processing of this grain.

The extraction of cellulose could be one possible application for this residue. At the moment there is a great demand of different types of cellulose such as: amorphous, crystalline and nano-cellulose, this last one having the same interesting properties as the other celluloses. Because the SBH has a particularly open structure, obtaining these types of cellulose is easier from a technological point of view.

On the other hand, the polysaccharides that are present in the SBH can be extracted and hydrolyzed to their constituent monomers to be used in the production of second-generation ethanol. In this case, the extraction and hydrolysis methods are being optimized in order to reduce process costs and increase their performance.

Due to its great polysaccharides content, SBH can also be used as a carbon source for solid and liquid state fermentation for the production of various enzymes, especially fungal cellulases and xylanase.

Finally, Soybean hull is a great source of numerous proteins with biotechnological and industrial applications. For example, the SBP is an enzyme present in this residue and it presents a very high activity. It is used in water treatment and also in the development of kits for clinical exams, while obtaining it from SBH would significantly increase the added value of this waste.

It is for all the reasons expressed above, that we believe it is of great importance to study and revalue the wastes present on our continent, to be able to reduce environmental pollution and the number of residues discarded into the environment, and also develop new processes that allow to add value to them and thereby encourage the industry on our continent.

REFERENCES

Adsul, M., Bastawde, K., Varma, A., Gokhale, D. Strain improvement of *Penicillium janthinellum* NCIM 1171 for increased cellulase production. *Bioresource Technology*, 2007, 98, 1467–1473.

Al Loman, A., Ju, L.-K. Soybean carbohydrate as fermentation feedstock for production of biofuels and value-added chemicals. *Process Biochemistry*, 2016, 51, 1046–1057.

Al Loman, A., Ju, L.-K. Enzyme-based processing of soybean carbohydrate: Recent developments and future prospects. *Enzyme and Microbial Technology*, 2017, 106, 35–47.

Arami, M., Limaee, N.Y., Mahmoodi, N.M., Tabrizi, N.S. Equilibrium and kinetics studies for the adsorption of direct and acid dyes from aqueous solution by soy meal hull. *Journal of Hazardous Materials*, 2006, 135, 171–179.

Benkő, Z., Andersson, A., Szengyel, Z., Gáspár, M., Réczey, K., Stålbrand, H. Heat extraction of corn fiber hemicellulose, *Applied Biochemistry and Biotecnology*, 2007, 137, 253–265.

Bracco, L.F., Levin, G.J., Navarro del Cañizo, A.A., Wolman, F.J., Miranda, M.V., Cascone, O. Simultaneous purification and immobilization of soybean hull peroxidase with a dye attached to chitosan mini-spheres. *Biocatalysis and Biotransformation*, 2017, 35, 306–314.

Brijwani, K., Oberoi, H.S., Vadlani, P.V. Production of a cellulolytic enzyme system in mixed-culture solid-state fermentation of soybean hulls supplemented with wheat bran. *Process Biochemistry*, 2010, 45, 120–128. DOI: https://doi.org/10.1016/j.procbio.2009.08.015.

Brodeur, G., Yau, E., Badal, K., Collier, J., Ramachandran, K.B., Ramakrishnan, S. Chemical and physicochemical pretreatment of lignocellulosic biomass: A review. *Enzyme Research*, 2011, Article ID 787532, 2011, 17 pages.

Camiscia, P., Giordano, E.D., Brassesco, M.E., Fuciños, P., Pastrana, L., Cerqueira, M., Picó, G.A., Valetti, N.W. Comparison of soybean hull pre-treatments to obtain cellulose and chemical derivatives: Physical chemistry characterization. *Carbohydrate Polymers*, 2018, 198, 601–610.

Corapcioglu, M., Huang, C. The adsorption of heavy metals onto hydrous activated carbon. *Water Research*, 1987, 21, 1031–1044.

Corredor, D.Y., Sun, X.S., Salazar, J.M., Hohn, K.L., Wang, D. Enzymatic hydrolysis of soybean hulls using dilute acid and modified steam-explosion pretreatments. *Journal of Biobased Materials and Bioenergy*, 2008, 2, 43–50.

Crini, G. Recent developments in polysaccharide-based materials used as adsorbents in wastewater treatment. *Progress in Polymer Science*, 2005, 30, 38–70.

Charerntanyarak, L. Heavy metals removal by chemical coagulation and precipitation. *Water Science and Technology*, 1999, 39, 135.

da Silva Menezes, B., Rossi, D.M., Ayub, M.A.Z. Screening of filamentous fungi to produce xylanase and xylooligosaccharides in submerged and solid-state cultivations on rice husk, soybean hull, and spent malt as substrates. *World Journal of Microbiology and Biotechnology*, 2017, 33, 58.

de Morais Teixeira, E., Corrêa, A.C., Manzoli, A., de Lima Leite, F., de Oliveira, C.R., Mattoso, L.H.C. Cellulose nanofibers from white and naturally colored cotton fibers. *Cellulose*, 2010, 17, 595–606. DOI: 10.1007/s10570-010-9403-0.

Demirbas, A. Heavy metal adsorption onto agro-based waste materials: A review. *Journal of Hazardous Materials*, 2008, 157, 220–229.

Deshpande, M.V., Eriksson, K.-E., Pettersson, L.G. An assay for selective determination of exo-1, 4,-β-glucanases in a mixture of cellulolytic enzymes. *Analytical Biochemistry*, 1984, 138, 481–487.

Ferrer, A., Salas, C., Rojas, O.J. Physical, thermal, chemical and rheological characterization of cellulosic microfibrils and microparticles produced from soybean hulls. *Industrial Crops and Products*, 2016, 84, 337–343.

Gao, J., Weng, H., Zhu, D., Yuan, M., Guan, F., Xi, Y. Production and characterization of cellulolytic enzymes from the thermoacidophilic fungal *Aspergillus terreus* M11 under solid-state cultivation of corn stover. *Bioresource Technology*, 2008, 99, 7623–7629. DOI: https://doi.org/10.1016/j.biortech.2008.02.005.

Garg, V.K., Amita, M., Kumar, R., Gupta, R. Basic dye (methylene blue) removal from simulated wastewater by adsorption using indian rosewood sawdust: A timber industry waste. *Dyes and Pigments*, 2004, 63, 243–250. DOI: https://doi.org/10.1016/j.dyepig.2004.03.005.

Gijzen, M., Kuflu, K., Qutob, D., Chernys, J.T. A class I chitinase from soybean seed coat. *Journal of Experimental Botany*, 2001, 52, 2283–2289.

Guan, X., Yao, H. Optimization of viscozyme L-assisted extraction of oat bran protein using response surface methodology. *Food Chemistry*, 2008, 106, 345–351. DOI: https://doi.org/10.1016/j.foodchem.2007.05.041.

Hashem, M., Hauser, P., Smith, B. Reaction efficiency for cellulose cationization using 3-chloro-2-hydroxypropyl trimethyl ammonium chloride. *Textile Research Journal*, 2003, 73, 1017–1023.

Hilhorst, R., Dunnewind, B., Orsel, R., Stegeman, P., Van Vliet, T., Gruppen, H., Schols, H. Baking performance, rheology, and chemical composition of wheat dough and gluten affected by xylanase and oxidative enzymes. *Journal of Food Science*, 1999, 64, 808–813.

Hölker U., Höfer M., Lenz J. Biotechnological advantages of laboratory-scale solid-state fermentation with fungi. *Applied Microbiology and Biotechnology*, 2004, 64, 175–186.

Hsieh, C.-C., Hernández-Ledesma, B., Jeong, H.J., Park, J.H., Ben, O. Complementary roles in cancer prevention: Protease inhibitor makes the cancer preventive peptide lunasin bioavailable. *PLoS ONE*, 2010, 5(1), e8890.

Jin-fa, C. Study on biosorption of cr(vi), cu(ii) by a new low-cost adsorbent. *Water Saving Irrigation*, 2013, 11.

Johar, N., Ahmad, I., Dufresne, A. Extraction, preparation and characterization of cellulose fibres and nanocrystals from rice husk. *Industrial Crops and Products*, 2012, 37, 93–99. DOI: https://doi.org/10.1016/j.indcrop.2011.12.016.

Julia, B.M., Belén, A.M., Georgina, B., Beatriz, F. Potential use of soybean hulls and waste paper as supports in SSF for cellulase production by *Aspergillus niger*. *Biocatalysis and Agricultural Biotechnology*, 2016, 6, 1–8.

Kornegay, E. Soybean hull digestibility by sows and feeding value for growing-finishing swine. *Journal of Animal Science*, 1981, 53, 138–145.

Kunitz, M. Crystalline soybean trypsin inhibitor: II. General properties. *The Journal of General Physiology*, 1947, 30, 291–310.

Liu, C., Cheng, F., Sun, Y., Ma, H., Yang, X. Structure-function relationship of a novel PR-5 protein with antimicrobial activity from soy hulls. *Journal of Agricultural and Food Chemistry*, 2016a, 64, 948–959.

Liu, C., Cheng, F.F., Liu, X., Ma, H.Y., Yang, X.Q. Improved extraction of disulphide-rich bioactive proteins from soya hulls: Characterisation of a novel aspartic proteinase. *International Journal of Food Science & Technology*, 2016b, 51, 1509–1515.

Liu, H.-M., Wang, F.-Y., Liu, Y.-L. Hot-compressed water extraction of polysaccharides from soy hulls. *Food Chemistry*, 2016c, 202, 104–109.

Maran, J.P., Manikandan, S., Thirugnanasambandham, K., Nivetha, C.V., Dinesh, R. Box-Behnken design based statistical modeling for ultrasound-assisted extraction of corn silk polysaccharide. *Carbohydrate Polymers*, 2013, 92, 604–611.

Marshall, W.E., Wartelle, L., Boler, D., Johns, M., Toles, C. Enhanced metal adsorption by soybean hulls modified with citric acid. *Bioresource Technology*, 1999, 69, 263–268.

Martelli-Tosi, M., Torricillas, M.d.S., Martins, M.A., Assis, O.B.G.d., Tapia-Blácido, D.R. Using commercial enzymes to produce cellulose nanofibers from soybean straw. *Journal of Nanomaterials*, 2016, Article ID 8106814, 2016, 10 pages.

Matheis, G., Whitaker, J.R. Peroxidase-catalyzed cross linking of proteins. *Journal of Protein Chemistry*, 1984, 3, 35–48.

McIntosh, S., Vancov, T. Optimisation of dilute alkaline pretreatment for enzymatic saccharification of wheat straw. *Biomass and Bioenergy*, 2011, 35, 3094–3103.

Mielenz, J.R. Ethanol production from biomass: Technology and commercialization status. *Current Opinion in Microbiology*, 2001, 4, 324–329.

Mielenz, J.R., Bardsley, J.S., Wyman, C.E. Fermentation of soybean hulls to ethanol while preserving protein value. *Bioresource Technology*, 2009, 100, 3532–3539.

Neris, J.B., Luzardo, F.H.M., da Silva, E.G.P., Velasco, F.G. Evaluation of adsorption processes of metal ions in multi-element aqueous systems by lignocellulosic adsorbents applying different isotherms: A critical review. *Chemical Engineering Journal*, 2018, 357, 404–420.

Neto, W.P.F., Silvério, H.A., Dantas, N.O., Pasquini, D. Extraction and characterization of cellulose nanocrystals from agro-industrial residue–Soy hulls. *Industrial Crops and Products*, 2013, 42, 480–488.

Ng, H.-M., Sin, L.T., Tee, T.-T., Bee, S.-T., Hui, D., Low, C.-Y., Rahmat, A. Extraction of cellulose nanocrystals from plant sources for application as reinforcing agent in polymers. *Composites Part B: Engineering*, 2015, 75, 176–200.

Orzua, M.C., Mussatto, S.I., Contreras-Esquivel, J.C., Rodriguez, R., de la Garza, H., Teixeira, J.A., Aguilar, C.N. Exploitation of agro industrial wastes as immobilization carrier for solid-state fermentation. *Industrial Crops and Products*, 2009, 30, 24–27.

Panwar, N., Kaushik, S., Kothari, S. Role of renewable energy sources in environmental protection: a review. *Renewable and Sustainable Energy Reviews*, 2011, 15, 1513–1524.

Phinichka, N., Kaenthong, S. Regenerated cellulose from high alpha cellulose pulp of steam-exploded sugarcane bagasse. *Journal of Materials Research and Technology*, 2018, 7, 55–65.

Prado, H.J., Matulewicz, M.C. Cationization of polysaccharides: A path to greener derivatives with many industrial applications. *European Polymer Journal*, 2014, 52, 53–75.

Preece, K., Hooshyar, N., Zuidam, N. Whole soybean protein extraction processes: A review. *Innovative Food Science & Emerging Technologies*, 2017, 43, 163–172.

Purkait, M.K., Maiti, A., Dasgupta, S., De, S. Removal of congo red using activated carbon and its regeneration. *Journal of Hazardous Materials*, 2007, 145, 287–295.

Qing, Q., Guo, Q., Zhou, L., Gao, X., Lu, X., Zhang, Y. Comparison of alkaline and acid pretreatments for enzymatic hydrolysis of soybean hull and soybean straw to produce fermentable sugars. *Industrial Crops and Products*, 2017, 109, 391–397.

Rabinovich, M., Melnick, M., Bolobova, A. The structure and mechanism of action of cellulolytic enzymes. *Biochemistry*, 2002, 67, 850–871.

Rafatullah, M., Sulaiman, O., Hashim, R., Ahmad, A. Adsorption of methylene blue on low-cost adsorbents: a review. *Journal of Hazardous Materials*, 2010, 177, 70–80.

Roy, D., Greenlaw, P.N., Shane, B.S. Adsorption of heavy metals by green algae and ground rice hulls. *Journal of Environmental Science & Health*, 1993, Part A 28, 37–50.

Sessa, D.J., Wolf, W.J. Bowman–Birk inhibitors in soybean seed coats. *Industrial Crops and Products*, 2001, 14, 73–83. DOI: https://doi.org/10.1016/S0926-6690(00)00090-X.

Sessa, D.J. Processing of soybean hulls to enhance the distribution and extraction of value-added proteins. *Journal of the Science of Food and Agriculture*, 2004, 84, 75–82.

Steevensz, A., Madur, S., Al-Ansari, M.M., Taylor, K.E., Bewtra, J.K., Biswas, N. A simple lab-scale extraction of soybean hull peroxidase shows wide variation among cultivars. *Industrial Crops and Products*, 2013, 48, 13–18.

Sud, D., Mahajan, G., Kaur, M. Agricultural waste material as potential adsorbent for sequestering heavy metal ions from aqueous solutions—A review. *Bioresource Technology*, 2008, 99, 6017–6027.

Sun, J.X., Sun, X.F., Zhao, H., Sun, R.C. Isolation and characterization of cellulose from sugarcane bagasse. *Polymer Degradation and Stability*, 2004, 84, 331–339. DOI: https://doi.org/10.1016/j.polymdegradstab.2004.02.008.

Taddia, A., Boggione, M.J., Tubio, G. Screening of different agroindustrial by-products for industrial enzymes production by fermentation processes. *International Journal of Food Science & Technology*, 2018, 54, 1027–1035.

Tharanathan, R.N., Kittur, F.S. Chitin—The undisputed biomolecule of great potential. *Critical Reviews in Food Science and Nutrition*, 2003, 43(1), 61–87.

Wang, X.-S., Qin, Y. Equilibrium sorption isotherms for of Cu^{2+} on rice bran. *Process Biochemistry*, 2005, 40, 677–680.

Ziemiński, K., Romanowska, I., Kowalska, M. Enzymatic pretreatment of lignocellulosic wastes to improve biogas production. *Waste Management*, 2012, 32, 1131–1137.

Chapter 9

Section B:
Uses of Renewable Feedstocks

José Fernando Solanilla-Duque*,
Margarita del Rosario Salazar-Sánchez
and Héctor Samuel Villada-Castillo

Departamento de agroindustria,
Facultad Ciencias Agrarias Universidad del Cauca, Sede Las Guacas,
Calle 5 # 4–70, A.A:190002, Popayán, Cauca, Colombia.

INTRODUCTION

This chapter has been structured in such a way that the basic concepts and trends related to the use of Renewable Raw Materials (RRM) can be understood in a general way, since these sources can be obtained from plants, animals, microorganisms (Fig. 9.1) and as a result of the use of biological systems, there are in most cases the reduction of emissions of carbon dioxide, methane or greenhouse gases, and the increase of organic matter, water, among others.

This is why the biological or chemical conversion of many of the complex materials provided by this biomass, into chemical products or energy, have required the development of technologies that integrate the fractionation of the different matrices. On the other hand, different researches and technological developments are taking place to establish chemical and biochemical routes to optimize these processes, among which are the aqueous treatment in supercritical conditions (SCW), ultrasound, gasification, Aqueous Phase Reforming (APR), pyrolysis, steam explosion, enzymatic hydrolysis, etc.

---

*For Correspondence: jsolanilla@unicauca.edu.co; Phone: (+57) (2) 8245976-79 ext. 101-1401.

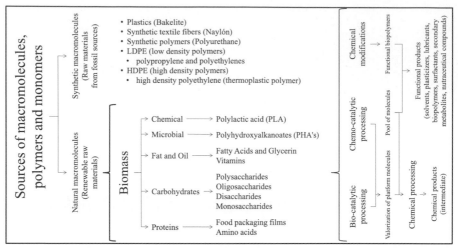

Figure 9.1   Scheme sources macromolecules, polymers and monomers.

The use of renewable sources, due to their richness in essential substances such as lipids, carbohydrates, proteins and other natural products, and also their chemical or biological conversion, are of great interest to the chemical industry (biofuels, agrochemicals, surfactants, lubricants, solvents), textile (fibers), pharmaceutical (metabolites), food (proteins, carbohydrates, lipids), manufacturing (polymers), among other applications, and in terms of reducing negative effects on the environment, the use of renewable raw materials has implied an advance in the biodegradability studies of the compounds developed (Table 9.1)

Table 9.1   Main applications of compounds obtained through green chemistry

Compound	Agri-food sector	Chem-ical sector	Pharma-ceutical sector	Textile sector	Manufac-turing sector
Essential oils (mint, basilic acid, etc.)	X				
Organic acids (coffee, cinnamic, oxalic, ferulic, acetic, etc.)		X			
Idol alkaloids, Tropane, Isoquinolic, Furanquinolic			X	X	
Polyacrylonitrile or poly-acidacrylic starch					X
Amino acids: L-Glutamic, L-Lysine, L-Threonine, L-Tryptophan	X	X	X		
Anti-inflammatories			X		
Antileucimics and antitumors (Camptotecina, telicarpina, solamarina, cefalotassina, harringtonina, etc.)			X		
Antimicrobials			X		
Aromas, spices	X			X	
Biofuel (Bioethanol, diesel)		X			

Table 9.1   .... (contd...)

Table 9.1   Main applications of compounds obtained through green chemistry (*contd...*)

Compound	Agri-food sector	Chem-ical sector	Pharma-ceutical sector	Textile sector	Manufac-turing sector
Cardiotonics			X		
Cellulose, hemicellulose				X	
Compounds of three carbons: Lactic Acid, 1, 2-propanoidol, 1, 3-propanoidol, 1, 3-Hydroxypropionic, Acetone, Propionic Acid	X	X	X	X	X
Four-carbon compounds: Succinic Acid, Butanol, 1,4-butanediol, 3-Hydroxybutyrolactone, 3-Hydroxybutanone, Fumaric Acid, Malic Acid		X	X		X
Five-Carbon Compounds: Furfural, Levulinic Acid, Itaconic Acid, Xylonic Acid, Isoprene		X	X	X	X
Compounds of six carbons: Hydroxymethylfurfural HMF, 2,5-Furandicarboxylic Acid, Citric Acid, Gluconic Acid, Sorbitol	X	X	X	X	X
Colorants	X			X	
Fuel	X				X
Benzoic derivatives (vanilic acid, parahydrohixobenzoic), Benzopyrone (Coumarin), benzoquinone (Ubiquinone, Tocorefol)		X			
Phenolic derivatives		X			
Dextrins		X			X
Diuretics			X		
Non-glucid sweeteners	X		X		
Enzymes	X	X	X		
Steroids		X			
Fiber				X	
Furan Chrome and furanocoumarin Furfural		X			
Gelling agents		X			
Glycerin	X	X			X
Glucids		X			
Glutathione		X			
Hemolytics			X		
Plant growth hormones and regulators	X				
Enzyme inhibitors (fibrinolytic proteinase inhibitors)			X		
Insecticides	X				
Insulin		X			
Lignin				X	

Table 9.1   .... (*contd...*)

Table 9.1   Main applications of compounds obtained through green chemistry (*contd...*)

Compound	Agri-food sector	Chem-ical sector	Pharma-ceutical sector	Textile sector	Manufac-turing sector
Lipids		X			
Maltoses					X
Nucleosides and nucleotides		X			
Smells (Pink, jasmine, etc.)	X				
Opiaceae (codeine, Thebaine, papaverine, etc.)			X		
Peptides and proteins	X	X			
Fodder				X	X
Pigments (Chlorophyll, carotenoids, xanthoilas, flavonoids, tannins, etc.)	X				
Resin, oleoresins (colophony, turpentine, etc.)				X	
Sapogenins, Steroid saponins (diosgenin, stigmasterol, sitosterol)		X			
Tannins				X	
Terpenes and terpenoids		X			
Vasoconstrictors			X		
Vitamins (ascorbic acid, thiamine, vitamin K)		X	X		

CHEMICAL INDUSTRY

Biofuels

World transport consumes 60% of liquid fuels derived from petroleum, which generate greenhouse gas emissions such as carbon monoxide (CO) by 70% and carbon dioxide (CO_2) by 19%, as well as methane (CH_4), nitrous oxide (N_2O), hydrofluorocarbons, perfluorocarbons, among others (Balat, 2011; Ireland and Clausen, 2019). These contribute to environmental problems such as melting glaciers, the death of coral reefs, rising sea levels, droughts, among other consequences (Laine, 2009).

The U.S. oil production has decreased from 1973 by 4.1 to 2.6 billion barrels in 2006 (Shafiee and Topal, 2008; Fantazzini et al., 2011). However, according to the EIA (Energy Information Administration) of the United States for 2019 production will remain on the rise to reach 11.5 million barrels per day, this being a reference framework of world behavior. The relative stabilization of the market for this fuel in the 1960s has generated little interest in alternative fuels to this day.

For this reason, biofuels from crops rich in biomass have been proposed as a viable alternative to petroleum derivatives due to their sustainability, availability, reduction of greenhouse gases, among others (Cobos et al., 2007). In this sense, bioethanol, due to its high octane number, is a biofuel widely used in the transport sector (Limayem and Ricke, 2012), producing lower emissions and preventing early ignition (Celik, 2008). The negative effects of these biofuels are associated with the increase in food prices on the international market (Von Braun, 2008), due to the fact that they are related to the production of these biofuels, as the expansion of land for this type of crops generates the risk in food security and loss of natural ecosystems (Fargione et al., 2008).

Bioethanol can be produced by bioconversion of pentose obtained from second-generation feedstocks, resulting in cost-effective ethanol production (Kuhad et al., 2011). However, there is currently a trend in the production of bioethanol from sugars and starches obtained by enzymatic means, as a substitute for 10–20% of gasoline (Zabed, 2019; Buckeridge et al., 2019).

On the other hand, the production and use of bioethanol from lignocellulosic materials have been compared with more efficient production routes, such as saccharo-ethanol. In an alternative environment, the industry proposes new strategic approaches to implement the comparative advantages of sustainable lignocellulosic ethanol (Gnansounou, 2010; Tyagi et al., 2019).

Biological and thermochemical methods can be be used to produce biofuels. To this end, pretreatments are the technological factor for obtaining bioethanol from cellulosic materials, since they define the degree and cost with which the carbohydrates of cellulose and hemicellulose can be converted into bioethanol (Dalena et al., 2019).

Table 9.2   Different pre-treatments for lignocellulosic materials

Mechanical pre-treatment		
Method	*Description*	
Mechanical shredding	Grinding Particle Size Reduction (40 mesh) Minimal effect on hydrolysis yields	Chang and Holtzapple, 2000
Thermal pre-treatment		
Steam Explosion	Material is subjected to $T = 160$–$260°C$, Direct injection of saturated steam is used, the time between 1 and 10 minutes.	Duff and Murray, 1996
High-temperature liquid water (LHW)	Biomass undergoes $T = 170$–$230°C$ Time 46 minutes The reactor is pressurized to keep the water liquid. Application in lignocellulosic materials	Laser et al., 2002
Physical-chemical pre-treatments		
Ammonia Fiber Explosion Process (AFEP)	Material impregnated with liquid ammonia (1 to 2 kg ammonia/kg dry biomass) $T = 90°C$ Time 30 minutes The material is subjected to rapid decompression.	Sun and Cheng, 2002
Explosion with CO_2	The molecule forms carbonic acid benefiting hydrolysis. Dosage: 4 kg of CO_2/kg of fiber Pressure: 5.62 Mpa, carried out with high pressure Temperatures up to 200°C Time: variable	Purl and Mamers, 1983

Source: Adapted from Bernier and Ricón, 2012

Accessible technologies exist for bioethanol production using rice, wheat straw, maize straw, and sugarcane bagasse as agricultural waste (Swain et al., 2019). But to solve the technological obstacles of the conversion process, new pre-treatment sciences and technologies will have to be applied (Table 9.2), which are more efficient when releasing cellulose and hemicellulose from its molecular complex with lignin (Arora et al., 2019). Alternative procedures for saccharification of complex polymers, such as enzymatic saccharification, have been proposed. And finally, the implementation of recombinant microbial strains capable of simultaneous xylose and glucose fermentation (Sarkar et al., 2011).

Another alternative is pretreatment with the steam explosion in lignocellulosic materials (Negro et al., 2003). This can cause changes in the atomic ratios of oxygen/carbon and hydrogen/carbon, as well as the crystallinity index of the cellulose and the soluble organic content when performing such pretreatment (Momayez et al., 2019). This treatment can be preceded by washing, grinding and enzymatic loads (Soares et al., 2011). From the pre-treated lignocellulosic materials of these processes, an amount of glucose is obtained that can affect by the elimination of hemicellulose and lignin. However, grinding these materials does not influence glucose production by enzymatic hydrolysis (Wyman et al., 2017).

Other methods have been the enzymatic saccharification of lignocellulosic materials after being pre-treated with supercritical carbon dioxide, the glucose yields with this pre-treatment are around 72%. However, 20% more glucose yield is achieved by combining the supercritical CO_2 explosion in alkaline medium (Santos et al., 2011).

Agrichemical

The production of functional foods, with organic character, free of toxic residues, with high nutritional value and which during their production do not cause alterations to natural resources, are a priority in the implementation of food and nutritional security systems.

Different technologies for agrochemical production have emerged, which are focused on the implementation of strategies to minimize environmental impact and establish clean agriculture. Such a strategy has resulted in a serious threat to humanity due to the indiscriminate use of synthetic chemical compounds that have caused ecological imbalance and depletion of natural resources. The damage caused to crops that lead to low yields has been due to the use of agrochemical supplements for pest eradication and in the production of soil nutrients (Malusá and Vassilev, 2014). The cost of agrochemicals has increased considerably in recent years. For this purpose, the energy required to produce 1 kg of fertilizer is 11.2 kWh for nitrogen (N), 1.1 kWh for phosphorus (P) and 1 kWh for potash (Saritha and Tollamadugu, 2019). Therefore, in organic agriculture it becomes inevitable to overcome the constraints.

An important element for the development of clean agriculture is the production and use of botanical agrochemicals, which, when applied to crops, present comparative advantages over other synthetic products (Rojas and Luque, 2012); a strategy that must be adopted in rural areas to achieve sustainable development. These agrochemicals are considered biodegradable, economical, easy to use and have a low level of toxicity.

In the case of agrochemicals, green chemistry seeks to offer prevention rather than remediation. Accordingly, the relationship between agrochemicals and the environment is analyzed not only from the point of view of the effects of the same on the pest species, soil nutrient deficiency, elimination of undesirable plants or other organisms, but also from the point of view of the formulation and production of the chemical agent. In this way, this perception conforms to the criteria of green chemistry, contextualized in pursuit of pollution prevention. Agrochemicals for crop protection and prevention are described as the following:

— **Insecticides**: These are for protection against insects or plague, eliminating them or impeding their action; controlling them below their level of activity.
— **Fungicides**: Protection against fungi and their actions can be protective and eradicating. The first, prevents or inhibits growth and the second, eliminates the plague.
— **Herbicides**: Eliminating undesirable plants, can be selective and non-selective.
— **Bio pesticides**: They are protected and ecologically obtained from natural substances such as plants, animals, bacteria and certain minerals.
— Others (fumigants, rodenticides, plant growth regulators, etc.): fumigants and rodenticides protect crops during storage. Regulators control or modify the growth process of plants. They are frequently used in cotton, rice and fruit crops.

The use of new environmentally friendly agrochemicals, with the use precautions, indicated in each case, has had a strong impact on a highly competitive agricultural production (Collins, 2001). Thus, in 2012 the FAO recommended the implementation of "good agricultural practices" such as:

a. Make a prior diagnosis of the crop before using agrochemicals,
b. Crop rotation,
c. Implement an integrated pest management program,
d. Do not overuse the same active ingredient,
e. Keeping records of periodic applications,
f. Use of biological control to minimize the use of agrochemicals.

The interest in the use of plant extracts for this purpose has increased considerably, with promising results from *in vitro* and *in vivo* research with plant species from different ecological environments and abundance in nature (Xu et al., 2014). The study of secondary metabolites present in extracts of diverse plants with biological activity as a biocide can be useful for phytosanitary control in different crops (Sharaby and Al Dhafar, 2019). Depending on the time of implementation, the resources involved and the implementation risks associated with agrochemicals, green chemistry strategies can be classified into three types:

Near-term implementation strategies: industry must optimize its commercial practices by developing zero discharge solutions and the means to reduce COD (Chemical Oxygen Demand) levels in wastewater generated by the use of agrochemicals (Díez et al., 2019; del Carmen et al., 2018).

Medium-term implementation strategies: The industry should seek to change its chemistry to maximize the efficiency of solvent consumption, focusing on eco-friendly technologies such as microwave chemistry involving the use of radiation for chemical reactions, solvent recovery, implementation of alternative solvents, use of isolated enzymes or complete cells for synthetic transformation (biocatalysis), among others (Sheldon and Brady, 2018; Clarke et al., 2018). The objective is to reduce the amount of waste generated, improve the efficiency of its materials and to reduce the costs involved.

Long-term implementation strategies: The trend is to shift dependence from petroleum products to renewable resources and biomass as feedstock for environmentally friendly biopesticides or agrochemicals. Therefore, the industry must undertake innovation processes in the search for green routes of chemical synthesis. These strategies focus, for example, on the use of living pathogenic organisms for pest control, among them we find biofungicides (*Trichoderma*) (Faruk and Rahman, 2018), bioherbicides (*Phytopthora*) (Mathur and Gehlot, 2018) and bioinsecticides (*Bacillus thuringiensis*) (Zghal et al., 2018), which provide a solution to pest problems of different types of crops, both ecologically and effectively. On the other hand, at the world level, the market for organic food has increased and the demand for residue-free phytosanitary products has increased, due to the growing concern for the safety and toxicity of traditional pesticides (Watson, 2018; Fortunati et al., 2018; Upholt, 2018). This leads to the search for tangible benefits that affect cost-effectiveness in terms of the number of applications, low residual effect and biodegradation properties and control of resurgence of pests with less damage to beneficial pests (Röös et al., 2018).

An example of these strategies could be the application of fermented fique juice as disease control in different crops. It was found that fique juice (*Furcraea gigantea* Vent.) contains saponins and alkaloids, secondary metabolites responsible for antifungal activity (Gokarneshan and Velumani, 2018). This type of product may have biofungicide incidence on plant pathogens such as *Colletotrichum gloeosporoides, Sclerotinia sclerotiorum, Fusarium* spp. and *Phytophthora infestans* (Arango, 2013). According to Santander et al., 2014, the biocidal action of fique juice against *C. gloeosporioides*, which causes anthracnose in tree tomato (*Solanum betaceum*) (Vallad et al., 2018), has a maximum inhibition of the pathogen at concentrations of 100,000 µg mL^{-1}.

Surfactants

Surfactants are tensioactive coadjuvantes or active surface agents, are defined as a substance capable of reducing the surface tension on the two-phase contact surface, i.e., reducing the surface tension (or interfacial tension) between two liquids or between a solid and a liquid (Prud'homme, 2017; Waldhoff and Spilker, 2016).

These chemical products are for final consumption, sold directly to the public, including detergents, soaps, and other household cleaning products, which also have applications in the food industry, cosmetics, textiles, agriculture and hydraulic fracking.

They currently have an approximate production of 16 million MT/year with an annual growth of 4.7% (Global Surfactant Industry, 2018). For this reason, there is a great concern to obtain surfactants that are environmentally safe and effective which are capable of cleaning almost any surface from sensitive skin to large industrial plants. The surfactants that stand out in the world panorama are:

a. Linear alkylbenzene sulfonates (Linear alkylbenzene sulfonate)
b. Alcohol derivatives:
—Sulfates of fatty alcohols (PAS –Primary Alcohol Sulfate)
—Ethoxylated alcohols (EA)
—Sulfates of ethoxylated fatty alcohols (AES, Alcohol Ethoxy Sulfates)
c. Ethoxylated Alkylphenols (APE, alkylphenol ethoxylates)
d. The sulfonates of α-olefins and kinds of paraffin (AOS/PS, α-olefin and paraffin sulfonates).

The trend in surfactant production is towards the production of renewable raw materials of animal and plant origin, called oleophins. They conform to principle nine of Sustainable Chemistry or green chemistry that these chemicals must break down into harmless degradation products and do not persist in the environment. With this in mind, environmental regulations seek to make products less aggressive to the environment. But their big disadvantage is that their prices are higher than synthetic surfactants.

Surfactants are characterized by having a hydrophobic part in their structure, being soluble in non-polar oils and media, and a hydrophilic part, being soluble in water and other polar media (Fig. 9.2).

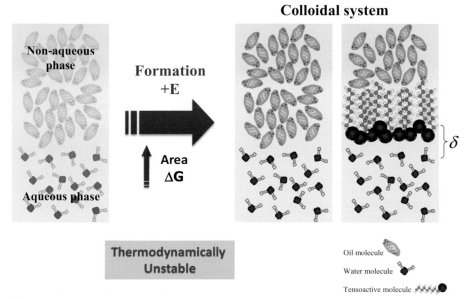

Figure 9.2 The activity of a surfactant dissolved in water. The colloidal system is formed as the free energy of Gibbs (ΔG) decreases and the surface area (A) (thermodynamically unstable system) decreases the surface tension at the interface (δ) of the water (73 mN/m).

The molecular structure of the surfactant is dissociated in the presence of an electrolyte and thus can be classified as ionic or non-ionic (neutral). In addition, depending on the nature of the charge, ions are classified as anionic, cationic and amphoteric.

Anionic surfactants: These are compounds that when dissociated in aqueous solutions give rise to a negatively charged ion. These ions are functional groups that are usually of the carboxylate ($-COO^-$), sulfate ($-SO_4^-$), sulfonate ($-SO_3^-$) or phosphate ($-O^-PO_3$) type. They are used in the formulation of detergents for domestic and industrial use and are the ones with the highest production volume.

Cationic surfactants: They act in the same way, in this case, the ion is positively charged which decreases the surface tension (Fig. 9.3). These compounds have a chain of 8 to 25 carbon atoms, derived from fatty acids or a petrochemical derivative and a positively charged nitrogen atom, constituted by a long chain of quaternary ammonium salts. They are efficient as bactericides, germicides, algicides as they eliminate or limit the growth of microorganisms.

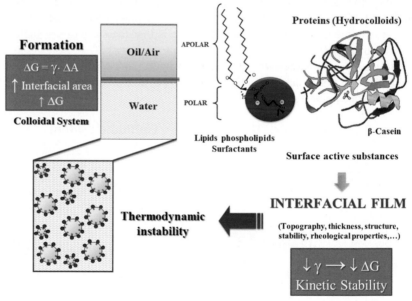

Figure 9.3   Scheme of an interfacial film formation in an oil drop. The colloidal systems formed are kinetically stable giving rise to different properties due to the presence of micellar structures.

Non-ionic surfactants: they are not dissociated into ions, they have an ether ($-O-$) function or hydroxyl ($OH-$) group as a polar part and their apolar part is an aliphatic chain. They are derived from polyoxyethylene and polyoxypropylene, are also derived from sorbitan and fatty alkanolamides, among others. They are stable, do not ionize in water, do not form salts with metallic ions and are effective in waters of soft or hard nature.

Surfactant capacity: According to Griffin (1949), this is determined by calculating the HLB value (Hydrophilic-Lipophilic Balance); this parameter helps to understand the relationship between the nature of the surfactant and the properties as a surfactant and emulsifier. This concept consists in attributing a certain value to the emulsifying agents from data relative to the stability of an emulsion. For example, this method calculates HLB for non-ionic surfactants as follows:

$$HLB = 20\frac{M_h}{M} \qquad \text{(Eq. 1)}$$

where M_h is the molecular mass of the hydrophilic part of the molecule, and M represents the molecular mass of the entire molecule.

The amphiphilic nature of surfactants can be expressed in terms of an empirical scale of HLB values. A high HLB value (10 to 18) indicates that a substance has a hydrophilic character and is suitable for forming Oil-Water (O/W) emulsions while substances with low HLB (3 to 8) are lipophilic and suitable for aqueous-oil (W/O) emulsions (Table 9.3). Different formulas have been established for calculating HLB, based on the composition of surfactants, but they can also be determined experimentally. The optimal HLB value for forming an emulsion depends to some extent on the nature of the system.

Table 9.3 Empirical scale of HLB values (Hydrophilic-Lipophilic Balance)

Characteristic behaviors related to water		HLB	Relation		Functions	
			Hydrophilic	Lipophilic		
Hydrophobic	There is no dispersion	0 / 1	0	100	Not reported (NR)	Emulsification W/O
	Little dispersion	2 / 3	10	90	Antifoaming agent	
		4 / 5	20	80	NR	
		6 / 7	30	70	Wetting agent	
	Milky dispersion with vigorous agitation	8 / 9	40	60	Wetting agent	
Hydrophilic	Stable milky dispersion	10 / 11	50	50	NR	Emulsification O/W
	Transparent, translucent and clear dispersion	12 / 13	60	40	Cleaning agent or detergents	
		14 / 15	70	30	Cleaning agent or detergents	
	Colloidal solution or solution	16 / 17	80	20	Solubilizing agent	
		18 / 19	90	10	Solubilizing agent	
		20	100	0	NR	

Biodegradability of surfactants: Although there are a large number of applications and advantages of surfactants at an industrial, economic and sanitary level, it has been necessary to demonstrate that there are correlations between the degree of biodegradability of a surfactant and its length and chain structure (linear or branched) (Pérez-García et al., 1996). Such cases, are the non-ionic surfactants that have a degree of biodegradation because the hydrophobic part of the molecule is linear and does not have an aromatic ring in the hydrophobic part (Torres et al., 2005).

Another factor affecting the biodegradability of these compounds is their toxicity. It should be noted that the load has a strong impact on the toxicity of a surfactant. With this in mind, the most toxic are the cationic surfactants and have been used as antimicrobials, while the anionic and biosurfactants have less toxicity. Among those reported are Tween (Polyoxyethylene sorbitan) and Span (Tiehm, 1994).

Accordingly, we can emphasize that the use of renewable raw materials associated with the use and production of surfactants, contributes to the reduction of CO_2 emissions, and also contributes to minimizing the environmental impact of these chemicals.

FOOD SECTOR

Modern food production is a consequence of the lifestyle changes that have taken place during this century. At the beginning of the 19th century all food was prepared in individual houses, often using raw materials derived from agriculture, practised at the family level. The industrialization of our society has created a demand for processed products. With progress in food technology and the use of modern transport and storage systems, food today can be produced on a large scale and distributed over long distances to the consumer, in the space of days or even hours. Consumer requirements have also changed. Food is not considered only as a source of energy to be used in daily survival. Today, nutritional aspects related to health, variety, and safety are factors of great importance, which is why "Green Chemistry" has become part of this new era of harnessing biomolecules for a biologically and functionally active diet.

Proteins

Proteins are important structural components of many foods that are also used as ingredients due to their nutritional value and physical-chemical properties. For 25 years, great interest has been given to the modification of edible proteins (of animal and vegetable origin) in order to improve their functional properties. The main objectives that are pursued with the modification of proteins are:

- Increase surface hydrophobicity.
- Increase the hydrophilic character (potentiation of solubility).
- To increase the adsorption capacity on the interfaces.
- Decrease surface or interfacial tension.
- Optimize the distribution of hydrophobic and hydrophilic domains.
- Improve the viscoelasticity of the adsorbed film on the O/W and A/W interfaces.

- To increase the charge density on the aqueous phase.
- Increase the flexibility of the interface.
- Provide specific attributes in functional foods.

The methods used for the modification of proteins by non-contaminating processes, and their consequences on the formulation, can be summarized as follows:

Physical methods

Heat: Increase hydrophobicity without affecting solubility.
Changes in pH: Modification of the load and surface and functional properties.

Chemical methods

Acylation and succinylation of the amino group: Increases net negative charge, increases electrostatic repulsions, esterification of the carbonyl group, increases net positive charge, increases the value of surface properties, phosphorylation of the amino group, increases emulsifier activity, increases protein hydration, improves steric repulsion.

— **Disamidation**: improves functional properties, increases hydrophobicity.
— **Lipophilization**: improves functional properties.
— **Glycosylation**: improves protein-solvent interactions. Increases steric stabilization.
— **Reducing alkylation of the amino group**: improves emulsifying capacity.
— **Reduction of disulfide bonds**: improves flexibility and rapid denaturation of the protein on the interface.

Enzymatic modifications

Partial hydrolysis with trypsin: Increases the number of groups loaded. Decreases molecular weight. Polymerization with transglutaminase. Modification of the molecular configuration and improvement of functional properties: addition of hydrophobic ligands.

Obtaining Protein Isolates and Hydrolysates

To obtain a protein isolate involves a series of steps aimed at eliminating or reducing non-protein components to achieve a final product with 80–90% protein (Vioque et al., 2001). In the first stage, the proteins are solubilized to separate them from the rest of the non-soluble compounds, the extract obtained contains, in addition to the proteins, the rest of the soluble compounds of the protein concentrate. The second stage aims at the concentration of proteins and especially their purification against other compounds (Belén et al., 2007).

A protein hydrolysate is formed by peptides, which are small sequences of inactive amino acids within the protein but which can be released after hydrolysis, catalyzed by chemical agents or enzymes (Vioque et al., 2000). Hydrolysis of proteins and control in the degree of hydrolysis are carried out for several reasons, including improving the nutritional properties and texture of foods, increasing or decreasing the solubility of protein, achieving better emulsion and foaming properties, reducing or eliminating unpleasant odors. Most protein hydrolysates have the ability to reduce interfacial tension between phases, and therefore may be able to form and stabilize emulsions and foams. Proteins, while decreasing interfacial tension, can form a continuous film at the interface through intermolecular interactions (Rodríguez-Hueso et al., 2012).

Protein hydrolysates are of paramount importance in the food industry as functional ingredients. Fish protein hydrolysates have proved useful in the formulation of substitutes as an important supplement in product formulation, also showing potential as antioxidants (Belén et al., 2007). For the monitoring and control of protein hydrolysis it is necessary to evaluate the degree of hydrolysis, which corresponds to the percentage measure of a dissolving compound (Guadix et al., 2000).

Proteins as Sources of Bioactive Peptides

Proteins are long chains of amino acids that form polypeptides, and have mainly a structural function. They are composed as follows: 50–55% carbon, 20–23% oxygen, 6–7% hydrogen, 0.2–0.3% sulfur and are an important source of 12–19% hydrogen that can be assimilated by the body. (Lorenzo et al., 2018).

Proteins are compounds with functions that intervene in the growth, the correct functioning of the organism, support the organism and are powerful antibodies that act on infections or external agents. Additionally these macronutrients have a functionality with added value, since they are deposited bioactive peptides that are good for health.

For this reason, an interest has been generated in the use of peptides derived from dietary proteins as intervention agents against chronic human diseases as well as for the general well-being of the organism (Garrido et al., 2018).

Bioactive peptides are a small part of the proteins that act positively on the biological functions of the organism, produced from the enzymatic hydrolysis of these proteins. These compounds are formed by peptide chains of 2 to 20 residues of amino acids that are inactive but can be released by digestion although this process is not as effective as the separation by enzymatic hydrolysis previously performed before consumption (Kisioglu and Nergiz, 2018).

Peptides can be derived from both animal and plant food sources and provide several specific beneficial alternatives for the proper functioning of the human organism. Some bioactive peptides extracted from animal sources such as: milk, egg, blood plasma, fish muscle and plant origin such as: soy, chickpea, beans, among others, have demonstrated antihypertensive capabilities, generating great curiosity at an industrial level since these peptides could be included in food products converting them into foods with functional characteristics that provide health benefits (Kaur et al., 2018).

This is why the hydrolysis process has gained importance today, given that from chemical (acid or alkali) or biological (enzymes) degradation processes on the protein, not only is a high content of essential amino acids guaranteed, but it has also been possible to obtain smaller peptide units, to which functional, bioactive and nutritional benefits are attributed, especially as an emulsifier, texture extender and antioxidant, which is because of the increase in solubility due to the increase in hydrophilic groups that favors its use in food formulations (He et al., 2013). Regarding the importance of the availability of amino acids in peptide units, its structure should be taken into account, highlighting that they have an amino group (NH_2) and a carboxyl group (COOH) attached to a central carbon atom, in addition each amino acid has unique properties conferred to the polarity that characterizes it (Krochta, 2002). These peptides can actively participate in chemical reactions through covalent (bisulfide bridge) and non-covalent (ionic bonds, hydrogen bonds and van der Waals forces) interactions, as well as hydrophobic interactions between apolar groups of amino acid chains (Kokini et al., 1994).

The peptides of greatest interest are those to which antimicrobial activity is attributed, due to the interaction they establish with the cytoplasmic membrane of microorganisms (Powers and Hancock, 2003). On the other hand, peptides also possess antifungal characteristics (Jang et al., 2006) and antiviral characteristics, the latter associated with cationic peptides (Albiol-Matanic and Castilla, 2004; Mohan et al., 2010).

Thus, the inclusion of peptides in the formation and production of PCs and CRs requires the implementation of three sequential stages: (i) rupture of intermolecular bonds of the protein by means of physical or chemical reactions that allow its structure to unfold and increase the surface area of contact with the other compounds of the matrix; (ii) organization and orientation of the polymer chains present according to the free radicals existing in the peptides, until reaching a desired conformation; (iii) formation of new intermolecular bonds and interactions to form the film or coating (Jerez et al., 2007).

The kinetics of protein adsorption, an important phenomenon in obtaining emulsions that form edible films (PC) and edible coatings (RC), have focused their interest on determining the rate of adsorption and the amount of protein adsorbed at the interface (Andrade, 1985). Thus, protein adsorption is understood to be the net result of interactions between system components, including the gaseous, liquid or solid interface, governed by intermolecular interactions between protein-interface, solvent-interface or solvent solvent (Fraaije, et al., 1991; Haynes and Norde, 1994; Guzey and McClements, 2006; Norde and Lyklema, 1991). It has been pointed out that protein molecules can be strongly adsorbed at the interface at different contact points, according to the orientation of the polypeptide chain, the number of residues or the segments in contact at the interfacial level, which determines the molecular flexibility of this chain and its affinity to the dissolution medium. Consequently, the polypeptide chains at the interface can adopt different arrangements: (i) **Row (train)**: refers to the amino acid segments that are in direct contact with the interface; (ii) **Loop**: corresponds to the polypeptide segments that are suspended within the phase, between the rows; (iii) **Tail**: made up of the terminal segments of the polypeptide chain that are frequently immersed in the aqueous phase, because they are usually residual groups NR and CR that are charged to neutral pH.

The structure acquired by the polypeptide chains at the interface depends on the flexibility of the chain and its affinity to the medium, as well as the surface pressure at which the protein film is found. Therefore, it is more likely that the protein unit will be denatured at low surface pressure than at high pressure, so that it is feasible that most segments of polypeptide chains can be attached to the interface in row configurations (Davies, 1953). Otherwise, if the surface pressure is high, there will be less chance of forming rows at the interface, than adopting loop and tail configurations. It should be noted that the process of protein adsorption at the interface takes place according to the following mechanism: (i) Diffusion from the dissolution sinusoidal to the interface; (ii) Adsorption, penetration and deployment of the protein at the interface; (iii) Reordering of the protein molecules that have been adsorbed at the interface (MacRitchie, 1989).

Carbohydrates

The most abundant renewable raw materials are carbohydrates from the available biomass. From this biomass cellulose, starch, chitin and chitosan stand out as the most abundant (75%), as organic matter. Among the most common applications are the production of polymeric materials with chemical transformations, as raw materials for obtaining fuels, in obtaining monomers of economic interest for the development of materials for industry in general. The main monomers that can be obtained from these are monosaccharides such as glucose (fermentation processes), xylose (production of furfuryl alcohol) and arabinose (food applications), from cellulose, starch and hemicellulose that through bioconversion processes, chemical modifications or a combination of both technologies can become basic chemical products (Yarema, 2005; Ito and Matsuo, 2010).

Currently, there is an interest in identifying chemical products from sugars of this kind of biomass (Lipták, 2018), adapting technologies of the traditional petrochemical industry, basic chemical products have been identified called building blocks, with structural versatility to obtain co-products and their respective derivatives. For such an effect, the biomass is hydrolyzed which are obtained by means of processes of fermentation, chemical transformation (catalytic cyclodehydration, processes of oxidation and reduction) or pyrolysis (Kunkes et al., 2008; Huber et al., 2005), hexose (alcohols, acetic acid, lactic acid acetone, hydroxymethylfurfural (HMF), levulinic acid, glycolic acid, polyols, etc), pentoses (furfural, polyols: xitol, xylose, etc.) and lignocelluloses (phenolic derivatives, hydrocarbons, levoglucosenone, etc.) that can be used as solvents, monomers for polymer production and intermediates in the production of other substances (Comba et al., 2018; Werpy and Petersen, 2004; Marques et al., 2018; Gallezot, 2012).

The potential of these products has been studied and evaluated. Recent technological advances have significantly improved their production in biorefineries. Some of these potential products have lost interest (aspartic, glutamic, glucaric, itaconic), while others have aroused interest in the scientific community and industry in general, for the commercial opportunities offered by these products such as ethanol, furfural, 5-hydroxymethylfurfural (HMF), isoprene, bio hydrocarbons or lactic acid (Holladay, 2007).

The main component of lignocellulosic biomass is lignin, a raw material with multiple applications. Lignin is formed by dehydration of sugars creating aromatic compounds. The world production of lignin from the paper industry is 50×106 MT/year, similar to the production of plastics. It is used as a substituted polymer for phenol-formaldehyde resins. Another product obtained from lignin is lignin sulfate which is used as an ion exchange resin. One of its characteristic properties is its adhesive character (Farrán et al., 2015).

To extract the maximum amount of lignin from the Lignol process, consists of bringing the lignin in contact, in a first stage, with a mixture of ethanol/water at 50% at temperatures of 200°C and pressures of 400 psi (Zhou et al., 2018). Their high purity, low molecular weight, and abundance of reagent groups are of great importance for the production of adhesives, lignin-based fibers, products such as vanillin and complex mixtures of quinines, phenols and catechols (Humbird et al., 2011). Pyrolysis and gasification followed by reforming, as well as water treatments such as Aqueous Phase Reforming (APR) or supercritical water (SCW), are possibly the most promising technologies for the production of chemicals and fuels from lignocellulosic biomass (Wyman, 2018; Mosier et al., 2005).

There are three routes to hydrolyze polysaccharides: acid hydrolysis, enzymatic and autohydrolysis. The latter has been studied just a little, but is the most suitable for the fractionation of lignocellulosic biomass. As mentioned earlier, water can be applied at high temperatures that hydrolyze the acetyl to acetic acid groups, this being the catalyst that totally or partially solubilizes hemicellulose. It consists of two stages, pre-hydrolysis of hemicelluloses and hydrolysis of cellulose, in some cases it can be combined with an agent that solubilizes/degrades the lignin leaving the cellulose in a solid phase or by means of an organic solvent for the delignification of the material (Nguyen et al., 2000).

Another polymer of interest as a renewable raw material is chitosan, which is the second most abundant polysaccharide after cellulose. It is used as it is or after small transformations in biopolymers, in applications such as biomedicine (biodegradable sutures, controlled drug release), agriculture and livestock (seed coating, controlled release of fertilizers, animal feed additives) and in the industry (catalyst, support of enzymatic systems).

Chitosan is a natural polymer obtained from chitin, it is one of the most abundant biopolymers in nature after cellulose, converting them into important renewable resources, as they are the most abundant organic compounds on earth. Cellulose reinforces the cell wall of plants while chitosan is the resistant exoskeleton of arthropods: crustaceans, insects, arachnids and is one of the main components of the walls of fungi (Caro, 2011). However, the most important source of chitosan, at an industrial level, is chitin, which, through a process of chemical or enzymatic deacetylation, has allowed it to be produced on a large scale (Lárez Velásquez, 2006).

In 1894, the term chitosan was named by Hoppe-Seyler, based on studies by Rouget, who discovered it in 1859, finding that treating chitin with a hot solution of potassium hydroxide produces a product soluble in organic acids. This "modified chitin," as he called it, turned violet in diluted solutions of iodide and acid, while chitin took on a green color (López et al., 2018). The study of the functional properties of this biopolymer has promoted its use over the years in several different fields such as agriculture, medicine, the food industry and pharmaceuticals.

When partial de-acetylation of chitin is generated, different types of chitosan are produced, with ranges of de-acetylation degrees and different molecular weights. The difference in the properties of these polymers is noted by the different solubility they present in aqueous media. The degree of deacetylation of chitosan is in the approximate range of 70–95%, while in the case of chitin it is around 5–15% (Villar-Chavero et al., 2018). In terms of molecular weight, commercial type chitosan can vary between 100 and 800 kDa 2004 (Baxter et al., 1992; 2005). The chemical structure of this polysaccharide is linear, obtained by extensive deacetylation of chitin, after replacing acetamide groups with amino groups. It is also known as a copolymer composed of two types of structural units, N-acetyl-D-glucosamine, and D-glucosamine linked together by glycosidic bonds of the type β (1→4), which are distributed randomly and vary in proportion along the chain (Lárez Velásquez, 2006).

Chitin consists of 2-acetamide-2-deoxy-β-D-glucose units joined by links β-(1→4), which when treated by a complete deacetylation reaction produces a totally soluble material in an acidic medium known as chitane, which theoretically should be 100% deacetylated; however, when deacetylation of chitin is incomplete or eliminates at least 50% of its acetyl groups, it becomes chitosan β(1-4) 2-acetamido-2-desoxy-D-glucose and β(1-4) 2-amino-2-desoxy-D-glucose, generating chitosan with different properties, since it depends on the reaction conditions (Rinaudo 2006).

The source and method of production determine the composition of the chitosan chains and their size. For this reason, the degree of acetylation and the average molecular weight are parameters of obligatory knowledge for the characterization of this polymer and for its use in different applications. There are also other chemical-physical characteristics such as its solubility, crystallinity, ashes among others, which must be considered when using the polymer in a specific application (López et al., 2018).

Obtaining chitosan: The extraction of chitosan can be performed by chemical or enzymatic methods, but most of the techniques developed use chemical processes of hydrolysis of the protein and removal of inorganic matter, because chitin is found in its natural sources associated with proteins, pigments and inorganic salts so, to be used, these compounds must be separated (Liman et al., 2011). These processes usually use large amounts of water and energy, which often generate corrosive wastes. Currently, enzymatic treatments are being studied as a promising alternative (López et al., 2018) with which processes have been achieved that use isolated enzymes or enzymatic extracts and microbiological fermentations, but still without the efficiency of chemical treatments, in which solutions of hydrochloric acid, sodium or potassium hydroxide, acetic acid, among others, intervene.

On a commercial level, chitosan is a compound of great interest, since they are obtained from natural sources, are biocompatible, biodegradable and non-toxic in comparison with synthetic materials (Rabea et al., 2003).

Its first applications were in water and effluent treatment, due to its ability to sequester transition and post-transition metallic ions, which are useful in the decontamination of industrial wastewater, it also works as a flocculant due to its polycyclic character, for this reason, it can be used in various industries. In addition, the reactivity conferred by its amino and hydroxide groups (NH_2 and OH) allows the preparation of derivatives that greatly expand its field of action.

The evolution of the uses and applications of this polymer is very significant and its trend is growing year by year, it has been possible to identify an enormous number of applications covering fields as varied as food, medicine, agriculture, cosmetics, pharmacy, and environmental control, among others (Hernández et al., 2012).

A polysaccharide of interest to both the food and polymer industries is starch due to its biodegradability and abundance (Yu et al., 2006). Among the calories consumed by humans, about 70–80% comes from starch (Karr et al., 2000).

Starch consists of two chemically distinguishable polysaccharides: amylose and amylopectin. Although amylose and amylopectin are composed of D-glucopyranose molecules, the differences between these two polymers result in large differences in functional properties. The other minor components in rice starch are lipids and proteins, and calcium, potassium, magnesium and sodium in the ionic form (Vandeputte and Delcour, 2004).

The rigid structure of the granules consists of concentric layers of amylose and amylopectin, which remain unchanged during grinding, processing and the production of commercial starches. They have two refractive indices, so when irradiated with polarized light they develop the typical "malt cross"; this is due to crystalline zones of amylose molecules ordered in parallel through hydrogen bridges, as well as amorphous zones caused mainly by amylopectin, which are not associated with each other or with amylose (Badui, 2006).

Starches have many applications in the food, beverage and polymer industries as multifunctional and easy-to-use compounds (Taggart, 2004). The most important properties to consider when determining the use of starch in food processing and other industrial applications include physico-chemical: gelatinization and retrogradation; and functional: solubility, swelling, water absorption, syneresis and rheological behavior of its pastes and gels (Wang and White, 1994).

Starches with low protein content make them feasible to use in the production of glucose syrups. Others, due to their high gelatinization temperatures, could be used in products that require high temperatures (polymers). Those that have greater swelling power are used for the production of items that require water retention, such as meat, jellies, etc. Those that present greater clarity can be applied in confectionery products. The firmness and elasticity, as well as the high stability to refrigeration and freezing of starches indicate that they can be used as thickening and stabilizing agents in different systems (Hernández et al., 2008).

Physically or chemically modified starches can replace fats by improving product stability, creaminess and moisture retention. The different products provide different physical characteristics (degree of spreading, thermal reversibility, gel stability, resistance to heat, acids and shear, etc.) to the product. A general positive aspect of all these products is the cost, wide availability and use of conventional procedures for storage and transport. However, they can be affected by thermal shock (freezing or heat), acid and shear. In addition, depending on the type of substitute it may mask the flavor or add its own flavor (which in certain cases may be positive) (Yarema, 2005).

On the other hand, others include maltodextrins and dextrins, which are used at high concentrations providing body, viscosity and the feeling of satiety to the food. With a 25% solution a net fat reduction of 8 kcal/g is achieved. At high concentrations it can brown the food subjected to heat during the process. This is not a problem for baked goods, but limits its application in desserts, creamy sauces, soups, etc. (Ito and Matsuo, 2010).

Lipids

The renewable raw materials most used by the food and chemical industry are oils and fats (Biermann et al., 2000). For this reason, the health sector recommends the use of vegetable oils because of their many advantages for human health. They are environmentally friendly due to their high degree of biodegradation, currently their industrial applications are focused on the paint sector, food and industrial coatings, automotive industry, among other uses, due to their contribution to the reduction of emissions of volatile organic compounds, significantly reducing toxic effects to the environment and health of the population (Bockisch, 2015). One of its most interesting applications is its use in the form of biodiesel as a renewable energy source (Singh and Singh, 2010), but it is also of importantance as raw materials for chemical products. In fact, 14% of their production is used to obtain chemical substances (Hill, 2007).

It was earlier mentioned that the procurement processes of carbohydrate and protein derivatives must be environmentally friendly and that in food processing offer nutritional, nutraceutical and technological functionality; in the case of production and use of oils and fats these products must substantially reduce calories rather than complete elimination of fat (Board, 2002). This group may include lipids with emulsifier function (mono and diglycerides), Medium Chain Triglycerides (MCTs), the so-called structured lipids and synthetic fats themselves (Hartel et al., 2018).

Lipids are a group of compounds of heterogeneous structure. They are soluble in organic compounds, major constituents of adipose tissue, are formed by esters of glycerol and saturated or unsaturated fatty acids, along with minor components. Similarly, within their chemical structure they are made up of C, H and O, as well as N, S and P (Hartel et al., 2018; Bockisch, 2015). They are classified into saponifiable and unsaponifiable lipids. This group also includes a wide variety of substances with different chemical characteristics, but with common physical properties, such as:

— Insoluble or not very soluble in water.
— Soluble in organic (non-polar) solvents such as chloroform, ether, alcohol, benzene, acetone.
— Are bipolar or amphipathic.
— Form micelles.
— Melting point.
— Low density.

— Its composition depends on:
— The nature of fatty acids linked to glycerin
— The proportion in which they find themselves
— Degree of esterification
— Isomeria cis - trans
— Esterification
— Saponification

They are obtained from seeds by hydrolysis containing fatty acids and glycerin, or transesterification. The transformations of industrial interest by reaction of double bonds of the alkyl chain are epoxidation, ozonolysis, metathesis or dimerization, among others. Highlights include the development of biodegradable polymeric materials and their composites, as well as the production of biodegradable solvents (Mohanty, et al., 2002). In this way, fatty acids and their derivatives are raw materials for the production of lubricants and surfactants (medium-sized fatty acids), resins (epoxides), linoleum and acrylates (polyols), polyesters and polyurethanes (dioles), solvents (glycerol carbonates), dicarboxylic acids, polycarboxylic acids, polyols with primary alcohols and polyamides (aldehydes) (Hartel et al., 2018). Other products include stabilizers, plasticizers, adhesives, paints and coatings, pharmaceuticals, etc. (Ibarra et al., 2018).

Biodiesel is another product of industrial interest, composed of methyl esters of fatty acids is the transesterification of oils from different raw materials of both animal and vegetable origin, with methanol and usually catalyzed with bases. This product is susceptible to oxidative and hydrolytic processes when stored (Lusas et al., 2017).

Glycerin has a high functionalization that makes it susceptible to oxidation, reduction, halogenation, etherification, esterification, as well as microbiological and enzymatic bioconversions. In addition, it can be used for fuel production. Among the production of monomers derived from glycerin are: propylene glycol, acrylic acid, polyethylene esters, glycerine carbonate, epichlorohydrin or glycidol. On the other hand, microorganisms can transform glycerin, as a carbon source, into the metabolic production of different chemicals, including succinic acid (Shibata et al., 2018).

Lipids with emulgent function

Mono- and diglycerides, which have been used for about 50 years, due to their emulsifying properties, are suitable when added at low concentrations (0.5%). Given the HLB of these emulsifiers, they are suitable for forming Water-in-Oil (W/O) emulsions (see earlier). In fact, the use of emulsifiers achieves the partial substitution of fat by water and through this formula successful formulations have been achieved (Martin et al., 2018). These compounds can be used to replace all or part of the fat content in numerous applications and produce a finished product with sensory properties similar to those of the original product. A mixture of such ingredients can combine the functional properties of each of the ingredients separately (Ponphaiboon et al., 2018).

The most frequently introduced substitutes within this section are the so-called N-Flate (a mixture of mono and diglycerides with modified starch and guar gum), which has a caloric value of 5.1 kcal/g, and Dur-Lo (a mixture of mono and diglycerides of plant origin) (Ognean et al., 2006).

Medium-chain triglycerides

They belong to a group of fats with low caloric content. These products have been available in the market for approximately two decades and are obtained by esterification of glycerine with caprylic and capric acids from coconut or palm oils. These compounds do not require the involvement of lipases, are absorbed directly into the bloodstream, are rapidly metabolized and have a caloric value of 8 kcal/g. Among them, medium chain fatty acids with chain lengths of 8 to 14 carbons can be used for the production of detergents, soaps, lubricants and biofuels (Bansal et al., 2018).

Structured lipids

They are obtained by interestification between MCT and Long Chain Triglycerides (LGT). In this process, polyunsaturated fats and oils are subjected to enzymatic hydrolysis instead of performing a hydrogenation process (chemical reaction, redox type, whose final result is the addition of hydrogen (H2) to another compound), which allows trans fats that are being subjected to prohibitions and regulations for their harmful effects on health, to promote the increase of those fats of the LDL type and decrease of the HDL and minimize problems of a cardiovascular nature. This treatment offers possibilities of interest, both from the nutritional point of view (introduction of essential fatty acids in the molecule), and from that of obtaining fats with a low calorie content. These lipids show advantages in comparison with the use of mixtures of TMC and TLC when used in third generation emulsions in parenteral nutrition. A typical example is that known as "Caprenine" (Procter & Gamble) (Samateh et al., 2018), a triglyceride with approximately 50% long chain saturated fatty acids (ac. behenic) and the rest, a mixture of caprylic and capric acids with random distribution in the positions of the glyceride. This produces a fat with plastic properties similar to those of cocoa butter but with a much lower caloric value (5 kcal/g) due to the fact that behenic acid, and also its monoglyceride, are absorbed very weakly, which means that only 29% of the ingested caprenine is used. It is intended to be used to replace cocoa butter in pastry products. It is digested, absorbed and metabolized by the same mechanisms as any other triglyceride (Kleiner and Akoh, 2018; Ghosh, 2018; Venugopal et al., 2018; Engelmann et al., 2018).

Synthetic greases

These substances are designed in order to obtain substitute products with zero or very low caloric content. The strategies used so far to obtain non-absorbable synthetic fats can be summarized as follows:

— Replace glycerin with other alcohols or fatty acids with different acids in order to achieve a steric impediment on the ester bond.
— Use polycarboxylic acids esterified with long chain alcohols.
— Reduce the ester to the corresponding ether.

The first product of this type was patented by Procter & Gamble under the name "Olestra". It consists of a mixture of hexa, hepta and octaesters of sucrose with a long chain of fatty acids. The number of ester groups means that absorption by the body is less than 5%, so it is considered an achaloric product. Due to its great stability, it can be used in frying and other thermal treatments. Due to its rheological characteristics, texture, flavor and stability against conservation, it is known as "pseudo-fat". Its use in the USA as a food additive is pending final approval. Numerous studies indicate that it is neither toxic, nor carcinogenic, nor mutagenic and does not produce toxins (Kavadia et al., 2018).

FIBER SECTOR

Fibers

Vegetable fibers are important globally as they are a productive alternative with less environmental impact compared to synthetic fibers (Rodríguez et al., 2011). Among the fibers with the most presence, ancestral tradition and use is Fique (*Furcraea* spp).

In the world fique is known as jute, sisal, cavul or with its scientific name *Furcraea bedinghausii* of the family of Agavaceae. This is a species native to Tropical America, more precisely the Andean Region of Colombia and Venezuela, from where the species spread to the east coast of Brazil and all the Antilles. Its cultivation is characterized by producing a natural fiber of multiple uses, among which is the packaging industry, the manufacture of hammocks, ropes, espadrilles, sacks, among other diverse uses (Hoyos et al., 2019).

At the end of the 19th century, there was a notable increase in the demand and production of the fique fiber as a consequence of the beginning of coffee exports; thus, the eastern Antioquia region developed the production of sacks, ropes, and making of espadrilles and enjalmas. The demand was of such magnitude that it generated the need to import sacks, manufactured with a fiber of similar characteristics as jute. This led the state to create fiber processing companies, resulting in the import of specialized machinery for processing fique. Thus the interest for the cultivation of fique increased; among the reasons that different varieties of fique exist in Colombia are their high content of fiber, their conditions of optimal life, and that as such are characteristics of the fiber of fique (Ovalle-Serrano et al., 2018).

In Colombia the fique crop is a means of livelihood for more than 20,000 families throughout the country, in all these regions the importance of this crop has been growing, presenting significant advances in traditional production methods, resulting in higher quality raw material, widening the gap for research and development of new products. In addition, the benefit of fique has allowed farmers to form small cooperatives, which collect the production of fiber for marketing at a fair price, highlighting the industrial, economic and social importance for the regions cultivating fique (MADR and MAVDT, 2006; Muñoz et al., 2014).

PHARMACEUTICAL SECTOR

Metabolites

The metabolic action that some food constituents, nutrients and non-nutrients, may have in preventing diseases in the early or chronic stages, or in preventing diseases and disorders that occur at any time has attracted the attention of nutritionists, medical, pharmaceutical and food industries and the general public. The interrelationship between diet and some of the causes of death, including heart disease, cancer, trauma and diabetes, has received the highest levels of governmental support in developed countries and in science. The performance of the secondary metabolites present in many of the foods that we currently consume can be useful for the prevention or treatment of diseases, resulting in an important alternative to public health and medicine. Many of the non-nutrient constituents in foods (especially those derived from various foods such as fruits, vegetables, grains and some herbs and species) are of interest because they show some bioactivity and participate in some aspects of human metabolism (although the bioactivity they show is more related to a medicine than to nutrient components) (Chadwick and Whelan, 2008; Storey, 2005).

Interest in the Concept of Nutraceutical Products

Interest in health and well-being in developed countries has attracted citizens to the attention of diet. In addition, the enormous health costs in these countries have led to a reduction in hospital stays, out-of-hospital treatment and greater control of the economy of the medical and health system. Attention has focused on the high cost of treating chronic diseases (such as heart disease, hypertension and cancer), which could be significantly reduced in severity, postponed in their stages of development or even prevented in many individuals by simply changing their earlier living habits. For example, for almost all individuals, recommendations for a healthy diet to prevent disease include avoiding smoking, regulating exercise, maintaining body weight at an appropriate interval for height and age, eating a diet low in fat (especially saturated fat) and calories, a higher firer intake from fruits, cereals and vegetables, etc. (Hanson, 2003; Dixon, 2001; Chatzikonstantinou et al., 2018; Li and Lou, 2018).

Social, environmental, public health and economic aspects act as driving forces to make the concept of functional food very attractive. In the near future, consumers will demand not only foods with a good texture, flavor, etc., but also the ability to reduce or prevent chronic diseases.

Functional Products of Interest to Health

From the point of view of medical and public health, products that provide a health benefit should be developed for the population with real medical needs. These involve products that can be targeted to well-defined segments of the market, consisting of people who will see or feel a measurable benefit for themselves or with the help of a health professional (Tawfike et al., 2018). Some examples of disease prevention promoted by public bodies are:

— The addition of B vitamins to cereals and bakery products.
— The addition of iodine to salt.
— Fluoridation of water.

MANUFACTURING SECTOR

Polymers

Polymers provide an additional advantage derived from the use of renewable sources for their manufacture. From a global point of view, however, this is not always an advantage over conventional plastics. Life cycle analysis studies show a positive effect on the use of bioplastics when two specific environmental impacts are assessed: consumption of fossil sources and reduction of CO_2 emissions.

The use of agricultural resources in turn allows the possibility of a closed cycle waste management (from biomass to biomass). However, the benefits of using this type of management should be tested on a case-by-case basis according to evaluation criteria established through standardized LCA studies.

Among the polymers that are used in the production of compounds for the formation of coatings, bioplastics, biobased and components for their production (Andreeßen and Steinbüchel, 2019) are: polythioesteroides, amides, esters, polyesters, for the development of copolymer mixtures or biocomposites.

BIODEGRADATION

Biodegradation is a biochemical transformation of compounds into mineralization by microorganisms. Mineralization of organic compounds in aerobic conditions produce carbon dioxide and water and under anaerobic conditions methane and carbon dioxide. Abiotic hydrolysis, photo-oxidation and physical disintegration of the material can improve its biodegradation by increasing its surface area for microbial colonization or by reducing molecular weight (Palmisano and Pettigrew, 1992). Biodegradation has been defined as a change in surface properties or loss of mechanical strength (Lemm et al., 1981), assimilation by microorganisms (Potts et al., 1981), degradation by enzymes (Swift, 1994), main chain breakage and subsequent reduction in mean molecular weight of polymers (Ratner et al., 1988; Hergenrother et al., 1992). Degradation can occur by any of the above mechanisms, alone or in combination with each other (de Sousa et al., 2019).

The enzymatic action of microorganisms is one of the main mechanisms during the biodegradation process where materials can be broken down into carbon dioxide, water, methane or into inorganic components or biomass. Biodegradability is also defined as the propensity of a material to break down into its constituent molecules by natural processes (often by microbial digestion). Biological degradation is chemical in nature, but the sources of the chemicals they attack are microorganisms (Singh and Sharma, 2008).

Biodegradation can occur at different structural levels, i.e., molecular, macromolecular, microscopic and macroscopic, depending on the mechanism (Karan et al., 2019), it has been argued that the phenomenon for *in vivo* degradation cannot be equated with the term biodegradation, since biodegradation involves the active involvement of biological entities such as enzymes or microorganisms in the degradation process (Smith et al., 1987). It is, however, difficult to identify the involvement and role of biological species in *in vivo* degradation. Both hydrolytic and enzymatic processes may contribute to degradation in varying degrees during the different stages of the degradation process. Degradation can begin by hydrolysis and enzymatic degradation can dominate. Thus biodegradation includes all types of degradation that occur *in vivo* if the degradation is due to hydrolysis or metabolic processes. Therefore, biodegradation is also defined as the conversion of materials into intermediate or less complex end products by solubilization, simple hydrolysis or the action of biologically formed entities, such as enzymes and other products of the organism. Polymer molecules can, but do not necessarily break down to produce fragments, but the integrity of the material decreases in that process (Anderson and Domsch, 1989; Yang and Wang, 2019).

Conclusion

This contribution describes basic concepts around trends in the transformation and use of renewable raw materials (biomass) focused on green chemistry, of commercial interest for different economic sectors. These concepts have been focused to give clarity of the industrial potential of biomolecules obtained from different biomasses from which are obtained functional products of interest for industry manufacturing, food, pharmaceuticals, cosmetics, among others, favoring humanity by reducing the environmental impact on ecosystems.

REFERENCES

Albiol Matanic, V.C., Castilla, V. Antiviral activity of antimicrobial cationic peptides against Junin virus and herpes simplex virus. *International Journal of Antimicrobial Agents*, 2004, 23(4), 382–389.

Anderson, T.H., Domsch, K.H. Ratios of microbial biomass carbon to total organic carbon in arable soils. *Soil Biology and Biochemistry*, 1989, 21(4), 471–479.

Andrade, J.D. Surface and Interfacial Aspects of Biomedical Polymers, vol. 2. New York; Plenum Press, 1985.

Andreeßen, C., Steinbüchel, A. Recent developments in non-biodegradable biopolymers: Precursors, production processes, and future perspectives. *Applied Microbiology and Biotechnology*, 2019, 103(1), 143–157.

Arango, O. Evaluation of bio-products of fique (*furcraea gigantea*) in the control of potato late blight. *Biotecnología En el Sector Agropecuario y Agroindustrial*, 2013, 11(2), 29–36.

Arora, R., Sharma, N.K., Kumar, S., Sani, R.K. Lignocellulosic ethanol: Feedstocks and bioprocessing. *In*: Ray, R.C. and Ramachandran, S. (eds), Bioethanol Production from Food Crops: Sustainable Sources, Interventions, and Challenges. San Diego, USA: Academic Press; 2019; pp. 165–185.

Badui, S. Química de los alimentos. 4ta Edición. México: Pearson Educación, 2006.

Balat, M. Production of bioethanol from lignocellulosic materials via the biochemical pathway: A review. *Energy Conversion and Management*, 2011, 52(2), 858–875.

Bansal, S., Kim, H.J., Na, G., Hamilton, M.E., Cahoon, E.B., Lu, C., Durrett, T.P. Towards the synthetic design of camelina oil enriched in tailored acetyl-triacylglycerols with medium-chain fatty acids. *Journal of Experimental Botany*, 2018, 69(18), 4395–4402.

Baxter, A., Dillon, M., Taylor, K.A., Roberts, G.A. Improved method for ir determination of the degree of N-acetylation of chitosan. *International Journal of Biological Macromolecules*, 1992, 14(3), 166–169.

Baxter, S., Zivanovic, S., Weiss, J. Molecular weight and degree of acetylation of high-intensity ultrasonicated chitosan. *Food Hydrocolloids*, 2005, 19(5), 821–830.

Belén Camacho, D.R., Álvarez, M.J.M., García, D., Medina, C., Sidorovas, A. Caracterización de un hidrolizado proteico enzimático obtenido del pez caribe colorado (pygocentrus cariba Humboldt, 1821). Interciencia, 2007.

Bernier Oviedo, D.J., Rincón Moreno, J.A. Estudio e agentes inhibitorios en el proceso de fermentación en la obtención de etanol a partir de material lignocelulósico y amiláceo. Trabajo de grado. Universidad del Tolima, 2012. http://repository.ut.edu.co/handle/001/1077

Biermann, U., Friedt, W., Lang, S., Lühs, W., Machmüller, G., Metzger, J.O., Rüsch gen. Klaas, M., Schäfer, H., Schneider, M. New syntheses with oils and fats as renewable raw materials for the chemical industry. *Angewandte Chemie International Edition*, 2000, 39, 2206–24.

Board, N.I.I.R. Modern Technology of Oils, Fats & Its Derivatives. National Institute of Industrial Re, 2002.

Bockisch, M. (ed). Fats and Oils Handbook (Nahrungsfette und Öle). Elsevier, 2015.

Buckeridge, M.S., Grandis, A., Tavares, E.Q. Disassembling the glycomic code of sugarcane cell walls to improve second-generation bioethanol production. *In*: Ray, R.C. and Ramachandran, S. (eds), Bioethanol Production from Food Crops: Sustainable Sources, Interventions, and Challenges. San Diego, USA: Academic Press; 2019; pp. 31–43.

Caro, R.R. Caracterización y evaluación de formulaciones con quitosano como modulador de la cesión de fármacos (Doctoral dissertation, Universidad Complutense de Madrid). 2011.

Celik, M.B. Experimental determination of suitable ethanol-gasoline blend rate at high compression ratio for gasoline engine. *Applied Thermal Engineering*, 2008, 28(5), 396–404.

Chadwick, D.J., Whelan, J. (eds). Secondary Metabolites: Their Function and Evolution (Vol. 171). UK, John Wiley & Sons, 2008.

Chang, V.S., Holtzapple, M.T. Fundamental factors affecting biomass enzymatic reactivity. *In*: Finkelstein, M., Davison, B.H. (eds), Twenty-First Symposium on Biotechnology for Fuels and Chemicals. Applied Biochemistry and Biotechnology. Totowa, NJ; Humana Press; 2000; pp. 5 37.

Chatzikonstantinou, A.V., Chatziathanasiadou, M.V., Ravera, E., Fragai, M., Parigi, G., Gerothanassis, I.P., Luchinat, C., Stamatis, H., Tzakos, A.G. Enriching the biological space of natural products and charting drug metabolites, through real time biotransformation monitoring: The NMR tube bioreactor. *Biochimica et Biophysica Acta (BBA)-General Subjects*, 2018, 1862(1), 1–8.

Clarke, C.J., Tu, W.C., Levers, O., Bröhl, A., Hallett, J.P. Green and sustainable solvents in chemical processes. *Chemical Reviews*, 2018, 118(2), 747–800.

Cobos, O.F.H., Reyes, J.L.T., García, L.C.F. Biocombustibles y su aplicación en Colombia. *Scientia Et Technica*, 2007, 34, 171–176.

Collins, T. Toward sustainable chemistry. *Science*, 2001, 291(5501), 48–49.

Comba, M.B., Tsai, Y.H., Sarotti, A.M., Mangione, M.I., Suárez, A.G., Spanevello, R.A. Levoglucosenone and its new applications: Valorization of cellulose residues. *European Journal of Organic Chemistry*, 2018, 2018(5), 590–604.

Dalena, F., Senatore, A., Iulianelli, A., Di Paola, L., Basile, M., Basile, A. Ethanol from biomass: Future and perspectives. *In*: Basile, A., Iulianelli, A., Dalena, F., Veziroğlu, T.N. (eds), Ethanol: Science and Engineering. USA, Elsevier; 2019; pp. 25–59.

Davies, J. On the shapes of molecules of polyamino acids and proteins at interfaces. *Biochimica et biophysica acta*, 1953, 11, 165–177.

de Sousa, L.S., Garcia, M.A.S., Santos, E.C.P., do Nascimento Silva, J., de Castro, A.G., de Moura, C.V.R., de Moura, E.M. Study of the kinetic and thermodynamic parameters of the oxidative degradation process of biodiesel by the action of antioxidants using the Rancimat and PetroOXY methods. *Fuel*, 2019, 238, 198–207.

del Carmen Durán-Domínguez-de, M., Navarro-Frómeta, A.E., Bayona, J.M. Treatment of wastewater from livestock activities with artificial or constructed wetland josé marrugo-negrete, juan figueroa-sánchez, iván urango-cárdenas and germán enamorado-montes. *In*: Durán-Domínguez-de-Bazúa, M.del.C., Navarro-Frómeta, A.E., Bayona, J.M. (eds), Artificial or Constructed Wetlands. Boca Raton; CRC Press; 2018; pp. 186–201.

Díez, A.M., Sanromán, M.A., Pazos, M. New approaches on the agrochemicals degradation by UV oxidation processes. *Chemical Engineering Journal*, 2019, 376, 120026.

Dixon, R.A. Natural products and plant disease resistance. *Nature*, 2001, 411(6839), 843.

Duff, S.J., Murray, W.D. Bioconversion of forest products industry waste cellulosics to fuel ethanol: A review. *Bioresource Technology*, 1996, 55(1), 1–33.

Engelmann, J.I., Silva, P.P., Igansi, A.V., Pohndorf, R.S., Cadaval Jr, T.R.S., Crexi, V.T., Pinto, L.A.A. Structured lipids by swine lard interesterification with oil and esters from common carp viscera. *Journal of Food Process Engineering*, 2018, 41(4), e12679.

Fantazzini, D., Höök, M., Angelantoni, A. Global oil risks in the early 21st century. *Energy Policy*, 2011, 39(12), 7865–7873.

Fargione, J., Hill, J., Tilman, D., Polasky, S., Hawthorne, P. Land clearing and the biofuel carbon debt. *Science*, 2008, 319(5867), 1235–1238.

Farrán, A., Cai, C., Sandoval, M., Xu, Y., Liu, J., Hernáiz, M.J., Linhardt, R.J. Green solvents in carbohydrate chemistry: From raw materials to fine chemicals. *Chemical Reviews*, 2015, 115(14), 6811–6853.

Faruk, M.I., Rahman, M.L. Management of cauliflower seedling disease (*Sclerotium rolfsii*) in seedbed with different substrate based Trichoderma harzianum Bio-fungicides. *Bangladesh Journal of Agricultural Research*, 2018, 42(4), 609–620.

Fortunati, E., Mazzaglia, A., Balestra, G.M. Sustainable control strategies for plant protection and food packaging sectors by natural substances and novel nanotechnological approaches. *Journal of the Science of Food and Agriculture*, 2018, 99(3), 986–1000.

Fraaije, J., Norde, W., Lyklema, J. Interfacial thermodynamics of protein adsorption, ion coadsorption and ion binding in solution: II. Model interpretation of ion exchange in lysozyme chromatography. *Biophysical Chemistry*, 1991, 40(3), 317–327.

Gallezot, P. Conversion of biomass to selected chemical products. *Chemical Society Reviews*, 2012, 41(4), 1538–1558.

Garrido, T., Uranga, J., Guerrero, P., de la Caba, K. The potential of vegetal and animal proteins to develop more sustainable food packaging. *In*: Gutiérrez, T. (ed.), Polymers for Food Applications. Cham, Springer; 2018; pp. 25–59.

Ghosh, M. Studies on Preparation, Characterization of Structured Lipids and Their Possible Application in Food. 2018.

Global Surfactant Industry Analysis - Forecast, January, 2018

Gnansounou, E. Production and use of lignocellulosic bioethanol in Europe: Current situation and perspectives. *Bioresource Technology*, 2010, 101(13), 4842–4850.

Gokarneshan, N., Velumani, K. Significant trends in nano finishes for improvement of functional properties of fabrics. *In*: Yusuf, Md. (ed.), Handbook of Renewable Materials for Coloration and Finishing, Wiley; 2018; pp. 387–434.

Griffin, W.C. Classification of surface-active agents by "HLB". *Journal of the Society of Cosmetic Chemists*, 1949, 1(5), 311–326.

Guadix, A., Guadix, E.M., Páez, M.P., Gonzales, P., Camacho, F. Procesos tecnológicos y métodos de control en la hidrólisis de proteínas. *Ars Pharmaceutica*, 2000, 41(1), 79–89.

Guzey, D., McClements, D.J. Formation, stability and properties of multilayer emulsions for application in the food industry. *Advances in Colloid and Interface Science*, 2006, 128, 227–248.

Hanson, J.R. Natural Products: The Secondary Metabolites (Vol. 17). Royal Society of Chemistry. 2003.

Hartel, R.W., Joachim, H., Hofberger, R. Fats, oils and emulsifiers. *In*: Confectionery Science and Technology. Cham: Springer; 2018. (Chapter 4; pp. 85–124).

Haynes, C.A., Norde, W. Globular proteins at solid/liquid interfaces. *Colloids and Surfaces B: Biointerfaces*, 1994, 2(6), 517–566.

He, S., Franco, C., Zhang, W. Functions, applications and production of protein hydrolysates from fish processing coproducts (FPCP). *Food Research International*, 2013, 50, 289–297.

Hergenrother, R.W., Wabers, H.D., Cooper, S.L. The effect of chain extenders and stabilizers on the in-vivo stability of polyurethanes. *Journal of Applied Biomaterials*, 1992, 3, 17–22.

Hernández, M., Torruco, J., Guerrero, L., Betancour, D. Caracterización fisicoquímica de almidones de tubérculos cultivados en Yucatán. *Ciência e Tecnologia de Alimentos*, 2008, 725.

Hernández-Ochoa, L., Macias-Castañeda, C.A., Nevárez-Moorillón, G.V., Salas-Muñoz, E., Sandoval-Salas, F. Antimicrobial activity of chitosan-based films including spices' essential oils and functional extracts. *CyTA-Journal of Food*, 2012, 10(2), 85–91.

Hill, K. Industrial development and application of biobased oleochemicals. *Pure and Applied Chemistry*, 2007, 79(11), 1999–2011.

Holladay, J.E., White, J.F., Bozell, J.J., Johnson, D. Top value-added chemicals from biomass-Volume II—Results of screening for potential candidates from biorefinery lignin (No. PNNL-16983). Pacific Northwest National Lab (PNNL), Richland, WA (United States), 2007.

Hoyos, C.G., Zuluaga, R., Gañán, P., Pique, T.M., Vazquez, A. Cellulose nanofibrils extracted from fique fibers as bio-based cement additive. *Journal of Cleaner Production*, 2019, 235, 1540–1548.

Huber, G.W., Chheda, J.N., Barrett, C.J., Dumesic, J.A. Production of liquid alkanes by aqueous-phase processing of biomass-derived carbohydrates. *Science*, 2005, 308(5727), 1446–1450.

Humbird, D., Davis, R., Tao, L., Kinchin, C., Hsu, D., Aden, A., Schoen, P., Lukas, J., Olthof, B., Worley, M., Sexton, D., Dudgeon, D. Process design and economics for biochemical conversion of lignocellulosic biomass to ethanol: dilute acid pretreatment and enzymatic hydrolysis of corn stover (No. NREL/TP-5100-47764). National Renewable Energy Lab. (NREL), Golden, CO. (United States); 2011.

Ibarra, V.G., de Quirós, A.R.B., Losada, P.P., Sendón, R. Identification of intentionally and non-intentionally added substances in plastic packaging materials and their migration into food products. *Analytical and Bioanalytical Chemistry*, 2018, 410(16), 3789–3803.

Ireland, P., Clausen, D. Local action that changes the world: Fresh perspectives on climate change mitigation and adaptation from Australia. *In*: Letcher, T.M. (ed.), Managing Global Warming. San Diego, USA: Academic Press; 2019; pp. 769–782.

Ito, R., Matsuo, Y. Handbook of Carbohydrate Polymers. Nova Science Publishers, 2010.

Jang, W.S., Kim, H.K., Lee, K.Y., Kim, S.A., Han, Y.S., Lee, I.H. Antifungal activity of synthetic peptide derived from halocidin, antimicrobial peptide from the tunicate, *Halocynthia aurantium*. *FEBS Letters*, 2006, 580(5), 1490–1496.

Jerez, A., Partal, P., Martínez, I., Gallegos, C., Guerrero, A. Proteinbased bioplastics: Effect of thermomechanical processing. *Rheologica Acta*, 2007, 46(5), 711–720.

Karan, H., Funk, C., Grabert, M., Oey, M., Hankamer, B. Green bioplastics as part of a circular bioeconomy. *Trends in Plant Science*, 2019, 24(3), 237–249.

Karr, G.S., Cheng, E., Sun, X.S. Physical properties of strawboard as affected by processing parameters. *Industrial Crops and Products*, 2000, 12(1), 19–24.

Kaur, J., Mattu, H.S., Chatha, K., Randeva, H.S. Chemerin in human cardiovascular disease. *Vascular Pharmacology*, 2018, 110, 1–6.

Kavadia, M.R., Yadav, M.G., Odaneth, A.A., Lali, A.M. Synthesis of designer triglycerides by enzymatic acidolysis. *Biotechnology Reports*, 2018, 18: e00246..

Kisioglu, B., Nergiz-Unal, R. The powerful story against cardiovascular diseases: Dietary factors. *Food Reviews International*, 2018, 34(8), 713–745.

Kleiner, L., Akoh, C.C. Applications of structured lipids in selected food market segments and their evolving consumer demands. *In*: Bornscheuer, U.T. (ed.), Lipid Modification by Enzymes and Engineered Microbes. San Diego, USA: AOCS Press; 2018; pp. 179–202.

Kokini, J., Cocero, A., Madeka, H., De Graaf, E. The development of state diagrams for cereal proteins. *Trends in Food Science & Technology*, 1994, 5(9), 281–288.

Krochta, J.M. Proteins as raw materials for films and coatings: Definitions, current status, and opportunities. *In*: Gennadios, A. (ed.), Protein-Based Films and Coatings. Boca Raton; CRC Press; 2002; pp. 1–42.

Kuhad, R.C., Gupta, R., Khasa, Y.P., Singh, A., Zhang, Y.-H.P. Bioethanol production from pentose sugars: Current status and future prospects. *Renewable and Sustainable Energy Reviews*, 2011, 15(9), 4950–4962.

Kunkes, E.L., Simonetti, D.A., West, R.M., Serrano-Ruiz, J.C., Gärtner, C.A., Dumesic, J.A. Catalytic conversion of biomass to monofunctional hydrocarbons and targeted liquid fuel classes. *Science*, 2008, 322(5900), 417–421.

Laine, J. Ciento cincuenta años de combustión de hidrocarburos fósiles: las alternativas emergentes. *Ingeniería y Ciencia*, 2009, (10), 11–31.

Lárez Velásquez, C. Quitina y quitosano: materiales del pasado para el presente y el futuro. *Avances en Química*, 2006, 1(2).

Laser, M., Schulman, D., Allen, S.G., Lichwa, J., Antal Jr, M.J., Lynd, L.R. A comparison of liquid hot water and steam pretreatments of sugar cane bagasse for bioconversion to ethanol. *Bioresource Technology*, 2002, 81(1), 33–44.

Lemm, W., Krukenberg, T., Regier, G., Gerlach, K., Bucherl, E.S. Biodegradation of some biomaterials after subcutaneous implantation. *Proc Eur Soc Artif Org.*, 1981, 8, 71–5.

Li, G., Lou, H.X. Strategies to diversify natural products for drug discovery. *Medicinal Research Reviews*, 2018, 38(4), 1255–1294.

Limam, Z., Selmi, S., Sadok, S., El Abed, A. Extraction and characterization of chitin and chitosan from crustacean by-products: Biological and physicochemical properties. *African Journal of Biotechnology*, 2011, 10(4), 640–647.

Limayem, A., Ricke, S.C. Lignocellulosic biomass for bioethanol production: Current perspectives, potential issues and future prospects. *Progress in Energy and Combustion Science*, 2012.

Lipták, A. CRC Handbook of Oligosaccharides, vol. 3. CRC Press, 2018.

Lopez, C., Antelo, L.T., Franco-Uría, A., Alonso, A.A., Pérez-Martín, R. Chitin production from crustacean biomass: Sustainability assessment of chemical and enzymatic processes. *Journal of Cleaner Production*, 2018, 172, 4140–4151.

Lorenzo, J.M., Munekata, P.E., Gómez, B., Barba, F.J., Mora, L., Pérez-Santaescolástica, C., Toldrá, F. Bioactive peptides as natural antioxidants in food products–A review. *Trends in Food Science and Technology*, 2018, 79, 136–147.

Lusas, E.W., Riaz, M.N., Alam, M.S., Clough, R. Animal and vegetable fats, oils, and waxes. *In*: Kent J., Bommaraju T., Barnicki S. (eds), Handbook of Industrial Chemistry and Biotechnology. Cham: Springer; 2017; pp. 823–932.

MacRitchie, F. Protein adsorption/desorption at fluid interfaces. *Colloids and Surfaces*, 1989, 41, 2534.

Malusá, E., Vassilev, N., A contribution to set a legal framework for biofertilisers. *Applied Microbiology and Biotechnology*, 2014, 98, 6599–6607.

Marques, S., Moreno, A.D., Ballesteros, M., Gírio, F. Starch biomass for biofuels, biomaterials, and chemicals. *In*: Vaz Jr., S. (ed.), Biomass and Green Chemistry. Springer, Cham.; 2018; pp. 69–94.

Martin, M.J., Trujillo, L.A., Garcia, M.C., Alfaro, M.C., Muñoz, J. Effect of emulsifier HLB and stabilizer addition on the physical stability of thyme essential oil emulsions. *Journal of Dispersion Science and Technology*, 2018, 1–8.

Mathur, M., Gehlot, P. Recruit the plant pathogen for weed management: Bioherbicide a sustainable strategy. *In*: Gehlot, P., Singh, J. (eds), Fungi and their Role in Sustainable Development: Current Perspectives. Singapore; Springer; 2018; pp. 159–181.

Ministerio de Agricultura y Desarrollo Rural (MADR) y Ministerio de Ambiente, Vivienda y Desarrollo Territorial (MAVDT). Guía Ambiental del Subsector Fiquero. Panamericana Formas e Impresos, Bogotá. Ed. 2, 2006; pp. 6–13.

Mohan, K.V., Rao, S.S., Atreya, C.D. Antiviral activity of selected antimicrobial peptides against vaccinia virus. *Antiviral Research*, 2010, 86(3), 306–311.

Mohanty, A.K., Misra, M., Drzal, L.T. Sustainable bio-composites from renewable resources: Opportunities and challenges in the green materials world. *Journal of Polymers and the Environment*, 2002, 10(1–2), 19–26.

Momayez, F., Karimi, K., Taherzadeh, M.J. Energy recovery from industrial crop wastes by dry anaerobic digestion: A review. *Industrial Crops and Products*, 2019, 129, 673–687.

Mosier, N., Wyman, C., Dale, B., Elander, R., Lee, Y.Y., Holtzapple, M., Ladisch, M. Features of promising technologies for pretreatment of lignocellulosic biomass. *Bioresource Technology*, 2005, 96(6), 673–686.

Muñoz Velez, M.F., Idalgo Salazar, M.A., Mina Hernandez, J.H. Fibras de fique una alternativa para el reforzamiento de plásticos. influencia de la modificación superficial. *Biotecnología En El Sector Agropecuario Y Agroindustrial*, 2014, 12(2), 60–70.

Negro, M.J., Manzanares, P., Oliva, J.M., Ballesteros, I., Ballesteros, M. Changes in various physical/chemical parameters of Pinus pinaster wood after steam explosion pretreatment. *Biomass and Bioenergy*, 2003, 25(3), 301–308.

Nguyen, Q., Tucker, M.P., Keller, F.A., Eddy, F.P. Two stage dilute acid pre-treatment of softwoods. *Applied Biochemistry and Biotechnology*, 2000, 84-86, 561–576.

Norde, W., Lyklema, J. Why proteins prefer interfaces. *Journal of Biomaterials Science, Polymer Edition*, 1991, 2(3), 183–202.

Ognean, C.F., Darie, N., Ognean, M. Fat replacers: Review. *Journal of Agroalimentary Processes and Technologies*, 2006, 12(2), 433–442.

Ovalle-Serrano, S.A., Blanco-Tirado, C., Combariza, M.Y. Exploring the composition of raw and delignified Colombian fique fibers, tow and pulp. *Cellulose*, 2018, 25(1), 151–165.

Palmisano, A.C., Pettigrew, C.A. Biodegradability of plastics. *Bioscience*, 1992, 42(9), 680–5.

Perez Garcia, M., Romero Garcia, L.I., Quiroga Alonso, J.M., Sales Marquez, D. Influence of LAS (linear alkylbenzene sulphonates) on biodegradation kinetics. *Chemical and Biochemical Engineering Quarterly*, 1996, 10(2), 75–82.

Ponphaiboon, J., Limmatvapirat, S., Limmatvapirat, C. Influence of emulsifiers on physical properties of oil/water emulsions containing ostrich oil. *Key Engineering Materials*, 2018, 277, 592–596.

Potts, J.E., Clendinning, R.A., Ackart, W.B., Neigisch, W.D. The biodegradability of synthetic polymers. *In*: Guillet, J. (ed.), Polymers and Ecological Problems. Polymer science and Technology Series, vol. 3. New York, NY: Plenum Press; 1973; pp. 61–79.

Potts, D.E., Ahlert, R.C., Wang, S.S. A critical review of fouling of reverse osmosis membranes. *Desalination*, 1981, 36(3), 235–264.

Powers, J.P.S., Hancock, R.E.W. The relationship between peptide structure and antibacterial activity. *Peptides*, 2003, 24(11), 1681–1691.

Prud'homme, R. Foams: Theory: Measurements: Applications. Routledge, 2017.

Puri, V.P., Mamers, H. Explosive pretreatment of lignocellulosic residues with high-pressure carbon dioxide for the production of fermentation substrates. *Biotechnology and Bioengineering*, 1983, 25(12), 3149–3161.

Rabea, E.I., Badawy, M.E.T., Stevens, C.V., Smagghe, G., Steurbaut, W. Chitosan as antimicrobial agent: Applications and mode of action. *Biomacromolecules*, 2003, 4(6), 1457–1465.

Ratner, B.D., Gladhill, K.W., Horbett, T.A. Analysis of in vitro enzymatic and oxidative degradation of polyurethanes. *Journal of Biomedical Materials Research*, 1988, 22, 509–527.

Rinaudo, M. Chitin and chitosan: Properties and applications. *Progress in Polymer Science*, 2006, 31(7), 603–632.

Rodrigez Hueso, M., Villagomez Zavala, D., Lozano Valdes, B., Pedroza, R. Revista Mexicana de I ngeniería Q uímica. *Revista Mexicana de Ingeniería Química*, 2012, 11(1), 2343.

Rodríguez, E.M.M., Lupín, B., Lacaze, M.V., González, J. La producción sustentable de fibras textiles. ¿Una alternativa viable para Argentina? UNMDP; 2011; p. 136.

Rojas Salas, M.C., Luque Turriago, J.E. Biofungicida a partir del jugo de fique (*Furcraea* spp.) y evaluación de su efectividad sobre la gota (*Phytopthora infestans*) en el cultivo de papa (*Solanum tuberosum*). *Revista Educación En Ingeniería*, 2012, 7(13), 13–22. https://doi.org/10.26507/rei.v7n13.162

Röös, E., Mie, A., Wivstad, M., Salomon, E., Johansson, B., Gunnarsson, S., Wallenbeck, A., Hoffmann, R., Nilsson, U., Sundberg, C., Watson, C.A. Risks and opportunities of increasing yields in organic farming. A review. *Agronomy for Sustainable Development*, 2018, 38(2), 14.

Samateh, M., Sagiri, S.S., John, G. Molecular oleogels: Green approach in structuring vegetable oils. *In*: Marangoni, A.G., Garti, N. (eds), Edible Oleogels: Structure and Health Implications. London, UK: AOCS Press; 2018; pp. 415–438.

Santander, M., Cerón, L., Hurtado, A. Acción biocida del jugo de fique (*Furcraea gigantea* Vent.) sobre Colletotrichum gloeosporioides aislado de tomate de árbol (*Solanum betaceum* Cav.). *Agro Sur*, 2014, 42(2), 13–17.

Santos, A.L.F., Kawase, K.Y.F., Coelho, G.L.V. Enzymatic saccharification of lignocellulosic materials after treatment with supercritical carbon dioxide. *The Journal of Supercritical Fluids*, 2011, 56(3), 277–282.

Saritha, M., Tollamadugu, N.P. The status of research and application of biofertilizers and biopesticides: Global scenario. *In*: Buddolla, V. (ed.), Recent Developments in Applied Microbiology and Biochemistry. San Diego, USA: Academic Press; 2019; pp. 195–207.

Sarkar, N., Ghosh, S.K., Bannerjee, S., Aikat, K. Bioethanol production from agricultural wastes: An overview. *Renewable Energy*, 2011.

Shafiee, S., Topal, E. An econometrics view of worldwide fossil fuel consumption and the role of US. *Energy Policy*, 2008, 36(2), 775–786.

Sharaby, A., Al-Dhafar, Z.M. Some Natural Plant Extracts Having Biocide Activities against the American Bollworm *Helicoverpa armigera* (Lepidoptera: Noctuidae). *Advances in Entomology*, 2019, 07, 10–20.

Sheldon, R.A., Brady, D. The limits to biocatalysis: Pushing the envelope. *Chemical Communications*, 2018, 54(48), 6088–6104.

Shibata, Y., Koizumi, T., Yoshikawa, T. *U.S. Patent Application No. 15/779,978*, 2018.

Singh, B., Sharma, N. Mechanistic implications of plastic degradation. *Polymer Degradation and Stability*, 2008, 93(3), 561–584. doi: http://dx.doi.org/10.1016/j.polymde gradstab.2007.11.008

Singh, S.P., Singh, D. Biodiesel production through the use of different sources and characterization of oils and their esters as the substitute of diesel: a review. *Renewable and Sustainable Energy Reviews*, 2010, 14(1), 200–216.

Smith, R., Williams, D.F., Oliver, C. The biodegradation of poly(ether urethanes). *J Biomed Mater Res.*, 1987, 21, 1149–1166.

Soares, I.B., Travassos, J.A., Baudel, H.M., Benachour, M., Abreu, C.A.M. Effects of washing, milling and loading enzymes on the enzymatic hydrolysis of a steam pretreated sugarcane bagasse. *Industrial Crops and Products*, 2011, 33, 670–675.

Storey, K.B. (ed.). Functional Metabolism: Regulation and Adaptation. John Wiley & Sons, 2005.

Sun, Y., Cheng, J. Hydrolysis of lignocellulosic materials for ethanol production: A review. *Bioresource Technology*, 2002, 83(1), 1–11.

Swain, M.R., Singh, A., Sharma, A.K., Tuli, D.K. Bioethanol production from Rice-and wheat straw: An Overview. *In*: Ray, R.C. and Ramachandran, S. (eds), Bioethanol Production from Food Crops: Sustainable Sources, Interventions, and Challenges. San Diego, USA: Academic Press; 2019; pp. 213–231.

Swift, G. Biodegradable polymers in the environment: Are they really biodegradable? *In*: Gebelein, C.G., Carraher, C.E. (eds), Biotechnology and Bioactive Polymers. Boston, USA: Springer; 1994; pp. 161-168.

Taggart, P. Starch as an ingredient: Manufacture and applications. *In*: Eliasson, A.-C. (ed.), Starch in Food: Structure, Function and Applications. Cambridge; Woodhead Publishing; 2004; pp. 363–392.

Tawfike, A.F., Abbott, G., Young, L., Edrada-Ebel, R. Metabolomic-guided isolation of bioactive natural products from *Curvularia* sp., an endophytic fungus of *Terminalia laxiflora*. *Planta Medica*, 2018, 84(03), 182–190.

Tiehm, A. Degradation of polycyclic aromatic hydrocarbons in the presence of synthetic surfactants. *Applied and Environmental Microbiology*, 1994, 60(1), 258–263.

Torres, L.G., Rojas, N., Bautista, G., Iturbe, R. Effect of temperature, and surfactant's HLB and dose over the TPH-diesel biodegradation process in aged soils. *Process Biochemistry*, 2005, 40(10), 3296–3302.

Tyagi, S., Lee, K.J., Mulla, S.I., Garg, N., Chae, J.C. Production of bioethanol from sugarcane bagasse: Current approaches and perspectives. *In*: Shukla, P. (ed.), Applied Microbiology and Bioengineering. San Diego, USA: Academic Press; 2019; pp. 21–42.

Upholt, W.M. The regulation of pesticides. *In*: Mandava, N.B. (ed.), Handbook of Natural Pesticides: Methods. Boca Raton, USA: CRC Press; 2018; pp. 273–295.

Vallad, G.E., Messelink, G., Smith, H.A. Crop protection: Pest and disease management. *In*: Heuvelink, Ep. (ed.), Tomatoes, 2nd Ed. Boston, USA; CABI; 2018; Vol. 13, pp. 207–257.

Vandeputte, G., Delcour, J. From sucrose to starch granule to physical behaviour: A focus on rice starch. *Carbohydrate Polymers*, 2004, 58, 256–266.

Venugopal, H., Rao, H.R., Kumar, H.A. Structured lipids: A unique designer lipid. *Research & Reviews: Journal of Dairy Science and Technology*, 2018, 5(2), 30–39.

Villar-Chavero, M.M., Domínguez, J.C., Alonso, M.V., Oliet, M., Rodriguez, F. Thermal and kinetics of the degradation of chitosan with different deacetylation degrees under oxidizing atmosphere. *Thermochimica Acta*, 2018, 670, 18–26.

Vioque, J., SánchezVioque, R., Clemente, A., Pedroche, J., Yust, M.D.M., Millán, F. Péptidos bioactivos en proteínas de reserva. *Grasas y Aceites*, 2000, 51(5), 361–365.

Vioque, J., SánchezVioque, R., Pedroche, J., del Mar Yust, M., Millán, F. Obtención y aplicaciones de concentrados y aislados protéicos. *Grasas Y Aceites*, 2001, 52(2), 127–131.

Von Braun, J. Food and Financial Crises: Implications for Agriculture and the Poor. International Food Policy Research Institute (IFPRI), *Food Policy Report No. 20*, Washington, DC, 2008.

Waldhoff, H., Spilker, R. (eds). Handbook of Detergents, Part C: Analysis (Vol. 123). Boca Raton, USA: CRC Press, 2016.

Wang, L., White, P. Structure and properties of amylose, amylopectin and intermediate materials of oat starches. *Cereal Chemistry*, 1994, 71, 263–268.

Watson, D. Pesticides and Agriculture: Profit, Politics and Policy. Burleigh Dodds Science Publishing, 2018.

Werpy, T., Petersen, G. Top value added chemicals from biomass: Volume 1–Results of screening for potential candidates from sugars and synthesis gas (No. DOE/GO-102004-1992). National Renewable Energy Lab., Golden, CO (US). Department of Energy, Washington, DC, 2004.

Wyman, C.E. Ethanol production from lignocellulosic biomass: Overview. *In*: Wyman, C.E. (ed.), Handbook on Bioethanol: Production and Utilization. New York: Routledge; 2018; pp. 1–18.

Wyman, C.E., Cai, C.M., Kumar, R. Bioethanol from lignocellulosic biomass. *In*: Meyers, R.A. (ed.), Encyclopedia of Sustainability Science and Technology. New York, USA: Springer, 2017; pp. 997–1022.

Xu, X.F., Lin, T., Yuan, S.K., Dai, D.J., Shi, H.J., Zhang, C.Q., Wang, H.D. Characterization of baseline sen-sitivity and resistance risk of *Colletotrichum gloeosporioides* complex isolates from strawberry and grape to two demethylation-inhibitor fungicides, prochloraz and tebuconazole. *Australasian Plant Pathology*, 2014, 43(6), 605–613.

Yang, G., Wang, J. Biohydrogen production by co-fermentation of sewage sludge and grass residue: Effect of various substrate concentrations. *Fuel*, 2019, 237, 1203–1208.

Yarema, K.J. Handbook of Carbohydrate Engineering. CRC Press, 2005.

Yu, L., Dean, K., Li, L. Polymer blends and composites from renewable resources. *Progress in Polymer Science*, 2006, 31, 576–602.

Zabed, H. Bioethanol production from high sugary corn genotypes by decreasing enzyme consumption. *In*: Islam, M.R. (ed.), Social Research Methodology and New Techniques in Analysis, Interpretation, and Writing. Pennsylvania, USA: IGI Global; 2019; pp. 216–240.

Zghal, R.Z., Ghedira, K., Elleuch, J., Kharrat, M., Tounsi, S. Genome sequence analysis of a novel *Bacillus thuringiensis* strain BLB406 active against *Aedes aegypti* larvae, a novel potential bioinsecticide. *International Journal of Biological Macromolecules*, 2018, 116, 1153–1162.

Zhou, Z., Lei, F., Li, P., Jiang, J. Lignocellulosic biomass to biofuels and biochemicals: A comprehensive review with a focus on ethanol organosolv pretreatment technology. *Biotechnology and Bioengineering*, 2018, 115(11), 2683–2702.

Chapter 10

Reduce Derivatives

Revathi Kottappara[1], Shajesh Palantavida[2],
and Baiju Kizhakkekilikoodayil Vijayan[1*]

[1]Department of Chemistry/Nanoscience, Kannur University, Swami Anandha
Theertha Campus, Payyannur, Edat P.O., Kerala-670 327, India.
[2]Centre for Nano and Materials Science, Jain University,
Jakkasandra, Ramanagaram, Karnataka, India.

INTRODUCTION

One of the key principles of green chemistry is to reduce or avoid derivatization in a chemical reaction. Derivatization necessitates the use of extra energy and reagents and will lead to extra waste generation in the synthesis. It also includes the use of protecting or deprotecting agents and any short term modification of the physical and chemical process. Blocking, protection and deprotection of functional groups has been part of standard organic synthetic methodology for decades and has made feasible the synthesis of numerous complex molecules. But this approach comes at a price, added synthetic steps, possible decrease in yield and a reduction in the atom economy of the process.

The use of protecting groups in a reaction introduces chemoselectivity in the reaction as shown in the Scheme 10.1. The parent compound contains two reactive sites with similar reactivity. The desired product is obtained when the Reagent (R) reacts with only one of the sites, say 'A'. Any reaction at site 'B' will lead to an unwanted product.

*For Correspondence: baijuvijayan@kannuruniv.ac.in, baijuvijayan@gmail.com

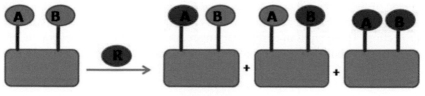

Scheme 10.1

Chemoselectivity is achieved when site 'B' is blocked by a protecting/blocking group and the subsequent reaction with the reactant 'R' can only occur at the desired site. The protecting group is removed after the reaction to regenerate site 'B', Scheme 10.2.

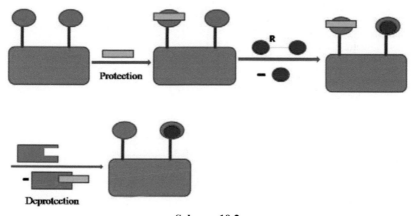

Scheme 10.2

Two additional steps, namely protection and deprotection, are required to bring out one change in this synthesis strategy.

A common example is the use of $LiAlH_4$ for the hydrogenation of esters to alcohols. $LiAlH_4$ is highly reactive and will respond to any carbonyl group present in the ester. The general strategy to selectively reduce the ester group in the presence of another carbonyl group is to convert the carbonyl group into an acetal, which is unreactive to hydrides. Acetal acts as the protecting group of the carbonyl. After the reduction using hydrides, the acetal is converted back to the original carbonyl group using aqueous acid, the deprotection step. In the absence of the protecting step, the reaction will yield diols. The acetal protection of a ketone during reduction of an ester, Vs. reduction to a diol when unprotected is shown in Scheme 10.3.

Similarly amine groups can be protected by converting them into the benzyl derivative. After the completion of the reaction, the NH_2 group can be regenerated by the cleavage of benzyl moiety. Selective reduction of the nitrile group in the presence of an amine using the protection strategy is given in Scheme 10.4.

Another example is the protection of alcohol groups by conversion to benzyl ether to achieve selective oxidation on another part of the molecule. After the completion of the oxidation process, the deprotection can be effected by the cleavage of the benzyl ether. Apart from additional steps, this strategy also requires handling of benzyl chloride which is very hazardous and becomes a waste after deprotection. These types of derivatization are usually common in several industrial processes involved in the synthesis of fine chemicals, drugs, pesticides and dyes.

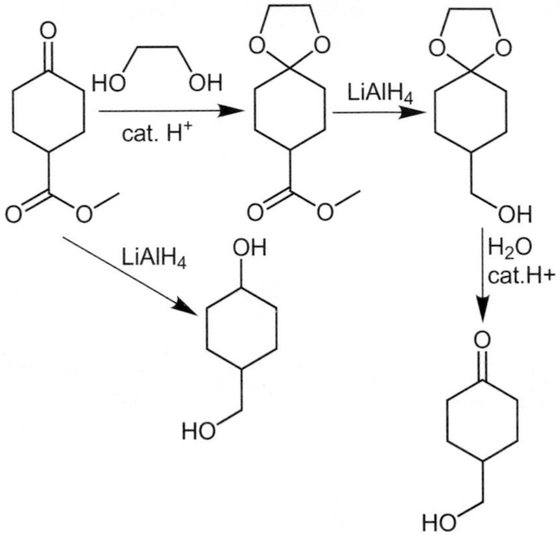

Scheme 10.3

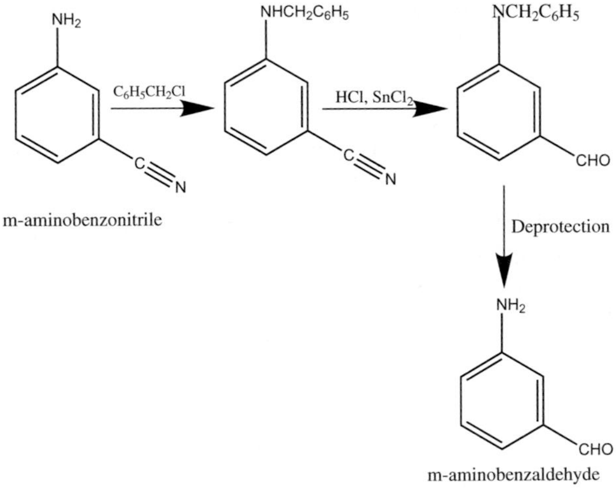

Scheme 10.4

Protecting Groups in Synthesis

The choice of the protecting group is a key factor behind the successful realization of a synthetic process. The overall efficiency and length of the synthetic process is substantially influenced by the choice of the protecting group. The antitumor activity of natural products containing 1,5-epoxybenzazocine, FR 900482 and FR66979 (H) triggered attempts to the total synthesis of these molecules and their analogues (McClure and Danishefsky, 1993; Schkeryantz and Danishefsky, 1995) (Scheme 10.5).

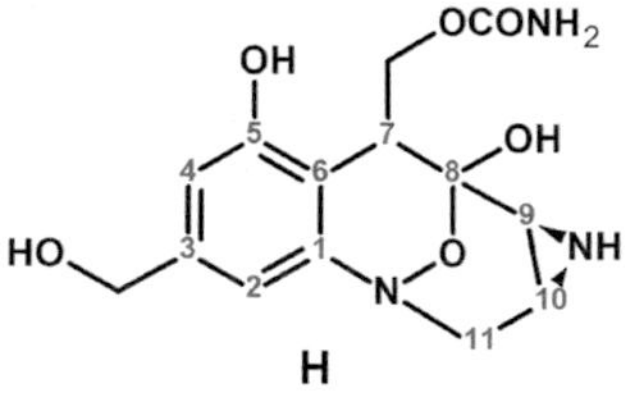

Scheme 10.5

A very attractive synthetic proposal for the molecules was the scheme establishing the benzoxazine ring system (D) by an intramolecular arylation of fused aziridine bearing seco systems (C). The primary advantage of this approach was that the cyclization substrates could be obtained through the Diels Alder addition of appropriate dienes (B) and dienophiles (A) (Scheme 10.6).

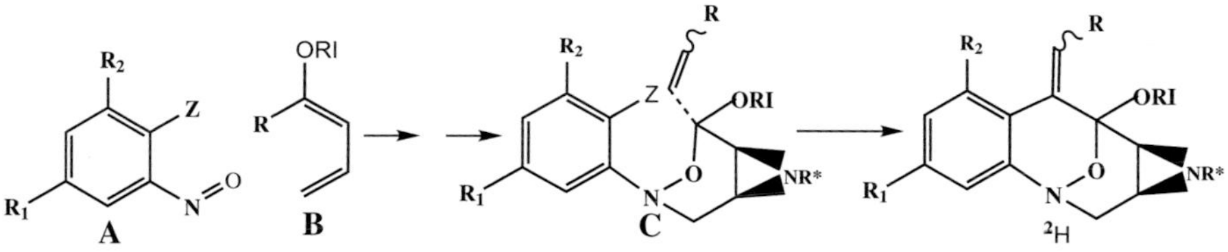

Scheme 10.6

The synthesis of (D) and further, epoxidation of the double bond at position 7 could be achieved for (E), though with some difficulty (Scheme 10.7). The further conversion to G was also possible. Among the final steps for the conversion to the natural product containing only hydroxyl groups, the ester group could be reduced to hydroxyl. But the hydrolysis of the ether protecting groups could not be achieved without the destruction of the aziridin ring. Hence the approach failed in the final deprotection step to deliver the product.

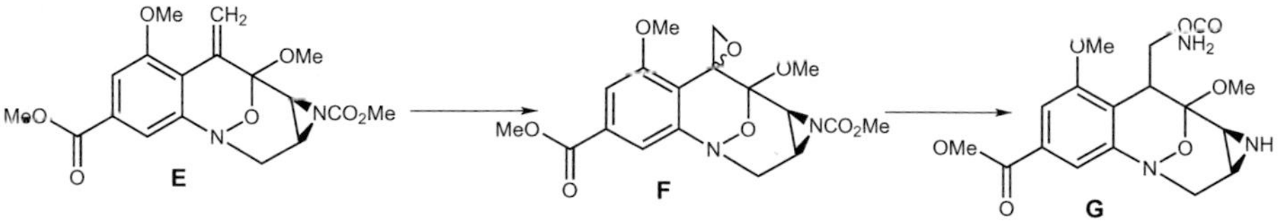

Scheme 10.7

Hence the major requirement of a successful synthesis, both in the greener as well as economic aspects seems to be the accurate retrosynthetic planning and distinct protecting group strategy that distinguishes the lability of the intermediate and the reagents. Table 10.1 shows some common protecting groups (Romanski et al., 2012).

Table 10.1 Some common protecting groups in organic synthesis and their deprotection method (Romanski et al., 2012)

	Protecting group	Deprotection method
Protection of alcohols	Benzoyl (Bz)	Removed by acid or base more stable than Ac group
	Tetrahydropyranyl(THP)	Removed by acid
	Trityl(triphenylmethyl, Tr)	Removed by acid and hydrogenolysis
	Trimethyl silyl(TMS)	Removed by acid or fluoride ion
	Acetyl(Ac)	Removed by acid or base
	ß-methoxyethoxymethyl ether (MEM)	Removed by acid
	Benzyl (Bn)	Removed by hydrogenolysis
	Methoxymethyl ether (MOM)	Removed by acid
	p-methoxybenzyl ether (PMB)	Removed by acid, oxidation or hydrogenolysis
	Methylthiomethyl ether	Removed by acid
Carbonyl protecting groups	Acetals and ketals	Removed by acid
	Dithianes	Removed by metal salts or oxidizing agents
	Acylals	Removed by Lewis acids
Carboxylic acid protecting groups	Methyl esters	Removed by acid or base
	Esters of 2, 6-disubstituted phenols like 2,6-dimethylphenol, 2,6-di-tert-butylphenol	Removed at room temperature by DBU-catalyzed methanolysis under high pressure conditions
	Benzyl esters	Removed by hydrogenolysis
	Oxazoline	Removed by strong hot acid
Amine protecting groups	p-methoxybenzyl carbonyl (Moz or MeOZ) group	Removed by hydrogenolysis
	Carbobenzyloxy(Cbz) group	Removed by hydrogenolysis
	Tert-Butyloxycarbonyl (BOC) group	Removed by concentrated strong acid like CF_3COOH
	Benzyl(Bn)	Removed by hydrogenolysis
	Carbamate group	Removed by acid and mild group
	p-methoxybenzyl (PMB)	Removed by hydrogenolysis
	Tosyl (Ts)	Removed by concentrated acid (HBr, H_2SO_4) and strong reducing agents
Phosphate protecting groups	Methyl(Me)	Removed by strong nucleophiles
	2-cyanoethyl	Removed by mild base

Use of Enzymes and Enzyme Labile Protecting Groups

Enzymes are biological catalysts that catalyze biochemical reactions with high chemo- and regioselectivity owing to their stereo-discriminating properties. Their extreme discrimination allows effective modification of specific sites, without disturbing the remaining sites in the molecule. Hence the use of enzyme to catalyze conversions can eliminate protecting and deprotecting steps in the reaction. A good example of this concept is the industrial synthesis of semisynthetic antibiotics including amoxicillin and ampicillin. The conventional approach of their synthesis starts by protecting R–H group of penicillin G with silyl ester (R = Si(Me)$_3$) which then reacts with phosphorous pentachloride at −40°C to form chlorimidate. The subsequent hydrolysis yields 6-aminopencillanic acid, a starting material for most semi-synthetic penicillins (Scheme 10.8) (Constable et al., 2007).

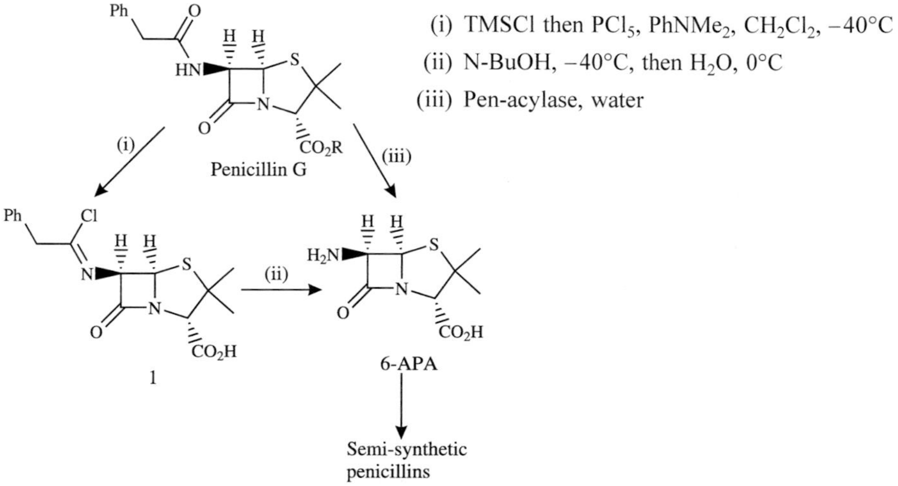

(i) TMSCl then PCl$_5$, PhNMe$_2$, CH$_2$Cl$_2$, −40°C

(ii) N-BuOH, −40°C, then H$_2$O, 0°C

(iii) Pen-acylase, water

Scheme 10.8

Following the green perspective, silyl protection step was eliminated by using pen-acylase enzyme that yields 6-APA directly in the aqueous phase and just above room temperature from penicillin G. In fact, every year several thousand metric tons of 6-APA is synthesized by this greener enzymatic approach.

Even though enzymes provide high selectivity in a reaction their use in synthetic chemistry is limited by their low stability, incompatibility to solvents and high cost. To overcome these limitations Cross-Linked Enzyme Crystals (CLECs) have been developed by Altus Biologics and they considerably increase the potential of enzymes in synthetic organic reactions (Mitchell et al., 1997; Anastas et al., 2000). Compared to conventional enzymes, CLECs are compatible with aqueous and organic solvents and can withstand a range of temperatures and pH. CLES-assisted synthesis of the antibiotic cephalexin avoids the N-protection step of D-phenylglycine (Anastas and Lankey, 2000) (Scheme 10.9).

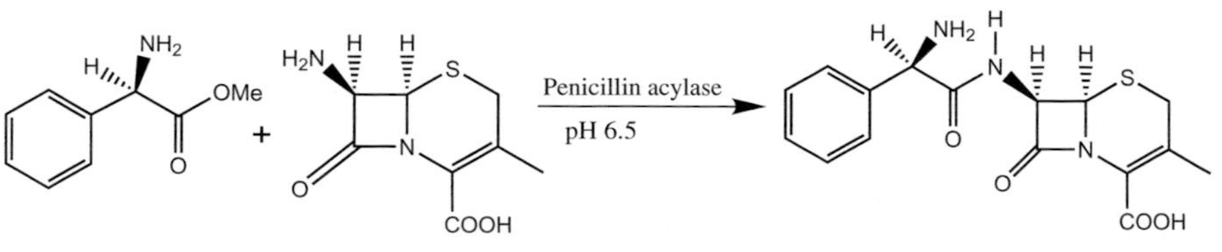

Scheme 10.9

The high regio- and enantio- selectivity of enzymatic synthesis have also found application in the area of polymer synthesis. Polyesters, polycarbonates, polysaccharides, polyurethanes, polyaromatics and vinyl polymers have been synthesized with the aid of enzymes like oxidoreductases, transferases and hydrolases (Matsumura et al., 2006; Gross et al., 2001a, b). For example, polyol-containing polyesters can be formed through the selective lipase-catalyzed condensation polymerization of diacids and reduced sugar polyols like sorbitol (Nakamura and Whited, 2003) Scheme 10.10.

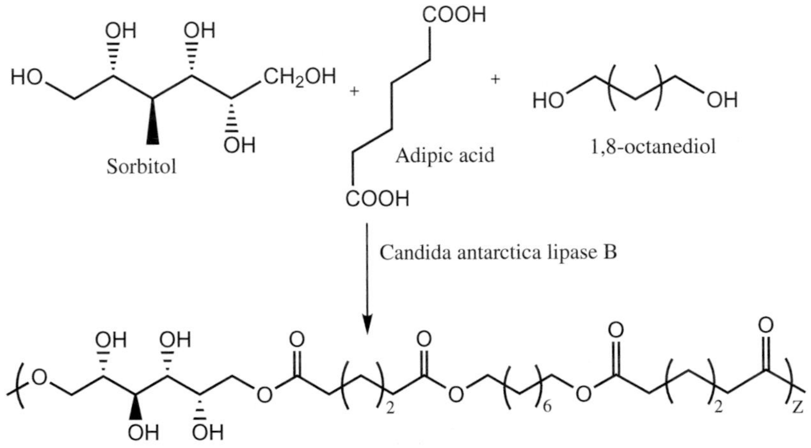

Scheme 10.10

Direct condensation of adipic acid and sorbitol in the bulk can be achieved at 90°C and 48 hours using immobilized lipase B from Candida Antartica. There is no need of any organic solvent, instead the monomers adipic acid, glycerol and sorbitol are solubilized in the binary or ternary mixture. Besides, the process does not require any preactivation of adipic acid. The condensation product, poly(sorbityl adipate) with an average molecular weight 10^4 (M_n) (1.7×10^4 (M_w)) can be obtained from this method. Here it is worth mentioning that the condensation reaction between sorbitol and glycerol proceeds with high regioselectivity The polyol monomers contain three or more hydroxy functional groups and the resulting polymer is expected to be highly cross-linked. But under enzymatic polymerization, only two of the hydroxyl groups are reactive and the high regioselectivity results in lightly branched polymers. The degree of branching changes with the stoichiometry of reaction time and monomer.

Another example is the ring opening polymerization of lactones and cyclic carbonates. Bisht et al. used enzymes lipase PS-30 and Novozyme-435 for the ring opening polymerization of ω-pentadecalactone and trimethylene carbonate at 70°C and obtained high molecular weight polymers with moderate dispersity (Bisht et al., 1997a, b). The enzyme catalysis offered improved propagation kinetics and molecular weight under mild conditions. Similarly, porcine pancreatic lipase has been reported to catalyze the ring opening polymerization of ε-caprolactone or trimethylene carbonate with ethyl glucopyranoside. The enzyme selectively promotes the polymerization reaction from the 6-hydroxyl position (Bisht et al., 1998). The reaction can be carried out under solvent less conditions since ethylglucopyranoside and monomers are in the liquid state (Ran et al., 2008).

Heavy Metal Salt Assisted Cleavage of Protecting Group

Another significant advance that needs mentioning is the catalytic deprotection. Noble metals have been found to activate protecting groups and cause deprotection when present in catalytic amounts. For example, the deprotection of allyl functional group can be carried out by using catalytic amounts of Rh^+, Ir^{2+} and Pd^0 complexes and even Pd-C (Scheme 10.11). This approach reduces wastage and eases the labor of the deprotection step. Scheme 10.9 shows the cleavage of allyl ethers, allyl esters, allyl urethanes and allyl carbonates by using heavy metal complexes.

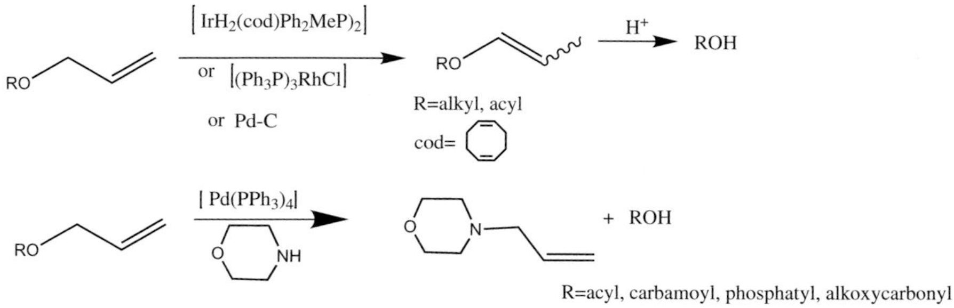

Scheme 10.11

Generation of Selectivity

Another approach commonly adopted by synthetic chemists is generating high selectivity in a reaction. When more than one reactive site is present in the reagent, the yield of the desired reaction can be raised by selectively directing the reaction to the preferred reactive site. Derivatizing the preferred reactive site such that it becomes more susceptible to the reacting species will induce selectivity in the reaction. A common example is directing nucleophilic substitution reactions by using halogen derivatives. The halogen atom imparts a high electropositive nature to the attached carbon, directing nucleophilic attack to the substituted carbon. Halogen, being a better leaving group, is eliminated in the product. Obviously, halogen derivatization results in the formation of unwanted halogen waste after the substitution reaction. Hence the comparative greenness of process cannot be estimated from the number of derivative steps alone.

Use of Photolabile Protecting Groups

Photochemical reactions usually involve milder conditions that are very well aligned with green principles. Photolabile protecting groups are useful in this regard. They help avoid the chemical conditions usually required when using conventional protecting groups. Photoremovable protecting groups possess a chromophore which is light sensitive but relatively stable to the reagents used in the synthetic process. One restriction is that the wavelength of the light used should be adsorbed only by the protecting group and not by other parts of the molecule. Photoremovable protecting groups must possess a short-lived excited state. A long-lived excited state will result in a decreased rate of cleavage reaction due to the increase in quenching processes. The successful demonstration of an efficient photo removable protecting group was first reported by Barltrop and coworkers. Benzyloxycarbonylglycine (A) can be converted to the free amino acid by ultraviolet irradiation (Barltrop and Schofield, 1962). Glycine (B) (75%) is formed along with phenylethylamine (C) and N-benzylglycine (D) as side products. Glycine can be isolated as its N-benzoyl derivative (E) (Scheme 10.12).

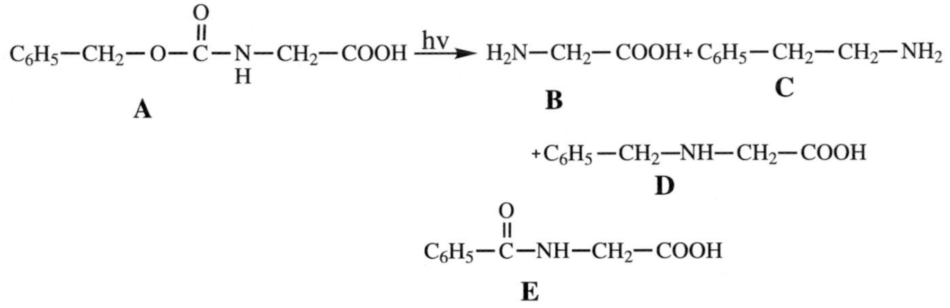
C_6H_5—CH_2—O—C(=O)—N(H)—CH_2—COOH
hv
H_2N—CH_2—COOH + C_6H_5—CH_2—CH_2—NH_2
A
B
C
+ C_6H_5—CH_2—NH—CH_2—COOH
D
C_6H_5—C(=O)—NH—CH_2—COOH
E

Scheme 10.12

Here N-benzylglycine formation occur by the interaction of benzylcarbenium ions with glycine and phenylethylamine is formed by the intramolecular N-and C-alkylation mechanism (Carpino, 1973). Chamberlin used 3,5-dimetoxybenzyloxycarbonyl group as a photolabile N-protecting group (Zimmerman and Sandel, 1963; Chamberlin, 1966). His studies reported that the reaction on 3, 5 dimethoxybenzyl p-nitrophenyl carbonate with amino acids gives N-protected derivatives which upon light irradiation in aqueous dioxan solution yields free amino acid with 3,5-dimethoxybenzyl alcohol as the side product (Scheme 10.13).

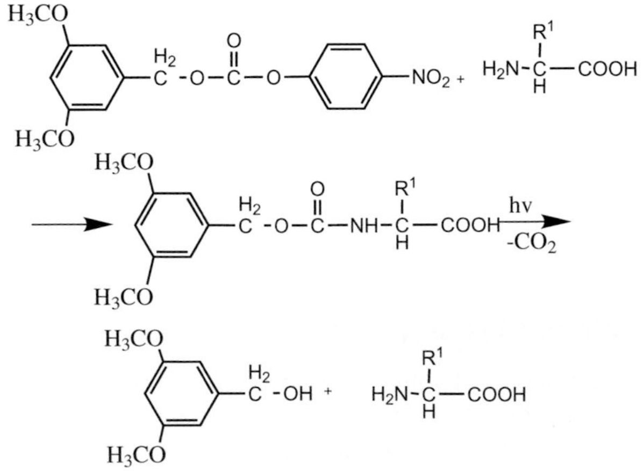
H_3CO
H_2
C-O-C-O
NO_2 +
H_2N-C—COOH
R^1
H
hv
-CO_2
C-O-C—NH-C—COOH
C-OH +

Scheme 10.13

Hence use of 3, 5-dimethoxybenzyloxycarbonyl group does not necessitate any use of chemical reagents or reduction reactions. The photolability of this protecting group is widely used in solid phase peptide synthesis for blocking and releasing amine groups (Shaaya and Sekeris, 1971). Other examples of photochemically removable protecting groups include 6-nitroveratryloxycarbonylchloride (NVOC-Cl) and 2-nitrobenzyloxycarbonyl chloride (NBOC-Cl) which can be prepared by reacting the parent alcohols with phosgene. Under Schotten-Baumann conditions these reagents react with amino acids, to yield NBOC- and NVOC-amino acids as blocking groups which can be easily removed by irradiation with light of wavelength longer than 320 nm. Here it is worth mentioning that, under these deblocking conditions, even tryptophan, the aminoacid which is most sensitive to light is not affected. The applications of a protecting group cover different areas of organic synthesis including synthesis of complex polyfunctional molecules (Amit et al., 1974; Zehavi and Patchornik, 1972; Pillai, 1980). However, the use of such a photolabile protecting group is yet to be frequent as the use of other protecting groups.

Chromatic Orthogonal Photolysis of Protecting Groups

One of the major bottlenecks in the use of protecting groups is orthogonality, which is defined as the possibility of selective removal of one group among a set of groups in a chronological order (Bochet, 2001). In 2002, Kessler successfully demonstrated a new protecting group strategy based on chromatic orthogonality which involved group differentiation using specific wavelengths of light (Blanc and Bochet, 2002). In synthetic organic chemistry, Solid-Phase Organic Synthesis (SPOS) has significant importance from a green perspective (Santini et al., 1998; Barany et al., 1987). SPOS involves attaching the preferred substrate to a solid support by means of a linker which can withstand the reagent used in the required synthesis. The final stage of the reaction involves detaching the linker using acids or bases or other harsh environments. A plausible alternative to such harsh conditions of acids and bases is the use of photolabile linkers (Lloyd-Williams et al., 1993; Patchornik et al., 1970; Peukert and Giese, 1998). Kessler and coworkers

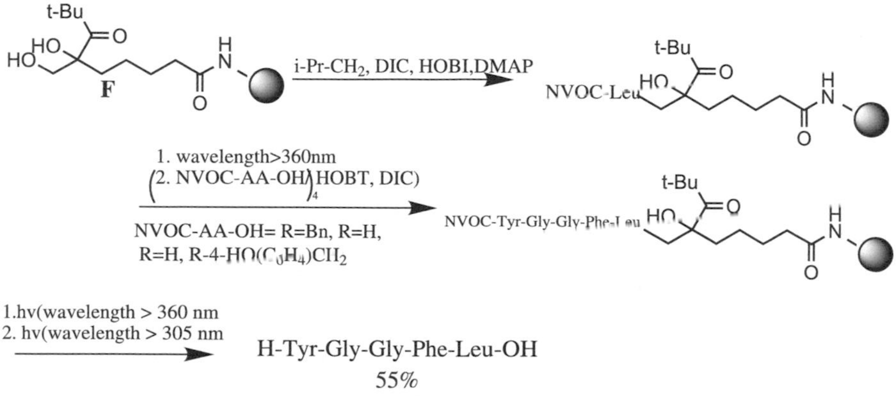

Scheme 10.14

demonstrated the synthesis of Leu-Enkephalin (H-Tyr-Gly-Gly-Phe-Leu-OH), and endorphin pentapeptide by using TentaGel-bound as the photosensitive linker (F). The sequential chromatic lability between this ter-butyl ketone derived linker (sensitive to 305 nm light) and a nitroveratryloxycarbonyl (NVOC) group (sensitive to 360 nm light) facilitated the synthesis of Leu-Enkephalin in an overall yield of 55% (Scheme 10.14). It is also worth mentioning that the whole synthesis was carried out in neutral medium and the proposed synthetic strategy avoided the use of common harsh deprotection reagents like trifluoroacetic acid or piperidine (Kessler et al., 2003).

Protecting Group as a Positive Ingredient in Chemical Reaction

According to green concepts, the use of a protecting group is undesirable. But if the use of a protecting group activates the upcoming transformation, a rethinking may be necessary. An example for such a protecting group is O-phenylenediamine, used in peptide synthesis (Schelhaas and Waldmann, 1996). As shown in Scheme 10.15, the peptide derivatives react with urethane groups to give 1-acylbensimidazoline-2-ones. Such activated compounds are capable of coupling directly with amino acid esters and peptide esters under basic conditions (Schelhaas and Waldmann, 1996).

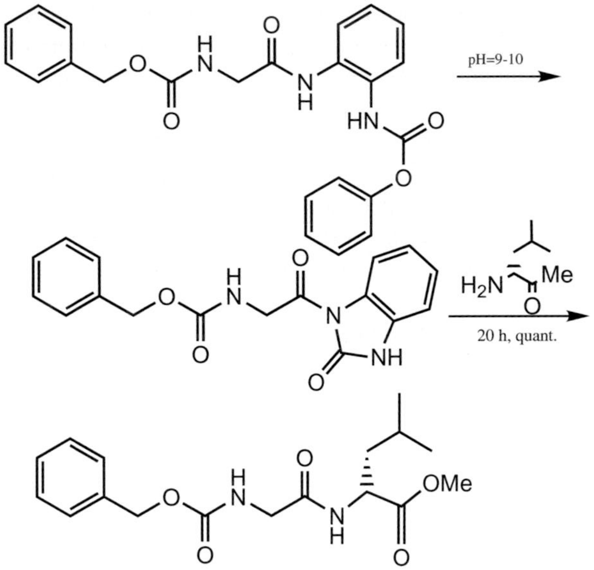

Scheme 10.15

Alternatives to Covalent Protecting Groups

Non-covalent derivatization

Non-Covalent Derivatization (NCD) is a possible alternative to traditional synthetic chemistry procedures which is based on "covalent" interactions between the reagent species. NCDs work by generating temporary variations in the precursor species by modifying the intermolecular interactions. The formation of NCDs occur by mere mixing of the starting material (co-former) with an auxiliary material, generally a solid state matrix and are held together by means of non-covalent forces under ambient conditions. The final product thus will be termed a non-covalent derivative.

Non-covalent derivatives (NCD) possess properties that significantly differ from those of the parent molecules. NCDs can be of two types: cocrystals and eutectics. Simply, a cocrystal can be defined as an NCD formed by the combination of stoichiometric amounts of two or more molecules held together by non-covalent interactions. A eutectic is also a homogeneous material formed from two solids, but it can be identified as a minimum in a phase diagram (Ågerstrand et al., 2015). If the enthalpy change predominates the entropy loss in the process, the result will be a cocrystal, or will be an eutectic (Bouissou-Schurtz et al., 2014; Stoler and Warner, 2015). Non-covalent derivatization can result in significant variations in the properties including solubility, melting point, stability and optical parameters of the parent molecule. Compared to the conventional techniques of covalent derivatization using protecting groups salt derivatives etc., the NCDs is an effective alternative for bringing out the same kind of property modifications in a species while being less toxic, generates lesser amount of waste and thus more environmentally benign (Gee and Green, 1998; Anastas and Eghbali, 2010).

The concept of non-covalent derivatization occurred in the early 1990s (Knapp, 1979; Parton et al., 1993). Non-covlent derivatization is a derivatization that relies on intermolecular interactions such as ionic, Van der Waals forces, hydrogen bonding or pi-pi interactions. Warner developed the innovative concept as a step towards minimizing the use of energy as well as materials during chemical transformations. An early and well-established system of non-covalent derivatization is given by the controlled diffusion and solubility of hydroquinone used in polaroid films (Cannon and Warner, 2002; Warner, 2006; Trakhtenberg and Warner, 2007). The major molecular component of a negative of the polaroid photograph is the organic developers. These organic developers are actually hydroquinone molecules. The major problem associated with the use of hydroquinone in the photographic system is auto-oxidation and premature diffusion. The speed of the photographic process relies upon the reactivity of the hydroquinone developers. However, the high reactivity of the hydroquinone may result in significant oxidation with uncontrolled vagrant oxygen in the system which may critically lower the hydroquinone concentration even before it reacts with silver ions. Under these conditions, hydroquinone oxidizes to form quinhydrone, a 1:1 complex of hydroquinone and quinine (Scheme 10.16).

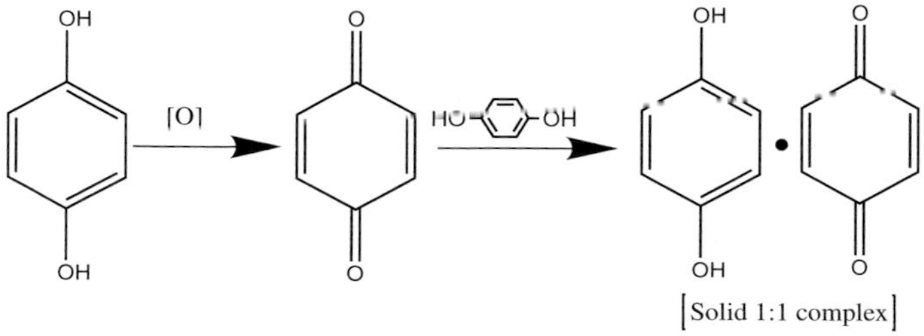

[Solid 1:1 complex]

Scheme 10.16

In addition to this, the inconsistent solubility behavior of the hydroquinone developers will lead to premature diffusion of the hydroquinone molecule. The conventional methods adopted for overcoming this problem was the covalent modification of hydroquinone, especially by protecting the hydroquinone oxygen with certain protecting groups (Guarrera et al., 1997; Blout and Simon, 1964). The most conducive protecting group used was base labile functionalized substituents, because of the presence of alkali present in the "pod" reagent. The hydroquinone oxygen was blocked by acyl hydrolysis and β-elimination mechanisms to prevent its oxidation. Thus, hydroquinone becomes oxidatively inert until its activation by the release of protecting group by base (Scheme 10.17).

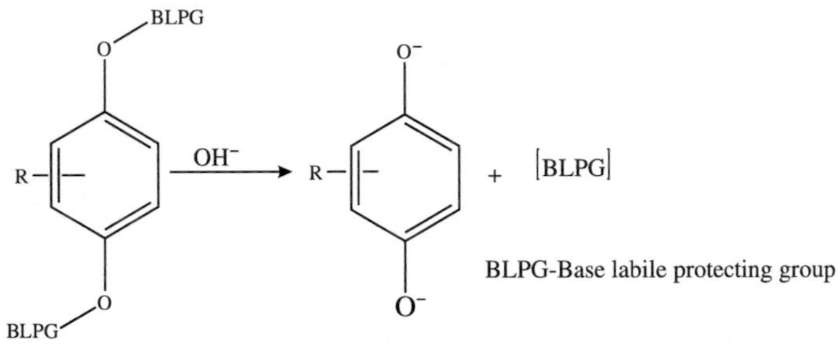

BLPG-Base labile protecting group

Scheme 10.17

The use of a protected hydroquinone candidate in a photographic system typically requires several additional steps including synthesis and purification, which leads to extra steps, wastage and hazardous waste generation. In order to design a photographic process in a greener way, researchers relied on supramolecular construction involving non-covalent interactions to control the chemical behavior (Guarrera et al., 1994). They developed a non-covalent protection between hydroquinones and bis-(N, N-dialkyl)terephthalimide in the form of a 1:1 cocrystal which is formed by both solvent recrystallization and mechanochemical process (Anastas and Eghbali, 2010). Hydroquinones protected by non-covalent interactions with bis-(N,N-dialkyl)terephthalamides is shown in Scheme 10.18.

The electron rich hydroquinone and electron deficient terephthalimide offers a π-stacking system which creates a base-labile attraction through tertiary amide-phenol hydrogen bond. The properties of the supramolecular non-covalent derivative between hydroquinone and terephthalimide are entirely different from the single components. The NCD thus formed are relatively water insoluble while hydroquinone has significant solubility. The insoluble behavior of the supramolecular derivative allows the easy processing of the photographic material in water. Besides, it is also noteworthy that since the solid complex is stabilized by hydrogen bonds, the hydroquinones are easily deprotonated when the pH of the film increases by the introduction of alkali from the "pod", they get readily solubilized and thus activated for use in the photographic system.

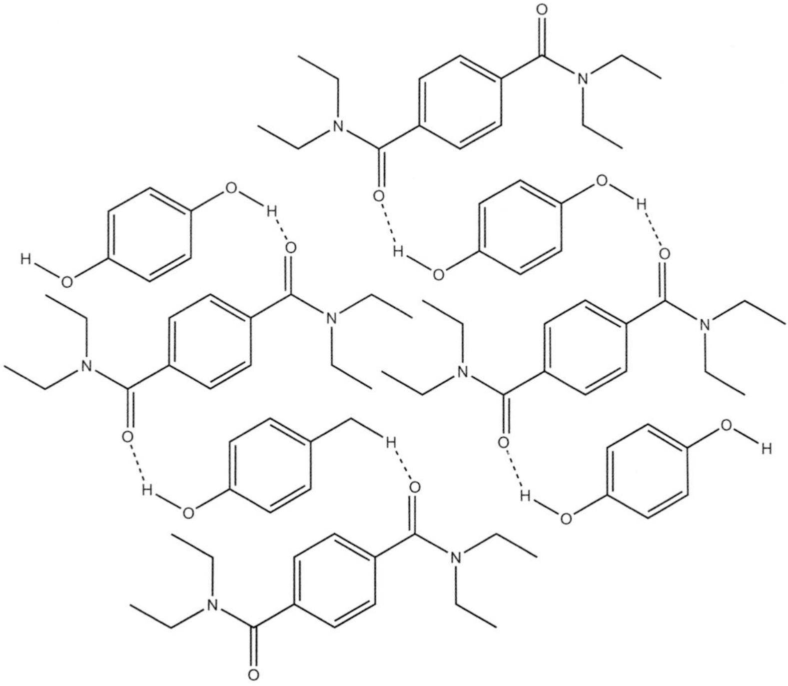

Scheme 10.18

Protecting Group Free Synthesis

The concept of a protecting group free synthesis was developed by Hardegger (Hardegger and Lohse, 1957; Kuhn and Kirschenlohr, 1956), in the muscarine synthesis. Later, two protecting group free synthesis of muscarin, were reported (Chan and Li, 1992; Chan et al., 1994; Dmochowska et al., 2009). However, none of those synthetic procedures appear any better by today's standards due to several functional group interconversions required in the process. Another approach reported on the muscarin synthesis by Chan was much shorter and free of protecting groups (Scheme 10.19)

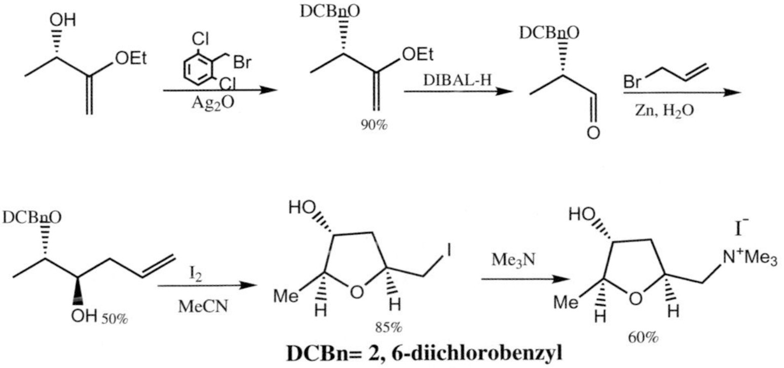

DCBn= 2, 6-diichlorobenzyl

Scheme 10.19

The synthesis starts with the benzoylation of a lactic acid precursor. The role of dichlorobenzyl moiety is to control the stereochemistry of the ring forming iodo-etherification reaction and is not to serve as a protecting group (Rychnovsky and Bartlett, 1981). Interestingly none of the above mentioned syntheses were developed as a protecting group free synthesis, but can be demonstrated as the first step towards the feasibility of a protecting group free synthesis.

The successful protecting group free synthesis of marine alkaloid, like hapalindole U has been demonstrated (Baran et al., 2007). The previously reported synthesis of the compound consisted of about 20 steps with the use of several protecting groups (Muratake and Natsume, 1990a, b). Baran et al. selected four representative members of a large class of natural products encompassing four different families. They then carried out the enantioselective total synthesis of the four, ambiguine H, hapalindole U, Wel-witindolinone A and fisherindole I members using schemes containing only seven to ten steps. Commercially available and inexpensive reagents were used on a preparative scale with deliberate exclusion of any protecting/blocking groups. By adopting the proposed total synthesis, they focused on greener guidelines which emphasized on atom economy (Trost, 1991; Hudlicky and Reed, 2009; Frey et al., 1999; Desiraju, 1995) and are given as follows:

- Strictly minimizing the oxidation-reduction reactions which do not result in C–C bonds (Hendrickson, 1975).
- Increasing the percentage of those steps which results in C-C bonds within the the total number of steps (Hendrickson, 1975; Corey and Cheng, 1996).
- Disconnections should be made to maximize convergency (Bertz, 1982).
- During the assembly of the molecular framework there should be a linear increase in the overall oxidation level of intermediates, except in cases where there is strategic benefit such as an asymmetric reduction (Baran et al., 2007).
- Wherever possible, during each step, maximum incorporation of tandem reaction to elicit maximum structural changes (Nicolaou et al., 2006).
- Minimizing the use of protecting groups in the synthesis by maximum utilization of the innate reactivity of each functional group (Hoffmann, 2006; Hoveyda et al., 1993).
- Planning of the synthetic process should open up a novel way to the invention of a new methodology based on the above mentioned criteria's (Wender and Miller, 2009).
- In case of target molecules of natural origin, there should be incorporation of biomimetic pathways (either known or proposed) in order to aid the above considerations (Nicolaou and Sorensen, 1996; Nicolaou and Snyder, 2004; Scholz and Winterfeldt, 2000; Eschenmoser, 1988; Heathcock, 1992).

Even though the above mentioned concepts existed earlier (Hoffmann, 2006; Nicolaou and Sorensen, 1996; Nicolaou and Snyder, 2004), the work focused on combining the principles for the total synthesis of the selected organic compounds. The protecting group free synthesis of ambiguine H (1) and hapalindole U (2) reported is shown in Scheme 10.20:

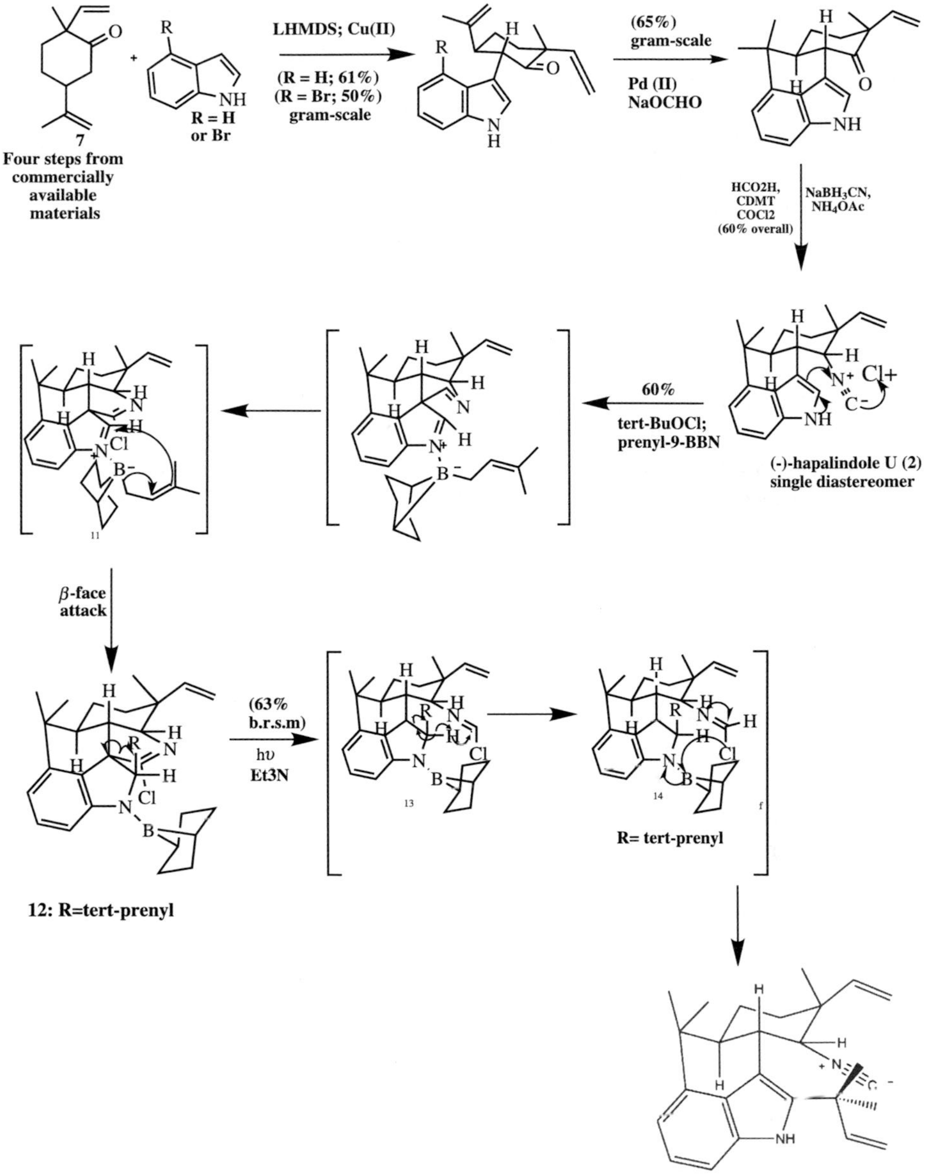

Scheme 10.20

Organolithium reagents are highly reactive to electrophilic functional groups like keto carbonyl groups (Schlosser, 2001; Whisler et al., 2004). There are reports on the generation of organolithium species in the presence of ketones but are generally quenched *in situ* by the ketone carbonyl group (Rutherford and Hawkins, 2007). Thus, in a chemical transformation, it is necessary to protect the ketone functional group if it is not involved in the desirable transformation, before an organolithium reaction. It is reliable to avoid such protection steps for greener synthesis. A notable report by Kim et al. demonstrated a flow-microreactor approach to the organolithium reactions without using the protecting step (Mason et al., 2007). In a flow microreactor, the reaction time is defined by the residence time between the reagent inlet and the quencher inlet. They successfully demonstrated the generation of aryl lithium species bearing ketone groups by iodine-lithium exchange reactions of the corresponding aryl iodides with mesityl lithium followed by a reaction with various electrophiles by strictly controlling the residence time to 0.003 seconds or less. They also reported the application of this method for the formal synthesis of Pauciflorol F, a natural product isolated from a stem bark. Even though the flow reactor concept is still in its infancy, it is undoubtedly a scalable process for industrial use (Kim et al., 2011).

Ge et al. demonstrated a successful protecting group free strategy for the total synthesis of (-)-lannotinide B in 10 steps and 23% yield by achieving excellent chemo and stereoselectivity (Ge et al., 2012). Another protecting group free synthesis was demonstrated by Hickmann et al. for a concise total synthesis of the marine oxylipins hybridalactone and three members of the ecklonialactone family (Hickmann et al., 2011). In 2010, an asymmetric total synthesis of (−)-englerin A in 15 steps with 8.1% yield was demonstrated by Zhou et al. without using any protecting groups (Zhou et al., 2010). Psychotrimine is a polymeric pyrroloindoline alkaloid containing N1–C3 union (compared to C3–C3 indole numbering) which is less common. It has been reported that the treatment of tryptophan derivative with N-iodosuccinimide and 2-iodoaniline yield pyrroloindoline in good yield by generating N1–C3 unit in a single step (Newhouse and Baran, 2008). Since a carbon from the carbamate functionality is incorporated into the natural product, this must be considered as a latent methylamine and not as a blocking group. From here, Psychotrimine can be obtained through a simple five step reaction of an iodide-selective Larock annulations followed by Buchwald-Goldberg-Ullmann reaction and Red-Al reduction. The significance of this protecting group free synthesis with respect to minimization of derivatives and step economy stems from the fact that the previous synthesis of Psychtrimine involved a 16-step process (Matsuda et al., 2008).

In 2008, Dangerfield and coworkers reported a protecting group free synthesis of 2,3-cis substituted hydroxypyrrolidines (Zuckermann and Kodadek, 2009). They proposed two novel synthetic methodologies. The first one involved the stereoselective synthesis of cyclic carbamates from olefinic amines and the second involved the formation of primary amines by a Vasella/reductive amination reaction in aqueous medium without using any protecting groups. Even though the deliberate exclusion of the protecting groups in this total synthetic reaction facilitated the development of new chemical reactions and utilized the intrinsic reactivity within organic compounds, this may fail in case of complex molecules. Undoubtedly, this suggests that excluding protecting groups from the retrosynthesis of the complex molecule might generate unforeseen chemical reactions in some cases which may critically affect the total synthesis (Omar et al., 2010). Thus care should be taken to choose a protecting group free synthesis for a total synthetic reaction.

Biogenesis–oriented Synthesis

Nature does not incorporate protecting groups in biosynthesis. Thus the significant aspect of biomimetic synthesis which closely follow nature is the avoidance of protecting groups (Yokoyama et al., 2004). This holds strictly for that synthetic part which is designed to be biomimetic and is usually the final step of the synthesis. A well-known example is the synthesis of elysiapyrone A involving a biomimetic cascadeof 8π and 6π electrocyclic reactions (Scheme 10.21).

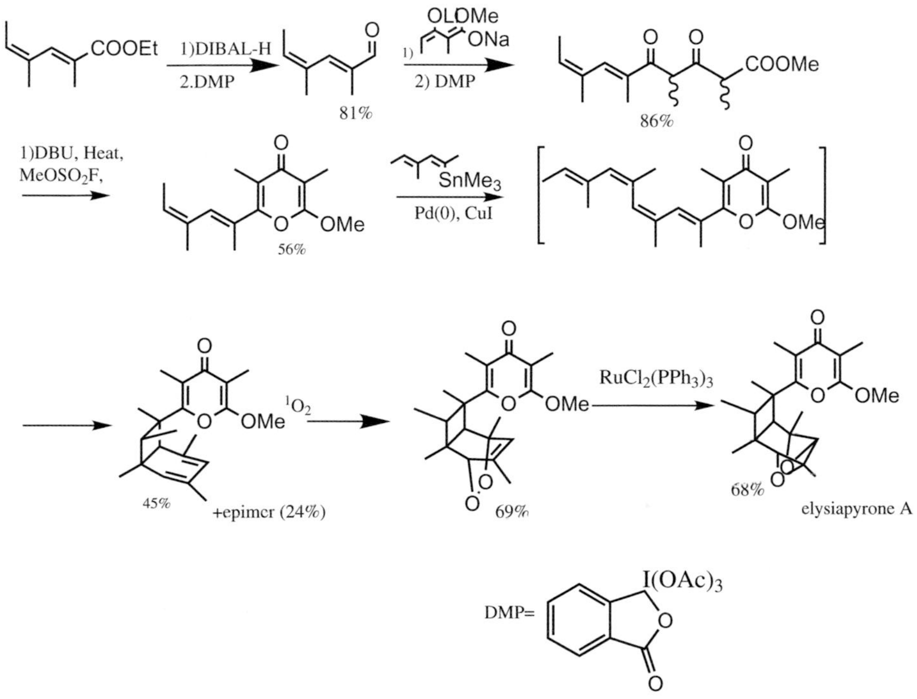

Scheme 10.21

There are reports on the synthesis of natural products following biomimetic techniques. Trost et al. have proposed high yield synthesis of (±)-allamcin, (±)-plumericin and allamandin by adopting biomimetic cascade as a key step (Trost et al., 1986). These examples demonstrates the fact that protecting group free synthesis is possible by mimicking nature (Young and Baran, 2009). Understanding the biosynthetic pathway of a natural product is the key to designing a biomimetic synthesis.

Order of Introduction of Functional Group

An alternative to the use of a protecting group in an organic synthesis is designing the order of introduction of functional groups into the molecule. The necessity of a protecting group can be greatly minimized/ avoided by first building the molecular skeleton followed by finishing it with the necessary functional groups during the synthesis. One of the best examples for demonstrating this approach is the synthesis of a membrarollin epimer shown in Scheme 10.22.

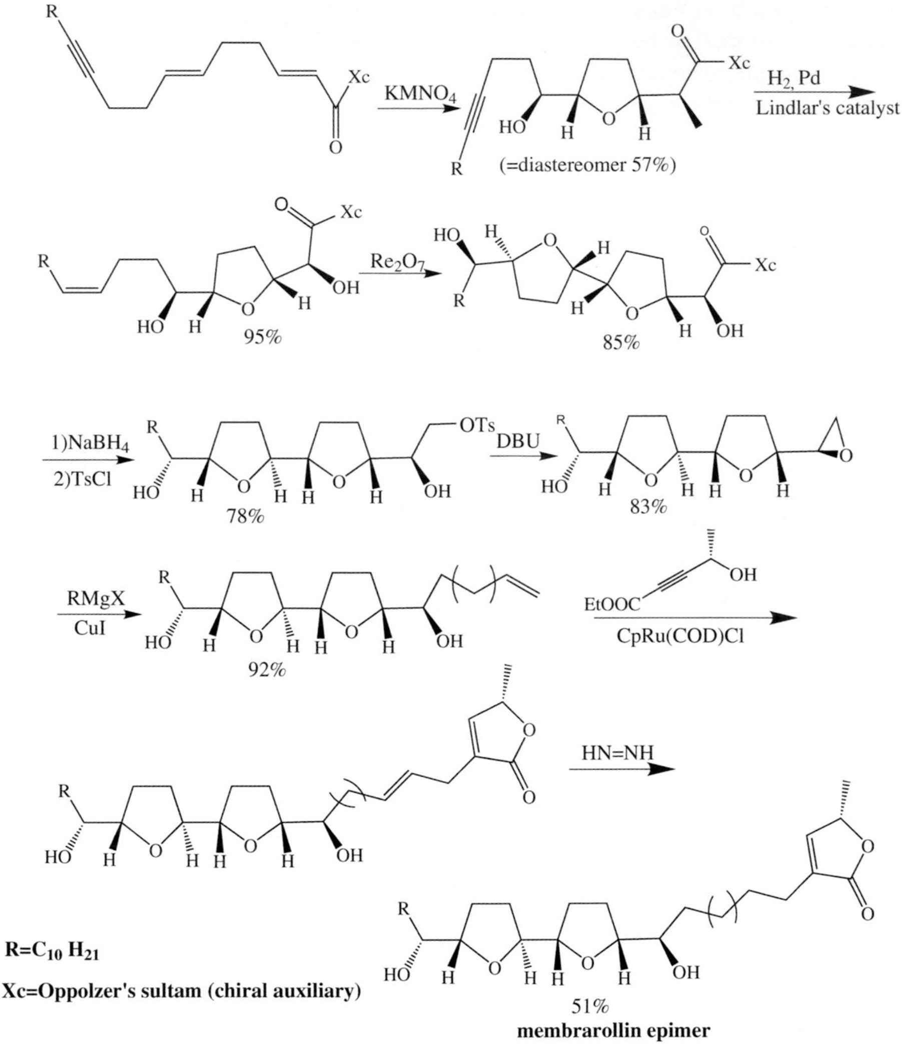

R=C$_{10}$H$_{21}$

Xc=Oppolzer's sultam (chiral auxiliary)

Scheme 10.22

The synthesis starts by assembling the molecular skeleton (not shown), with carbon-carbon double and triple bonds as profunctionality (Hu and Brown, 2005). This is followed by the generation tetrahydrofuran rings and alcohol functions through a series of functionalization reactions. The final stage of the synthesis involves a series of transition-metal-catalyzed processes, for the completion of the molecular skeleton, resulting in an overall avoidance of the protecting group. This undoubtfully suggests the fact that the variation in the order of introduction of functional groups can drastically minimize or even completely avoid the use of protecting groups in an organic synthesis. Thus this concept of designing the order of introduction of a functional group by first assembling the molecular skeleton in a reaction will indeed open up the realization of protecting group free synthesis (Fraunhoffer et al., 2005; Yoshimitsu et al., 1997; McFadden and Stoltz, 2006; Ding et al., 2006).

Conclusion

The percentage atom utilization in a chemical reaction is a critical factor that characterizes the greenness of the reaction. Minimizing the use of derivatives in a chemical synthesis can be achieved by avoiding the use of protecting groups which will result in an increase of atom economy on the reaction. Avoiding the use of protecting groups increases the atom economy of the synthesis. Enzyme catalysis, sequential introduction of functional groups and biomimicking approaches can help achieve this goal to a great extent. But one may have to be resigned to the fact that complete removal of protecting groups from a synthetic scheme, though ideal may not be feasible in all cases. Alternate approaches like non-covalent and photolabile protection/deprotection groups can be relied on to improve the greenness of a synthetic process. Functionalization to improve selectivity can be used to avoid additional deprotection steps and can further increase the eco-friendliness of a synthetic route. Careful choice of starting materials and well-designed sequential introduction of functional groups can significantly reduce dependency of a synthetic scheme on protecting groups. It may be worth noting that other aspects of a modification or derivatization step might determine the overall effectiveness of the synthetic route and indirectly contribute to a greener process.

REFERENCES

Ågerstrand, M., Berg, C., Björlenius, B., Breitholtz, M., Brunström, B., Fick, J., Gunnarsson, L., Larsson, D.G.J., Sumpter, J.P., Tysklind, M., and Rudén, C. Improving environmental risk assessment of human pharmaceuticals. *Environmental Science & Technology*, 2015, 49(9), 5336–5345.

Amit, B., Zehavi, U., Patchornik, A. Photosensitive protecting groups—A review. *Israel Journal of Chemistry*, 1974, 12(1–2), 103 113.

Anastas, P., Eghbali, N. Green chemistry: Principles and practice. *Chemical Society Reviews*, 2010, 39(1), 301–312.

Anastas, P.T., Lankey, R.L. Life cycle assessment and green chemistry: The yin and yang of industrial ecology. *Green Chemistry*, 2000, 2(6), 289–295.

Anastas, P.T., Barlett, L.B., Kirchhoff, M.M., Williamson, T.C. The role of catalysis in the design, development, and implementation of green chemistry. *Catalysis Today*, 2000, 55(1–2), 11–22.

Baran, P.S., Maimone, T.J., Richter, J.M. Total synthesis of marine natural products without using protecting groups. *Nature*, 2007, 446(7134), 404.

Barany, G., kneib-cordonier, N., Mullen, D.G. Solid-phase peptide synthesis: A silver anniversary report. *International Journal of Peptide and Protein Research*, 1987, 30(6), 705–739.

Barltrop, J., Schofield, P. Photosensitive protecting groups. *Tetrahedron Letters*, 1962, 3(16), 697–699.

Bertz, S.H., Convergence, molecular complexity, and synthetic analysis. *Journal of the American Chemical Society*, 1982, 104(21), 5801–5803.

Bisht, K.S., Svirkin, Y.Y., Henderson, L.A., Gross, R.A., Kapaln, D.L., Swift, G. Lipase-catalyzed ring-opening polymerization of trimethylene carbonate. *Macromolecules*, 1997a, 30(25), 7735–7742.

Bisht, K.S., Henderson, L.A. and Gross, R.A., Enzyme-catalyzed ring-opening polymerization of ω-pentadecalactone. *Macromolecules*, 1997b, 30(9), 2705–2711.

Bisht, K.S., Deng, F., Gross, R.A., Kaplan, D.L., Swift, G. Ethyl glucoside as a multifunctional initiator for enzyme-catalyzed regioselective lactone ring-opening polymerization. *Journal of the American Chemical Society*, 1998, 120(7), 1363–1367.

Blanc, A., C.G. Bochet, C.G. Wavelength-controlled orthogonal photolysis of protecting groups. *The Journal of Organic Chemistry*, 2002, 67(16), 5567–5577.

Blout, E.R., Simon, M.S. Photographic products, processes and compositins utilizing acyl hydroquinones. 1964, Google Patents.

Bochet, C.G. Orthogonal photolysis of protecting groups. *Angewandte Chemie International Edition*, 2001, 40(11), 2071–2073.

Bouissou-Schurtz, C., Houeto, P., Guerbet, M., Bachelot, M., Casellas, C., Mauclaire, A.C., Panetier, P., Delval, C. and Masset, D., Ecological risk assessment of the presence of pharmaceutical residues in a French national water survey. *Regulatory Toxicology and Pharmacology*, 2014, 69(3), 296–303.

Cannon, A.S., Warner, J.C. Noncovalent derivatization: Green chemistry applications of crystal engineering. *Crystal Growth & Design*, 2002, 2(4), 255–257.

Carpino, L.A., New amino-protecting groups in organic synthesis. *Accounts of Chemical Research*, 1973, 6(6), 191–198.

Chamberlin, J.W., Use of the 3, 5-dimethoxybenzyloxycarbonyl group as a photosensitive N-protecting group. *The Journal of Organic Chemistry*, 1966, 31(5), 1658–1660.

Chan, T., Li, C. A concise synthesis of (+)-muscarine. *Canadian Journal of Chemistry*, 1992, 70(11), 2726–2729.

Chan, T.H., Li, C.J., Lee, M.C., Wei, Z.Y. 1993 RU lemieux award lecture organometallic-type reactions in aqueous media—A new challenge in organic synthesis. *Canadian Journal of Chemistry*, 1994, 72(5), 1181–1192.

Constable, D.J.C., Dunn, P.J., Hayler, J.D., Humphrey, G.R., Leazer, J.L., Linderman, R.J., Lorenz, K., Manley, J., Pearlman, B.A., Wells, A., Zaks., A., Zhang, T.Y. Key green chemistry research areas—A perspective from pharmaceutical manufacturers. *Green Chemistry*, 2007, 9(5), 411–420.

Corey, E., Cheng, X.-M. The logic of chemical synthesis. *Journal of the American Chemical Society*, 1996, 118(43), 10678–10678.

Desiraju, G.R., Supramolecular synthons in crystal engineering—A new organic synthesis. *Angewandte Chemie International Edition in English*, 1995, 34(21), 2311–2327.

Ding, R., Katebzadeh, K., Roman, L., Bergquist, K., Lindstrom, U.M. Expanding the scope of lewis acid catalysis in water: Remarkable ligand acceleration of aqueous ytterbium triflate catalyzed Michael addition reactions. *The Journal of Organic Chemistry*, 2006, 71(1), 352–355.

Dmochowska, B., Skorupa, E., Świtecka, P., Sikorski, A., Łącka, I., Milewski, S., Wiśniewski, A. synthesis of some quaternary N-(1, 4-anhydro-5-deoxy-D, L-ribitol-5-yl) ammonium salts. *Journal of Carbohydrate Chemistry*, 2009, 28(4), 222–233.

Eschenmoser, A., Vitamin B12: Experiments concerning the origin of its molecular structure. *Angewandte Chemie International Edition in English*, 1988, 27(1), 5–39.

Fraunhoffer, K.J., Bachovchin, D.A., White, M.C. Hydrocarbon oxidation vs C−C bond-forming approaches for efficient syntheses of oxygenated molecules. *Organic Letters*, 2005, 7(2), 223–226.

Frey, D., Claeboe, C., Brammer, Jr., L. Toward a 'reagent-free'synthesis. *Green Chemistry*, 1999, 1(2), 57–59.

Ge, H.M., Zhang, L.D., Tan, R.X., Yao, Z.J. Protecting group-free total synthesis of (−)-lannotinidine B. *Journal of the American Chemical Society*, 2012, 134(30), 12323–12325.

Gee, J.P., Green, J.L. Chapter 4: Discourse analysis, learning, and social practice: A methodological study. *Review of Research in Education*, 1998, 23(1), 119–169.

Gross, R., Kalra, B., Kumar, A. Polyester and polycarbonate synthesis by in vitro enzyme catalysis. *Applied Microbiology and Biotechnology*, 2001a, 55(6), 655–660.

Gross, R.A., Kumar, A., Kalra, B. Polymer synthesis by in vitro enzyme catalysis. *Chemical Reviews*, 2001b, 101(7), 2097–2124.

Guarrera, D., Taylor, L.D., Warner, J.C. Molecular self-assembly in the solid state. The combined use of solid-state NMR and differential scanning calorimetry for the determination of phase constitution. *Chemistry of Materials*, 1994, 6(8), 1293–1296.

Guarrera, D.J., Kingsley, E., Taylor, L.D., Warner, J.C. Non-covalent derivatization: Diffusion control via molecular recognition and self assembly. Proceedings of the IS&T's 50th Annual Conference, The Physics and Chemistry of Imaging Systems. The Society for Imaging Science and Technology, 1997, 537–539.

Hardegger, E., Lohse, F. Über muscarin. 7. Mitteilung. Synthese und absolute konfiguration des muscarins. *Helvetica Chimica Acta*, 1957, 40(7), 2383–2389.

Heathcock, C.H. The enchanting alkaloids of yuzuriha. *Angewandte Chemie International Edition in English*, 1992, 31(6), 665–681.

Hendrickson, J.B., Systematic synthesis design. IV. Numerical codification of construction reactions. *Journal of the American Chemical Society*, 1975, 97(20), 5784–5800.

Hickmann, V., Kondoh, A., Gabor, B., Alcarazo, M., Fürstner, A., Catalysis-based and protecting-group-free total syntheses of the marine oxylipins hybridalactone and the Ecklonialactones A, B, and C. *Journal of the American Chemical Society*, 2011, 133(34), 13471–13480.

Hoffmann, R.W., Protecting-group-free synthesis. *Synthesis*, 2006, 2006(21), 3531–3541.

Hoveyda, A.H., D.A. Evans, and G.C. Fu, Substrate-directable chemical reactions. *Chemical Reviews*, 1993, 93(4), 1307–1370.

Hu, Y., Brown, R.C. A metal–oxo mediated approach to the synthesis of 21, 22-diepi-membrarollin. *Chemical Communications*, 2005(45), 5636–5637.

Hudlicky, T., Reed, J.W. Applications of biotransformations and biocatalysis to complexity generation in organic synthesis. *Chemical Society Reviews*, 2009, 38(11), 3117–3132.

Kessler, M., Glatthar, R., Giese, B., Bochet, C.G. Sequentially photocleavable protecting groups in solid-phase synthesis. *Organic Letters*, 2003, 5(8), 1179–1181.

Kim, H., Nagaki, A., Yoshida, J.-i. A flow-microreactor approach to protecting-group-free synthesis using organolithium compounds. *Nature Communications*, 2011, 2, 264.

Knapp, D.R., Handbook of Analytical Derivatization Reactions. John Wiley & Sons, 1979.

Kuhn, R., Kirschenlohr, W. 2-Amino-2-desoxy-zucker durch katalytische Halbhydrierung von Amino-, Arylamino-und Benzylamino-nitrilen; d-und l-Glucosamin. Aminozuckersynthesen II. *Justus Liebigs Annalen der Chemie*, 1956, 600(2), 115–125.

Lloyd-Williams, P., Albericio, F., Giralt, E. Convergent solid-phase peptide synthesis. *Tetrahedron*, 1993, 49(48), 11065–11133.

Mason, B.P., Price, K.E., Steinbacher, J.L., Bogdan, A.R., McQade, D.T. Greener approaches to organic synthesis using microreactor technology. *Chemical Reviews*, 2007, 107(6), 2300–2318.

Matsuda, Y., Kitajima, M., Takayama, H. First total synthesis of trimeric indole alkaloid, psychotrimine. *Organic Letters*, 2008, 10(1), 125–128.

Matsumura, S., Soeda, Y., Toshima, K. Perspectives for synthesis and production of polyurethanes and related polymers by enzymes directed toward green and sustainable chemistry. *Applied Microbiology and Biotechnology*, 2006, 70(1), 12.

McClure, K.F., Danishefsky, S.J. A novel Heck arylation reaction: Rapid access to congeners of FR 900482. *Journal of the American Chemical Society*, 1993, 115(14), 6094–6100.

McFadden, R.M., Stoltz, B.M. The catalytic enantioselective, protecting group-free total synthesis of (+)-dichroanone. *Journal of the American Chemical Society*, 2006, 128(24), 7738–7739.

Mitchell, A.D., Auletta, A.E., Clive, D., Kirby, P.E., Moore, M.M., Myhr, B.C., The L5178Y/tk+/− mouse lymphoma specific gene and chromosomal mutation assay: A phase III report of the US environmental protection agency Gene-Tox program1 This manuscript has been reviewed by the US Environmental Protection Agency Office of Toxic Substances, Pollution Prevention and Toxics, and the US Environmental Protection Agency National Health and Environmental Effects Research Laboratory and approved for publication. Approval does not signify that the contents necessarily reflect the views and policies of the Agency, nor does mention of trade names or commercial products constitute endorsement or recommendation for use. 1. *Mutation Research/Genetic Toxicology and Environmental Mutagenesis*, 1997, 394(1), 177–303.

Muratake, H., Natsume, M. Synthetic studies of marine alkaloids hapalindoles. Part I: Total synthesis of (±)-hapalindoles J and M. *Tetrahedron*, 1990a, 46(18), 6331–6342.

Muratake, H., Natsume, M. Synthetic studies of marine alkaloids hapalindoles. Part 2. Lithium aluminum hydride reduction of the electron-rich carbon-carbon double bond conjugated with the indole nucleus. *Tetrahedron*, 1990b, 46(18), 6343–6350.

Nakamura, C.E., Whited, G.M. Metabolic engineering for the microbial production of 1, 3-propanediol. *Current Opinion in Biotechnology*, 2003, 14(5), 454–459.

Newhouse, T., Baran, P.S. Total synthesis of (±)-psychotrimine. *Journal of the American Chemical Society*, 2008, 130(33), 10886–10887.

Nicolaou, K., Sorensen, E. Classics in Total Synthesis; VCH: New York, 1996. Google Scholar.

Nicolaou, K., Snyder, S.A. The essence of total synthesis. *Proceedings of the National Academy of Sciences*, 2004, 101(33), 11929–11936.

Nicolaou, K., Edmonds, D.J., Bulger, P.G. Cascade reactions in total synthesis. *Angewandte Chemie International Edition*, 2006, 45(43), 7134–7186.

Omar, N.Y., Rahman, N.A., Zain, S.M. An ONIOM study on the enantioselective diels–Alder reaction catalyzed by SiO_2–Immobilized chiral oxazaborolidinium cation. *Journal of Molecular Catalysis A*: *Chemical*, 2010, 333(1–2), 145–157.

Parton, W.J., Scurlock, J.M.O., Ojima, D.S., Gilmanov, T.G., Scholes, R.J., Schimel, D.S., Kirchner, T., Menaut, J.-C., Seastedt, T., Garcia Moya, E., Kamnalrut, A., Kinyamario, J.I. Observations and modeling of biomass and soil organic matter dynamics for the grassland biome worldwide. *Global Biogeochemical Cycles*, 1993, 7(4), 785–809.

Patchornik, A., Amit, B., Woodward, R. Photosensitive protecting groups. *Journal of the American Chemical Society*, 1970, 92(21), 6333–6335.

Peukert, S., Giese, B. The pivaloylglycol anchor group: a new platform for a photolabile linker in solid-phase synthesis. *The Journal of Organic Chemistry*, 1998, 63(24), 9045–9051.

Pillai, V.R., Photoremovable protecting groups in organic synthesis. *Synthesis*, 1980, 1980(01), 1–26.

Ran, N., Zhao, L., Chen, Z., Tao, J. Recent applications of biocatalysis in developing green chemistry for chemical synthesis at the industrial scale. *Green Chemistry*, 2008, 10(4), 361–372.

Romanski, J., Nowak, P., Kosinski, K., Jurczak, J., High-pressure transesterification of sterically hindered esters. *Tetrahedron Letters*, 2012, 53(39), 5287–5289.

Rutherford, J.L., Hawkins, J.M. Preparation of Novel Substituted Haloarene Compounds. Google Patents, 2007.

Rychnovsky, S.D., Bartlett, P.A. Stereocontrolled synthesis of cis-2, 5-disubstituted tetrahydrofurans and cis-and trans-linalyl oxides. *Journal of the American Chemical Society*, 1981, 103(13), 3963–3964.

Santini, R., Griffith, M.C., Qi, M. A measure of solvent effects on swelling of resins for solid phase organic synthesis. *Tetrahedron Letters*, 1998, 39(49), 8951–8954.

Schelhaas, M., Waldmann, H. Protecting group strategies in organic synthesis. *Angewandte Chemie International Edition in English*, 1996, 35(18), 2056–2083.

Schkeryantz, J.M., Danishefsky, S.J. Total Synthesis of (.+−.)-FR-900482. *Journal of the American Chemical Society*, 1995, 117(16), 4722–4723.

Schlosser, M., The organometallic approach to molecular diversity-halogens as helpers. *European Journal of Organic Chemistry*, 2001, 2001(21), 3975–3984.

Scholz, U., Winterfeldt, E. Biomimetic synthesis of alkaloids. *Natural Product Reports*, 2000, 17(4), 349–366.

Shaaya, E., Sekeris, C.E. Inhibitory effects of α-amanitin on RNA synthesis and induction of DOPA-decarboxylase by β-ecdysone. *FEBS Letters*, 1971, 16(4), 333–336.

Stoler, E., Warner, J.C. Non-covalent derivatives: Cocrystals and eutectics. *Molecules*, 2015, 20(8), 14833–14848.

Trakhtenberg, S., Warner, J.C. Green chemistry considerations in entropic control of materials and processes. *Chemical Reviews*, 2007, 107(6), 2174–2182.

Trost, B.M., Balkovec, J.M., Mao, M.K. A total synthesis of plumericin, allamcin and allamandin. Part 2. A biomimetic strategy. *Journal of the American Chemical Society*, 1986, 108(16), 4974–4983.

Trost, B.M., The atom economy—A search for synthetic efficiency *Science*, 1991, 254(5037), 1471–1477.

Warner, J.C., Entropic control in green chemistry and materials design. *Pure and Applied Chemistry*, 2006, 78(11), 2035–2041.

Wender, P.A., Miller, B.L. Synthesis at the molecular frontier. *Nature*, 2009, 460(7252), 197.

Whisler, M.C., MacNeil, S., Snieckus, V., Beak, P. Beyond thermodynamic acidity: A perspective on the complex-induced proximity effect (CIPE) in deprotonation reactions. *Angewandte Chemie International Edition*, 2004, 43(17), 2206–2225.

Yokoyama, Y., Hikawa, H., Mitsuhashi, M., Uyama, A., Hiroki, Y., Murakami, Y. Total synthesis without protection: Three-step synthesis of optically active clavicipitic acids by a biomimetic route. *European Journal of Organic Chemistry*, 2004, 2004(6), 1244–1253.

Yoshimitsu, T., Song, J.J., Wang, G.Q., Masamune, S. Application of newly developed anti-selective aldol methodology: Synthesis of C6−C13 and C19−C28 fragments of miyakolide. *The Journal of Organic Chemistry*, 1997, 62(26), 8978–8979.

Young, I.S., Baran, P.S. Protecting-group-free synthesis as an opportunity for invention. *Nature Chemistry*, 2009, 1(3), 193.

Zehavi, U., Patchornik, A. Light-sensitive glycosides. II. 2-nitrobenzyl 6-deoxy-. alpha.-L-mannopyranoside and 2-nitrobenzyl 6-deoxy-. beta.-L-galactopyranoside. *The Journal of Organic Chemistry*, 1972, 37(14), 2285–2288.

Zhou, Q., Chen, X., Ma, D. Asymmetric, protecting-group-free total synthesis of (−)-Englerin A. *Angewandte Chemie International Edition*, 2010, 49(20), 3513–3516.

Zimmerman, H.E., Sandel, V.R. Mechanistic organic photochemistry. II. 1, 2 Solvolytic photochemical reactions. *Journal of the American Chemical Society*, 1963, 85(7), 915–922.

Zuckermann, R.N., Kodadek, T. Peptoids as potential therapeutics. *Current Opinion in Molecular Therapeutics*, 2009, 11(3), 299–307.

Chapter 11

Catalysis

Fabiola N. de la Cruz[1]*, José Domingo Rivera-Ramírez[2], Julio López[3] and Miguel A. Vázquez[3]

[1]Departmento de Química Orgánica, Facultad de Ciencias Químicas, Universidad Autónoma de Coahuila, Coah., México.
[2]Departmento de Química, Centro Universitario de Ciencias Exactas e Ingenierias, Universidad de Guadalajara, Gda., México.
[3]Departmento de Química, División de Ciencias Naturales y Exactas, Universidad de Guanajuato, Gto., México.

INTRODUCTION

First a catalyst needs to be defined, this word is derived from the Greek words *kata* (down) and *lyein* (loosen), which means a catalyst is a substance (in small proportions) that accelerates a chemical reaction without being consumed or affected during the course of the reaction. It is a compound that is very important and appears in the steps of a reaction mechanism, but disappears in the overall chemical reaction; it is neither a reagent or a product.

Generally, the use of catalysts alter the reaction mechanism to reduce the activation energy and make new energy barriers (relative to the uncatalyzed reaction), making it easier for the reactants to overcome this new energy barrier and obtain the products.

There are many catalysts in use, it will depend on the product or mechanism that is being carried out to decide the type of catalyst that will be used (organic, organometallic or enzymatic). Catalysts can be in the same phase (homogeneous phase) as the chemical reactants or in a distinct phase (heterogeneous phase).

*For Correspondence: Fabiola N. de la Cruz, Departamento de Química Orgánica, Facultad de Ciencias Químicas, Universidad Autónoma de Coahuila, Blvd. Venustiano Carranza e Ing. J. Cárdenas Valdez S/N, 25280, Saltillo, Coah., México. Email: fcruz@uadec.edu.mx

ORGANOCATALYST

The synthesis reported by Wöhler in 1832 (Wöhler and von Liebig, 1832), where cyanide anion was present and was used as a catalyst in a condensation reaction to produce benzoin, was the first towards organocatalytic reactions. For a long time, the catalyst focused on the use of enzymes (biocatalysis) and synthetic metal complexes (metal catalysis), primarily geared to asymmetric synthesis. Besides the research of Wöhler, other works such as von Liebig, Bredig, Pracejus had a profound effect on the organocatalytic area path. Many decades later research contributions of List (List et al., 2000), Lerner, MacMillan, Hayashi, Jørgensen and Barbas groups to mention but a few, also led to its importance.

The term organocatalysis is related to use a substoichiometric quantity exclusive of small organic compounds to accelerate a chemical reaction (List, 2007). This area has been increased exponentially not only in organic chemistry, but also to other science areas. From the 1980's to the present day, more than 4242 records have been reported relating to organocatalysis, most of these records have focused on organic chemistry (Liu and Wang, 2017).

The impact of this synthetic tool is attributed to its ability to provide the transformation with high selectivity from simple carbon-carbon bonds formations to complex molecules with high efficiency, eco-compatible and environmentally benign.

There are essentially four types of organocatalysts (Fig. 11.1, panel 1): Lewis bases (Section A), Lewis acid (Section B), Brønsted bases (Section C) and Brønsted acid (Section D). However, it is dominated by the Lewis base.

Catalyst

The modes of activation of organic catalysts have been divided mainly in two ways:

a) **Covalent-based**: is a characteristic to covalently bind the substrate and this condition is reversible. In this respect, chiral or non-chiral primary or secondary amines are used, that produce enamines (Fig. 11.1, panel 2A) (Mukherjee et al., 2007) and iminium ion (Fig. 11.1, panel 2B) (Erkkilä et al., 2007) as intermediates, respectively. Also, the N-heterocyclic carbene (Enders et al., 2007) used as catalysts, enable an inverted reaction mainly to aldehydes groups (Fig. 11.1, panel 2C).

b) **Noncovalent-based**: Miscellaneous attractive interactions between the catalyst and subtract are the main characteristic. Within this are hydrogen-bonding activation (Taylor and Jacobsen, 2006), phase-transfer (Shirakawa and Maruoka, 2013; Craig et al., 2018), anion-binding (Reisman et al., 2008; Smajlagic et al., 2018) and Brønsted acid (Fig. 11.1, panel 3A-D, respectively) (Parmar et al., 2014).

It can be said that Brønsted acid, are divided from weak species such as: HAROLs (Dilek et al., 2018), silanediols (Schafer et al., 2013), BINOLs (Zhou, 2011; Parmar et al., 2017; Kikuchi et al., 2019), TADDOL's (Pellissier, 2008; Nguyen et al., 2018), squaramides (Held and Tsogoeva, 2016; Martínez-Crespo et al., 2018; Modrocká et al., 2018), (thio)ureas (Fan and Kass, 2017; Sonsona et al., 2018; Andrés et al., 2018), guanidine derivatives (Yuan et al., 2017; Shaabani et al., 2018), carboxylic acids (Min and Seidel, 2017) (Fig. 11.1, panel 1, section D), to

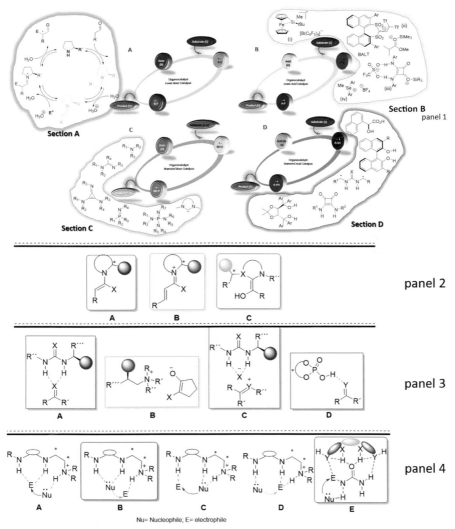

Figure 11.1 Activations modes and catalysts involves in organocatalyst.

strong characteristics as: binaphthyl phosphoric acids derivatives (Momiyama et al., 2015; Saito and Akiyama, 2016), sulfonimides (Galván et al., 2016; Dai et al., 2017). All of them, generally function through hydrogen bonding towards substrates. In this respect, there are also molecules with a bifunctional activity (Formica et al., 2018), typically hydrogen over nitrogen in the urea, thiourea, a similar functional group interacts with the electrophile and the additional groups as amine (Brønsted base) activate the nucleophile moiety. Structures such as proline (Liu and Wang, 2017), pipecolic acid (Mohapatra et al., 2015), cinchona-derivatives (Guo and Wong, 2017; Moczulski et al., 2018), iminophosphorane (Formica et al., 2018), ferrocenophane-amino derivatives (Yao et al., 2018), amine-thioureas (Fang and Wang, 2015), to mention a few are included within this characteristic. A different activation mode has been proposed to explain the selectivity of the reactions (Fig. 11.1), for example, when cinchona alkaloid-squaramide catalysts are used in the addition reactions, two modes of the transition state exist: Wynberg and Houk models (Fig. 11.1, panel 4A-E), the DFT studies showed the activation of mode B and the Houk´s model is appropriate to explain the selectivity (Guo and Wong, 2017). When the concept Networks of Cooperative Hydrogen Bonds (NCHB) was introduced, the main characteristic was a pure noncovalent interaction. The study of NCHB was carried out between arylideneureas with β-ketoesters to produce the Biginelli derivatives via Mannich addition using secondary aminophenol as a catalyst. The DFT analysis showed that the NCHB process has a lower energy barrier of the reaction in transition state compared with the uncatalyzed pathway (Lillo et al., 2016).

With respect to the Brønsted base, recently Brahmachari et al. described an update on the developments of the triethanolamine as a catalyst. It is also worth noting that the development of chiral organobases, occupy an extremely important position in asymmetric organocatalysis (Brahmachari et al., 2018). Cao and co-workers used tertiary amine, amidines, guanidines, cyclopropenimines and iminophosphoranes, etc. (Fig. 11.1, Section C) (Cao et al., 2018).

The Lewis acid uses neutral or cationic electron-deficient organic molecules, with the aim to active less reactive functional group. However, there are not many of these kinds of organic catalysts due the lack of sufficiently strong acidity. In this context, silicon-based catalysts have been used (Fig. 11.1, panel 1B (i)), this kind of silylium ion possesses high catalytic activity, though their synthesis is expensive and difficult as a few chiral structures have also been synthesized. When describing the chiral Lewis Acid, a strategy that has been used is an enantiopure anion, implementing the concept "asymmetric counteranion-directed catalysis" (ACDC) (Dumoulin and Masson, 2016). The Diels-Alder reactions were probed as a model between cinnamates and cyclopentadiene. The dinaphthyl-allyl-tetrasulfones (BALTs) where silyl ketene acetal formed ionic pair species highly activatied the Lewis acid catalyst to obtain enantiomeric ratios of up 97:3 and diastereomeric ratios > 20:1 (Fig. 11.1, panel 1B (ii)) (Gatzenmeier et al., 2016). Squaramides and silyl triflates interact to form a stable Lewis acid complex used in nucleophile addition (Fig. 11.1, panel 1B (iii)) (Banik et al., 2017). Other interesting organic Lewis acids are the trisubstituted selenonium salts, which are utilized for halogenations and aldol type reaction (Fig. 11.1, panel 1B (iv)) (He et al., 2018). Usually, the hydrogen bonding is the key to organocatalysis, taking its cue from this concept, hydrogen and halogen bonding share many similarities (Cavallo et al., 2016), for this reason, halogen interaction has been used as an organocatalyst (Bulfield and Huber, 2016).

When one uses a catalyst, which contains one or more of N-, O-, P-, or S-atoms, it is immediately associated it with Lewis bases. As depicted in Fig. 11.1, panel 1, section A, and panel 2A-C, the catalyst acts covalently to produce different intermediates depending on the substrates used. One of most important molecules in organocatalyst is the amino acid proline and its derivatives, due to it being inexpensive (Liu, 2017). However, the development of another catalyst that has that been carried out, e.g. the use of primary amines, generates an enamine with low nucleophilicity, which is a disadvantage compared with secondary amines. Nevertheless when this group is functionalized with another structure the activity is improved (Kaur et al., 2017), usually the use of diamines (1,2-; 1,3-; 1,4-) (Zlotin and Kochetkov, 2015; Ričko et al., 2016) or primary amine Cinchona-derived (Zhou et al., 2008; Lam and Houk, 2015) has been the strategy to increase the reactivity and selectivity. Other important catalysts have been the N-Heterocyclic Carbenes (NHCs) (Fig. 11.1, panel 2C) (Di Marco et al., 2016; Haghshenas et al., 2017), their application in Stetter reaction, Morita-Baylis-Hillman, Rauhut-Currier reactions, benzoin and esterification process are some of the major uses of this catalyst. The NHCs are structurally versatile due to their tuning physicochemical proprieties as well steric and electronic characteristic that this allows (Flanigan et al., 2015).

The activity of the different catalysts is summarized by the Barbas report (Cheng et al., 2014):

- **Proline-derived (2°amines)**: most commonly used for the activation of aldehydes.
- **Chiral phosphine catalysts**: used for the activation of allenes and Morita-Bayllis-Hillman carbonates.
- **Cinchona alkaloid-derived (1°amines)**: employed for the activation of ketones.
- **Bifunctional thiourea, urea, thiocarbamate and squaramide (3°amines)**: activate carbonyl groups by hydrogen bonding.
- **Binaphthyl phosphate**: employed to Diels-Alder reaction.
- **Binaphthyl phosphoric acid**: formation and activation of imines.
- **N-heterocyclic carbene catalysts**: activation of carbonyl groups.

Strategies and Synthetic Application

The organocatalyst has been used in domino reaction (Chauhan et al., 2017) as described Chanda and co-worker (Chanda and Zhao, 2018). The use of (S)-proline derivative (Scheme 11.1, panel 1(i)) are used to produce with high enantioselectivities pyrrolidine derivatives through Michael/Mannich (3+2) cyclization/Witting olefination sequence (Scheme 11.1, panel 1, structure A) (Chauhan et al., 2017). With a similar catalyst (Scheme 11.1, panel 1(ii)) the bicyclopyrazolidine has been obtained through a cascade vinylogous 1,6-Michael addition/1,4-proton shift/aza-Michael/hemiaminal formation sequence using 1,2-diaza-1,3-diene and 2-butenal derivatives (Scheme 11.1, panel 1, structure C) (Chauhan et al., 2017). Utilizing the sequence NHC-catalyzed (3+2) Michael/Michael/esterification via homoenolate/enolate intermediates produced tetrasubstituted cyclopentanes using α,β-unsaturated aldehydes and nitroallylic acetates (Scheme 11.1, panel 1(iii), structure D) (Shu et al., 2016). It was recommended that, spiro-oxindoles derivatives be obtained between α,β-unsaturated aldehydes and oxindolyl β,γ-unsaturated α-keto esters by domino Michael/intramolecular aldol/lactonization reactions process (Scheme 11.1, panel 1(iv), structure E) (Zhang et al., 2017). The quinine-derived phase transfer catalyst (Scheme 11.1, panel 1(v)) was applied to domino process to

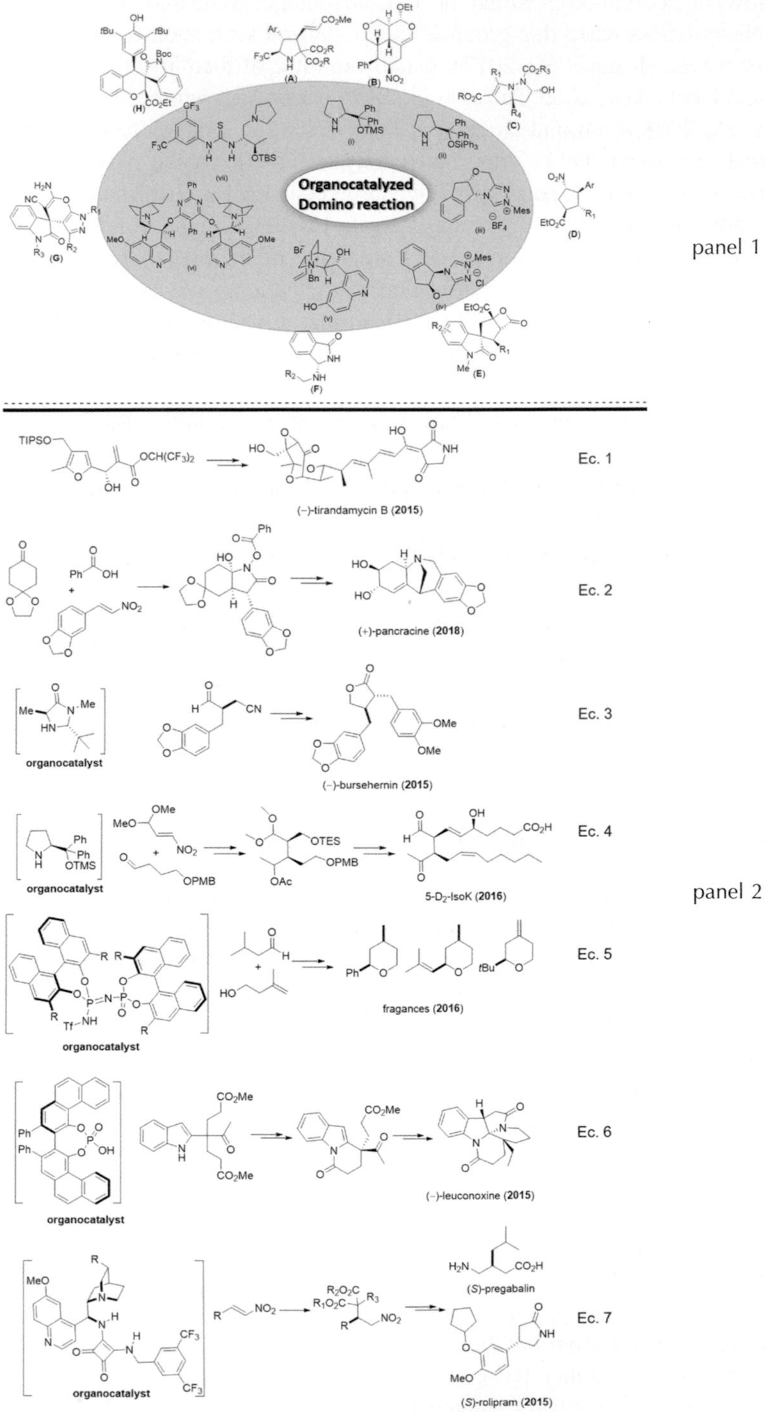

panel 1

panel 2

Scheme 11.1. Application of organocatalysis to: panel 1) different transformation; panel 2) biologically active compounds.

furnish 3-amino-substituted isoindolinones (Scheme 11.1, structure F) (Capobianco et al., 2016). By using DHQD2PYR (Scheme 11.1, panel 1(vi)) the spiro-indoline-pyrans could be obtained by means of domino Michael/cyclization reaction (Scheme 11.1, structure G) (Xie et al., 2016). The report of the synthesis of chromones derivatives (Scheme 11.1, structure H) by an asymmetric organocatalytic domino oxa-Michael/1,6-addition reaction of ortho-hydroxyphenyl-substituted para-quinone methides and isatin-derived enoates has been described. The dysfunctional thiourea (Scheme 11.1, panel 1(vii)) is used to yield the heterocyclic with >99% ee and >20:1 d.r. (Zhao et al., 2016) Enders and co-workers reported a stereoselective one-pot synthesis of functionalized complex tricyclic polyethers (Scheme 11.1, structure B). The structures have been achieved by one-pot quadruple reaction/hetero-Diels–Alder sequence using the combination of secondary amine (Scheme 11.1, panel 1(i)) and lanthanide catalysis (Dochain et al., 2016).

The aza-Morita-Baylis-Hillman reaction (Pellissier, 2017) is a clearly an example where the organocatalytic procedure is applied. This method synthetizes a key structure to obtain natural molecules, e.g. the synthesis of polyfunctionalized ester derivative (Scheme 11.1, panel 2, Ec. 1) using β-isocupreidine as a catalyst, this ester is an important intermediate to the enantioselective synthesis of (−)-tirandamycin B. The concise synthesis of (+)-pancracine was accomplished through of bicyclic octahydro-2H-indol-2-one as a key intermediate (Scheme 11.1, panel 2, Ec. 2) (de Gracia Retamosa et al., 2018). The latter was obtained by three component strategy using a ketone, a carboxylic acid and nitroalkane via cyclization reaction an unnatural proline as organocatalyts (Conde et al., 2012; Chen and Enders, 2018). A synergistic catalysis protocol has been reported by MacMillan (Welin et al., 2015), where photoredox and enamine catalysis allowed an enantioselective α-cyanoalkylation of aldehydes. The oxonitrile product could be applied to the total synthesis of the lignan natural product (−)-bursehernin (Scheme 11.1, panel 2, Ec. 3). Another interesting application of organocatalysis is described in the enantioselective synthesis of isoketal 5-D_2-IsoK, this is similar to prostaglandin. The latter has been achieved using the isoketals as a key core through an organocatalyzed Michael addition. Aldehyde and nitroolefin are transformed by Jørgensen-Hayashi´s catalyst to obtain high enantioselectivity (Scheme 11.1, panel 2, Ec. 4) (Candy et al., 2016). The Prins cyclization is used to form tetrahydropyran structures with high enantiomeric ratio (up to 96.5:3.5 er), the reaction was catalyzed by a highly acidic confined imino-imidodiphosphate (Scheme 11.1, panel 2, Ec. 5). This method produced diverse scents to gram-scale (Liu et al., 2016). The (−)-leuconoxine is a terpene indole alkaloid isolated from Leuconotis eugenifolius (Abe and Yamauchi, 1993). The structural challenge was addressed by Kawasaki T. and co-workers (Higuchi et al., 2015) from desymmetrization of prochiral diester using a chiral phosphoric acid catalyst as a key step of total synthesis (Scheme 11.1, panel 2, Ec. 6). Other derivatives such as (+)-melodinine E and (±)-mersicarpine can be furnished by this method. Protocols using water have been described for synthesis of inhibitor of phosphodiesterases IV (S)-rolipram (Zhu et al., 2001) and the potent antiepileptic (S)-pregabalin (Belliotti et al., 2005). The use of cinchona-based squaramide catalysts (Han et al., 2016) on water carried out in a proper noncovalent way to promoted enantioselective Michael addition of malonates to nitroolefins (Scheme 11.1, panel 2, Ec. 7) (Bae and Song, 2015).

USE OF ORGANOMETALLIC COMPOUND IN CATALYSIS

Since catalytic transformations replaced their original stoichiometric versions, several fields in science have benefited. In particular, metallic catalysis plays a central role in many synthetic pathways with outstanding applications. Thanks to the advances in algorithms and hardware in general, some studies based on density functional theory allow a better understanding of the phenomenon occurring in homogenous catalysis of organometallic species. The main differences compared with past research lies on the integration of solvent effects which are important for stabilization of charged fragments, transfer of electric charges and completion of the coordination sphere in unsaturated metals (Vidossich et al., 2016).

Understanding the chemistry of base metals (Fe, Co and Ni) in catalysis can be a big challenge parallel with the routes exhibited by noble metals (Ru, Rh, Ph, Os, Ir, Pt) due to the high spin electronic states commonly present in the first group but rarely associated to the second. Thus, several computational studies of the last decade have been dedicated to establish important guidelines related with the chemical behavior of noble metals (Sperger et al., 2015), but special calculations are needed for the rest of them because during the catalytic cycle, they must change spin states in a process called spin acceleration and those discrete stages are better followed through DFT approaches (Holland, 2015).

To expand the knowledge of typical d-block metals, some coordination complexes containing f-block metals have been structurally studied due to their interesting electronic properties. Between them, $ThCl_4(THF)_3$, $Cp*_2Sm-(\mu-Cl)_2MgCl(THF)_2$ and $NdCl_3$ proved certain activities for the polymerization of ε-caprolactone after being transformed into inclusion macrocyclic complexes under the action of catecholborane. However, research is still under progress to clarify the reason that only short chains of polycaprolactone can be obtained through this method and to find an explanation for the lack of catalytic activity of another complex bearing metal like uranium (Das et al., 2015).

At this point it is evident that metallic catalysis includes several variants not only in homogeneous but also heterogeneous processes. Despite the growing applications and better understanding experienced in the last years, supporting single metals atoms catalysis (Liu, 2017), here only those cases that mainly focus on the catalytic species interacting with the substrate only in a single phase will be described. Nevertheless, it is not possible to deny the benefits of heterogenous catalysts in terms of manipulation and cost, compared with the homogeneous ones. Such is the case of the cross-linked organometallic polymer P[5]-OCP, prepared from an imidazolium pillar[5]arene coordinated to Pd(II). During the experimental progression, the catalyst was tested as promoter of Suzuki-Miyaura couplings in aqueous media, exhibiting excellent results not only of catalytic activity, but also stability as well as reusability (Liu et al., 2018). At an industrial level, there are also challenges for this kind of catalysis, relating with the efficient conversion of small and simple materials into others with more attractive applications, for instance, the case for the set of methods whose main goal is the transformation of methane towards methanol, methyl bisulfate or any other functionalized hydrocarbon. To achieve this, metallic catalyst containing Mg, Cu, Fe, Hg and Pt have been tested with good results, each displaying specific advantages and drawbacks (Olivos-Suárez et al., 2016).

In general, heterogeneous catalysis can be often rationalized based on the concepts of surface organometallic chemistry, especially if new molecules must be designed and prepared to accomplish a specific performance with high standards (Pelletier and Basset, 2016). Between the most known metals suitable in this area, Ti, Zr, Hf, V, Ta, Cr, Mo, W, Re, Fe, Co and Ni can be mentioned, which constitute the active sites of groups like metallocarbenes (as well as the alkyls or hydrides substituted), metallocarbynes, metal imides, metal aziridines, metal alkyls and neutral or cationic metal hydrides. These families have been explored, for instance, in a set of polymerizations, depolymerizations and metathesis of hydrocarbons. Another important trend includes the manipulation of techniques and understanding of factors involved during the interaction of active organometallic compounds embedded over metal oxides. In this context, $[Cp^*(PMe_3)Ir(III)]$ has been established as a good activator of sp^2 and sp^3 C–H bonds (for instance, in the deuteration of methane) when supported in some acidic modified metal oxides. The same study, however, concludes that there is no direct correlation between the activity of the catalyst and the degree of the surficial acidity (Kaphan et al., 2018).

Regarding the activity of the catalysts, special attention needs to be given to those structures where a single or multiple metal-metal bond is present. Recent evidence has suggested their active implication in some catalytic scenarios previously attributed only to mononuclear scaffolds. In this way, their relevance is impacted since enzymatic transformations (for instance in [Ni-Fe]-hydrogenase) until redox processes (formation/breaking σ and π bonds) and stabilization of pre-catalytic mixtures (Pd(I) dimers for couplings, Cu(I)-nitrene dimers in amination reactions) or transient intermediates (Pd(III) dimers for C–H functionalization, Rh(I)/Rh(II) dimers in alkene hydroformylations), as well as intact dinuclear clusters (dirhodium for transfer reactions, diruthenium for propargylic substitutions, early/late bimetallic species for olefin hydrogenations, Pt/Ti species for allylic substitutions, Zr/Co for hydrosilylations and $Co_2(CO)_8$ for alkyne trimerizations) (Powers and Uyeda, 2017). The same structure-activity effect could be attested in paddlewheel structures like $Rh_2(esp)_2C(p\text{-MeOPh})_2$ and $BiRh(esp)_2C(p\text{-MeOPh})_2$ (esp $= \alpha,\alpha,\alpha',\alpha'$-tetra-methyl-1,3-benzenedipropoonate), where the replacement of one atom of Rh for Bi resulted in a more electrophilic carbene as the consequence of diminished π-back-bonding interactions. In a practical way, dirhodium species are useful in reactions involving decomposition of diazo-compounds, like cyclopropanations, while their heterobimetallic counterparts can exert an enhancement of reactivity of deactivated and small molecules, through a metallic remote control (Fig. 11.2) (Collins et al., 2018).

The exact nature of the ligands attached to the metal is another determining factor for the catalyst to exert its function. In this way, there is an interesting research line dealing with chiral cyclopentadienyl ligands in combination with metals like

Sc(III), Ir(III), Ru(II), Co(III), Rh(III) and Zr(IV), which are involved in asymmetric processes like C-H activation of hydroxamic acid derivatives, cycloisomerization of enynes to yield cyclopropanes and various cases of cycloadditions, just to mention a few (Newton et al., 2016). The substitution pattern of the ligands around the metal is also highly crucial. To exemplify this situation, polymerization of ethylene using Ni(II) salicylaldiminato species could be considered. Normally, non-substituted ligands yield a mixture of polymerized chains with non-negligible occurrence of ramifications due to β-eliminations; however, when electron-withdrawing groups are present, the ratio of linear polymers is maximized. Comparatively SF$_5$ showed better results than CF$_3$ in aqueous dispersions (Kenyon and Mecking, 2017). However organometallic catalytic pathways are not only under control of the metal and ligand, but also the counterion that are present as part of the catalyst. This situation is clearly exemplified using different gold species commonly used for total synthesis and asymmetric transformations, where the concepts of structure-activity take on great importance (Jia and Bandini, 2015).

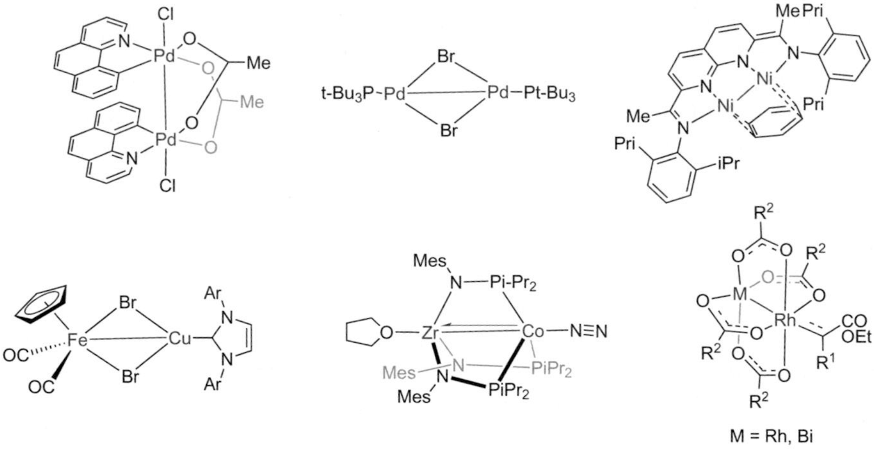

Figure 11.2   Examples of catalysts exhibiting homo and hetero M-M bonds.

Synergistic effects between different metals are also important to complete some catalytic cycles and can be possible in many organic reactions, as shown by the following four cases:

1. Diastereoselective synthesis of spiroketals through a [4+2] cycloaddition of *in situ* formed enol ethers and *o*-quinone methide precursors, in the presence of catalytic amounts of Ph$_3$PAuCl, AgOTf and Sc(OTf)$_3$. This method renders the products in a time lapse of only 5 minutes at room temperature, showing excellent yields. However, if the Au(I) catalyst is omitted, the expected reaction never takes place and if on the other hand the Sc(III) salt is avoided, the result is a messy mixture difficult to separate (Liang et al., 2017).

2. Synthesis of 1,2-diaryl ethanes through the Csp^3-Csp^3 coupling promoted by NiBr$_2$(glyme) in the presence of Zn and a diphospine-type ligand. This combination generates the active catalytic species Ni(0)L$_n$, which suffers two sequential oxidative additions for the substrate (benzyl ethers), regulated by the

presence of Zn, rendering a Ni(III) intermediary, responsible of releasing the final product by means of a reductive elimination. The main attraction of this method is the low-cost and the compatibility with several functional groups (Cao and Shi, 2017).

3. Dual catalyzed chain growth systems of ethylene using $(BiPy)_2FeEt$ are responsible for the reconstitution of the chain transfer agent $ZnEt_2$ [from Zn(oligomer)]. This catalysis occurs via Et/R exchange and the main product is released through a β-hydride elimination (Cariou and Shabaker, 2015).

4. *In situ* formation of a catalyst batch able to produce methanol from a mixture 3:1 H_2/CO_2, based on the initial interaction of $ZnEt_2$ and $Cu(stearate)_2$. After activation at room temperature a catalytic mixture of Cu(0) colloidal particles and ZnO is formed, which in turn drives the synthesis of the alcohol at 250°C and 50 bars (Brown et al., 2015).

The catalytic process is not always considered a single category, as the involved transformations can be placed at the edge of different principles or in other cases, of two and even more of them. One illustrative example is the ruthenium catalyzed metathesis of olefins, which, in combination with enzymatic oxidation, renders aryl epoxides enantioselectivelly in moderate yields (Denard et al., 2015). Another example was found in investigating the hydroformylation reactions of supramolecular complexes of cyclodextrins-triglycerides, where $Rh(CO)_2(acac)$ was used after being stabilized in aqueous media by coordination with the sodium salt of the trisulfonated triphenylphosphane, applying not very harsh conditions (6 hours, 80°C and 80 bar CO/H_2). This catalytic design could be reused several times and the best reported yield corresponds to the effect of a methylated β-cyclodextrin substituted at the position of the hydroxyl group in C2 (Vanbésien et al., 2015). Following the same concept of cooperative catalysis, the case of the C–C coupling between alkyl halides and tetralkyltin derivatives is found, which is regulated, at the same time, by the formation of supramolecular clusters and some Pt(II)/Pt(IV) species (Scheme 11.2, A) (Levin et al., 2016a).

In another context but related with synergistic pathways, we find the acylation of amides under mild conditions, implying a site-selective *N*-activation nucleophilic cycle and then, an efficient Suzuki-Miyaura coupling of boronic acids under the action of $Pd(0)L_n$ (Scheme 11.2, B). Since primary amides are considered as high potential in the pharmaceutical industry, the present method dealing with the functionalization of the C–N bond is attractive in this sector (Meng et al., 2016).

Since the last decade there has been a growing interest for the development of organic and metallic complexes with photoredox activity, as evidenced by the number of publications in the area, that increased 20 times just in the first half of this lapse. For example, along with the catalysis methods, photoredox and metallic strategies combine perfectly to offer an elegant alternative for the highly efficient decarboxylative cyanation to render chiral nitriles. According to reports, benzylic radicals and Cu(II) species are two key intermediates which emerged jointly during the photocatalytic stage, after $Ir(ppy)_3$, CuBr, TMSCN and a chiral bisoxazoline executed their action in the presence of an appropriate *N*-hydroxyphthalamide ester (Wang et al., 2017). On the other hand, some interesting examples of compounds with the ability to transform UV-Vis light into chemical energy via a single electron transfer process, include structures of metals in groups 8 and 9 bond to bispyridines, bispyrazines and phenylpyridines, like $Ru(bpy)_3^{2+}$, $Ru(bpz)_3^{2+}$, $Ir(ppy)_3$, $Ir(ppy)_2(dtbbpy)^+$ and $Ir[dF(CF_3)ppy]_2(dtbbpy)^+$ (Shaw et al., 2016). Metals such as Pt(II), Au(I), Cu(I), Ni(II) and Fe(II) have shown promising results related with photocatalysis for oxidative addition, reductive elimination, and trans-metalation processes (Levin et al., 2016b). Ir is considered a privileged metal in photocatalysis as evidenced by the numerous examples of applications in this area. Intense efforts have been done to improve the efficiency of molecules capable of oxidizing water under natural irradiation and fortunately, the system formed by $[Cp*Ir\{P(O)(OH)_2\}_3]Na$ anchored to $BiVO_4$ over a conducting surface and subsequently oxidized with IO_4^-, has overcome all expectations (Wang et al., 2018c).

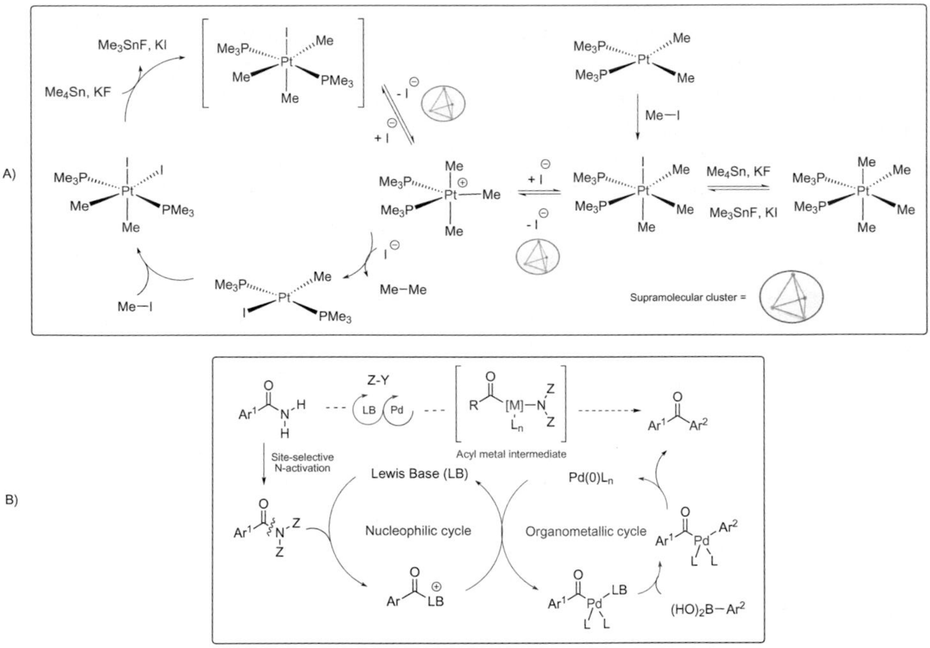

Scheme 11.2   Examples of cooperative catalysis.

Organometallic catalysis has been extensively exploited for the synthesis of porphyrins. Several palladium-mediated couplings are basic strategies to build these macrocycles by creation of C–C bonds, and other pathways employing palladium or iridium are now being used to functionalize the scaffolds via C–H activation; complementary, copper-assisted Huisgen cycloaddition is an important method to link porphyrin frameworks into a variety of functional compounds. Nevertheless, those metals are not the only ones implicated in the field of macrocycles, as depicted in the review made by Shinokubo et al. (Hiroto et al., 2017).

J. Bercaw, among many other researchers, dedicated his entire life to solve crucial problems in organometallic chemistry and catalysis. Between the most relevant contributions we can find the reduction of CO, some alkene polymerizations or β-H eliminations and the C–H and C–C activations through different zirconocenes, scandocenes, and niobocene complexes (Wolczanski and Chirik, 2015). M. Brookhart, on the other hand, drew important guidelines for the understanding olefin cyclopropanation through iron-containing carbenes, olefin polymerization using palladium and nickel-based catalysts as well as in C–H activation and functionalization using iridium species (Daugulis et al., 2016). This metal has also been investigated as a promising agent taking part of a catalyst for sp^2 C–H borylation of (hetero) arenes, in combination with N,B-type boryl anions. Its preparation was performed *in situ* through the reaction of a precatalyst (pyridintetraminodiborane) with $Ir(OMe)(COD)_2$ and its range of application included substrates with electronic excess character and high sterical hindrance (Wang et al., 2015). Ruthenium, on the other hand, was involved in the development of catalysts like $[RuCl_2(p\text{-cymene})]_2$, which is effective for the synthesis of indole rings through an electrochemical cyclization of aniline derivatives and alkynes. Remarkably, the catalyst is reconstituted using the electric current while hydrogen evolution takes place. This combination also works efficiently in water-based systems and is stable to air (Xu et al., 2018).

One of the concerns in catalysis deals with the development of species containing earth abundant metals capable to drive two-electron redox processes, commonly associated with noble metals. In this regard, $[(DPB)Fe]_2(N_2)$ and $(DPB)Co(N_2)$ (DPB = bis(o-diisopropylphosphinophenyl) phenylborane) have proved excellent activation of the E–H bond (E = C, N, O or S) in aromatic and heteroaromatic compounds. This cobalt complex also exhibited a remarkable potential in the hydrosilylation of several aldehydes and ketones (Nesbit et al., 2015).

In the case of manganese, diverse examples of C–H stoichiometric activations have been reported in the last 40 years, but their catalytic counterparts remain little explored. Even so, recent advances in the area confirmed that this rich metal can be used in combination with acids or bases to promote the activation of C–H frames in molecules bearing bonds such as C≡C, C=C, C=O, C=N and C≡N (taking part of allylations, alkylations, alkenylations, alkynilations, acylations and cyclizations), which results in low-priced and easier transformations because $MnR(CO)_5$ species (R=Me, Bn, Ph) are not required like in traditional stoichiometric reactions. While there are undeniable advantages in the use of manganese-based catalysts, there are other aspects which can be further improved, such as the development of their redox chemistry and their ability to be implicated in enantioselective C–H activations (Hu et al., 2018).

Those cases where metallic catalysis is applied to biphasic systems with chiral outcomings need special mentioning. On the base of this criterion, they can be sorted into two categories according to their manifested physicochemical properties:

a) Chiral metal catalysis in micellar systems (e.g. Au(I)-mediated intramolecular hydroxycarboxylation of allenic acids).

b) Non miscellar chiral metal catalysis, which can be represented by non-immobilized heterogeneous catalyst in water (e.g. asymmetric silyl conjugate addition assisted by Cu(II) complexes) or by water-soluble species (e.g. Ru(II)-catalyzed intramolecular cyclopropanation of trans allylic diazocompounds) (Guo et al., 2018).

On the other hand, among the wide field of catalysis, one category has especially emerged as a powerful tool for the synthesis and deconstruction of organic molecules. Reversible catalysis, also known as shuttle catalysis, has been exploited for the transfer of small groups or fragments in the course of isodesmic reactions. Between the most attractive applications we could find hydroacylation of alkenes promoted by $(PPh_3)_3RhCl$, retro-hydroformylation of aldehydes using $[Rh(COD)OMe]_2$, hydrocyanation of alkenes driven by $Ni(COD)_2$ in combination with DPEphos, hydromagnesiation of styrenes with $FeCl_2$ in presence of bis(imino) pyridine and silacyclopropanation using $Ph_3PAg(alkene)_3OTf$ (Bhawal and Morandi, 2016).

Table 11.1. Contributions of metallic *N*-heterocyclic carbenes to catalysis.

Metal	*Uses*	*Representative structure*
Pd	Suzuki, Heck, Negishi, Hiyama, Sonogashira and Buchwald-Hartwig couplings. Aerobic oxidations of alcohols and aminations.	
Ru	Aqueous isomerization of allylic alcohols. Olefin metathesis.	
Cu	Three components click reactions. Stereoselective anti-Markovnikov hydrothiolation of alkynes. Hydrosilylations.	
Au	Lactonization of acetylenic carboxylic acids. Cycloisomerizations of allenes. Alkyne hydration. Hydroarylation of arylakynes.	
Ir	Transfer hydrogenation from glycerol. Isomerization of allylic alcohols (in combination with zirconium-based MOFs).	
Rh	Hydrogenation. Hydrosilylation.	
W	Olefin metathesis.	
Ag	Three component coupling reactions of alkynes, aldehydes and amines.	

In addition to other features mentioned earlier, reusability is a prime factor when proposing a new catalyst. Methods inferring N-heterocyclic carbene complexes have been developed concerning this aspect and many examples in which these structures are applied following the principles of green chemistry are now described. This is remarkable because their differential complexes make them highly versatile (Wang et al., 2018b). In Table 11.1 some of the applications are summarized:

Paradoxically in some cases, better understanding of the behavior in the organometallic complexes, the more available knowledge to synthesize metal-free catalysts which can mimic the activity of the first. For instance, the investigations on $(Me_2AlCH_2PMe)_2$ interacting with $Ni(II)$ in the polymerization of phenylsylane and its conduct facing $Cp^*Rh^{(III)}(DMSO)(Me)_2$, have allowed the development of ambiphilic molecules without metallic centers to achieve diverse reactions like C–H and S–H borylations, H_2 activation, CO_2 hydroboration and hydrogenation, Csp^3–H bond activation, B–B bond formation, etc. (Fontaine and Rochette, 2018). From these examples, it is evident that many efforts carried out in the field of organometallic catalysis have allowed the possibilities to access complex molecular scaffolds using less time and preserving high standards of molecular economy. In spite of this, research continues with the aim to find combinations to provide both more effective and less expensive options.

BIOCATALYSIS

Interdisciplinary collaboration is both a necessary and fruitful attribute in science development. When a phenomenon is observed, measured and studied by different disciplines, large and intimate knowledge is obtained from it.

Examples of this are chemistry and enzymology, both have mutually benefited as they have extended their levels of collaboration, leading to a new area: biocatalysis. Incorporation of enzymology to chemistry has dissipated the idea that biological catalysis is holist and simplistic: when an enzyme is used in synthesis, greener (Wardencki et al., 2005), faster and efficient processes are obtained, while the use of materials and resources is minimum, a characteristic that it is hard to solve in classic protocols. On the other hand, when enzymology is carried on to its own work the reductionist and pragmatic foundations of chemistry, it was possible to understand the phenomenon as asymmetric induction, chemo-, regio- and enantioselectivity, physicochemical properties of enzymes, both hardness and softness of catalytic sites, enzymatic reaction mechanisms and catalytic promiscuity (Afagh and Yudin, 2010; Faber, 2011; Uritsky et al., 2016; Pleiss, 2017). In recent years evolution of proteins allowed the design and development of enzymes that catalyze non-natural reactions, and solvent engineering helped in understanding how to use enzymes in non-aqueous media.

A chronological model was proposed (Bornscheuer et al., 2012; Bornscheuer, 2017) to explain the effects of biocatalysis from the use of the first enzyme in synthesis 100 years ago (Rosenthaler, 1908) to their uses in the actual period. Such a model confirms that this is the third wave of biocatalysis. During the *first wave* animal or plant tissues extracts or microbial strains containing useful enzymatic activity for a given synthetic reaction were used, this wave is considered rudimentary since events as adverse reactions, selectivity and stability were poorly controlled (Gurung et al., 2013; Martínez-Cuesta et al., 2015).

The second wave came in the 1980s with gene technology and recombinant protein engineering (Young et al., 2012), which enabled the cloning and expression of the enzymes of interest in host microorganisms. This promoted the reduction in production cost of enzymes and increased their disposable amount. Thus, sited-directed mutagenesis to improve properties such as stability, chemo-, regio- and stereoselectivity started (Winter et al., 1982; Wilkinson et al., 1983; Estell et al., 1985). Handling features such as immobilization on inert resins were also incorporated (Bartlett and Cooper, 1993; Abian et al., 2004), and with this the use of enzymes in non-aqueous solvents was possible (Butler, 1979; Arnold, 1990; Dordick, 1992; Doukyu and Ogino, 2010). It was in this wave when the synthetic use of enzymes on industrial scales was increased (Singh et al., 2016).

The third wave started with advanced protein engineering in the 1900s, such as DNA shuffling and error-prone polymerase chain reaction in combination with high-throughput screening (directed evolution) methods (Bornscheuer and Pohl, 2001; Romero and Arnold, 2009), bioinformatics and computational modeling (van de Waterbeemd and Gifford, 2003; Xiao et al., 2014). Hence, characterizing, designing or expressing any enzyme, including understanding the mechanism of promiscuous enzymatic reactions can take few months instead of years.

Enzymes in Organic Synthesis

Currently the main aim of synthetic methods and products, such as polymers (Kadokawa and Kobayashi, 2010), fine chemicals, intermediates, and precursors of pharmaceutical actives (Margolin, 1993; Patel, 2013; Alcántara, 2017), flavorings, fragrances (Larios et al., 2004; Franssen et al., 2005; Dhake et al., 2013; Poornima and Preetha, 2017), and dyes (Polak and Jarosz-Wilkolazka, 2012; Wang et al., 2018a) can be obtained with enzymes at the laboratory or industrial level. The methods to carry them out include isolated enzymes, immobilized enzymes, combined enzymes to carry out cascade reactions (Ricca et al., 2011; Sperl and Sieber, 2018), and modified enzymes in microorganisms, which is called metabolomic engineering (Dixon et al., 1996; Fujii et al., 2005).

Due to similarities and in many cases the advantage of other catalysts and activators, the use of enzymes in organic syntheses is still increasing (Li et al., 2012b). Chemo- (García et al., 1994) end regioselectivity (Lavandera et al., 2004) of the enzymes allows directed planning of synthesis as well as diminution of protecting groups, which means fewer synthesis steps, less overall time reaction and waste reduction (Sweers and Wong, 1986; Bashir et al., 1995; Drauz and Waldmann, 2002; Illanes, 2008). The stereoselective properties of the enzymes allows the resolution of racemic mixtures and the asymmetric synthesis enables dispensing with chiral auxiliaries. A disadvantage is that it is not possible to obtain an enzyme with opposite chemo-, regio-, or stereoselectivity to a specific enzyme, which is possible with a non-enzymatic catalyst (Noyori et al., 1987; Brown, 2001; Christmann and Stefan, 2007; Mugford et al., 2008).

- Regarding the reaction medium, the mild conditions of the enzymes (20–40°C and pH≈7) diminish decomposition, isomerization and rearrangement in substrates and products, however, any variation can alter the process or the enzyme itself. An advantage is that, the enzymes work under the same conditions (except proteases) so that several enzymatic reactions can be carried out continously without isolating intermediates: artificial metabolism (García-Junceda, 2008; Erb et al., 2017). This property is limited when cofactors (heme, flavine or NAD[P]H) or organic energy sources are required, moreover, some enzymatic reactions are inhibited by the presence of substrates or products (Mordhorst et al., 2017). On the other hand, there are extremophile enzymes (Karan et al., 2012; Coker, 2016) or cases where it is possible to induce enzyme activity with alternative energy sources such as microwaves (Young et al., 2008).

Catalytic promiscuity (O'Brien and Herschlag, 1999; Penning and Jez, 2001; Walsh, 2001; Bornscheuer and Kazlauskas, 2004; Hult and Berglund, 2007), implies that the enzymes can catalyze a general type of reaction or several different reactions. It has been proposed that actually at least one enzymatic process for each known organic reaction exists (Sih et al., 1977) and that many of these reactions have mechanistic variants only observed in enzymes. Thus, many reported hydrolysis/synthesis of esters (Boland et al., 1991; Lau et al., 2004; Velonia et al., 2005; Laszlo et al., 2011), amides (Schmidt-Kastner and Egerer, 1984; García et al., 1993; de Zoete et al., 1996; Litjens et al., 1999; Conde et al., 2000; Baldessari and Mangone, 2001; Gill and Patel, 2006), lactones (Gutman et al., 1990), lactams (Taylor et al., 1990), ethers (Zhang and Poulter, 1993), anhydrides (Yamamoto et al., 1988), epoxides (Leak et al., 1992), and nitriles (Nagasawa and Yamada, 1989). Other reactions reported are alkane oxidation (Mansuy and Battoni, 1989; Bordeaux et al., 2011 and 2012), alcohols (Lemiére et al., 1985), aldehydes, sulfurs and sulfoxides (Phillips and May, 1981), as well as alkene epoxidation (Abdulmalek et al., 2014), aromatic hydroxylation and dihydroxylation (Boyd et al., 1991), and the Baeyer-Villiger oxidation of ketones (Walsh and Chen, 1988). On reduction reactions, there are those of aldehydes/ketones, alkenes and reductive aminations (Koszelewski et al., 2018). There are additions/eliminations of water (Findeis and Whitesides, 1987), ammonia (Akhtar et al., 1987), hydrogen cyanide (Effenberger et al., 1987), halogenation and dehalogenation (Neidleman and Geigert, 1986), Friedel-Crafts type alkylation (Stecher et al., 2009; Drienosvká et al., 2015), O- and N-dealkylations (Buist and Dimnik, 1986), carboxylations (Aresta et al., 1998), and decarboxylations (Ohta, 1999), as well as isomerizations (Schwab and Henderson, 1990), acyloin (Fuganti and Grasselli, 1988), and aldolic reaction (Toone et al., 1989), Michael addition (Kitazume et al., 1986) of amides to acrylonitiiles (Torre et al., 2004), aldehydes (Carlqvist et al., 2005) and esters (Dhake et al., 2010); of enolates to aldehydes, ketones and esters (Branneby et al., 2004; Svedendalhl et al., 2005); of peroxides to aldehydes (Svedendahl et al., 2008); of thiols to ketones and sulfoxides (Madalińska et al., 2012) in addition to double bonds (Lou et al., 2008), Setter (Pohl et al., 2002), Neff (Durchschein et al., 2010) and Diels-Alder (Oikawa et al., 1995; Pohnert, 2001; Williams, 2002; Bos et al., 2012) type reactions. Antibody (Braisted and Schultz, 1994) and enzyme (Aemissegger et al.,

2002; Luk et al., 2011) catalyzed Cope type rearrangements have been found as well as [3,3]-sigmatropic rearrangements (Abe et al., 1993; Ganem, 1996), and aziridinations (Farwell et al., 2015). Many enzymes carried out reactions that from a chemical point of view were not possible to achieve, some examples are the functionalization of non-activated positions as aliphatic carbon hydroxylation (Nagasawa and Yamada, 1989).

In recent years there have been reports of modified enzymes to catalyze non-natural reactions, to increase specificity, and reaction velocity of secondary reactions, to increase efficiency; catalytic site remodeling to increase stereospecifity (Renata et al., 2015), or avoid the inactivation phenomenon (Gershenson et al., 2000). These are currently known as Grubbs type reactions (Jeschek et al., 2016), metallic carbenoids formation to catalyze formation of N–H, C–H or O–H bonds (Baumann et al., 2007), cyclopropanations (Coelho et al., 2013 and 2014; Heel et al., 2014), insertions of N–H to athydildiazoketones and insertion of nitrenes metaloenzyme catalyzed (Arnold, 2015). Nucleophilic aromatic substitutions (Seffernick and Wackett, 2001), aliphatic chains desaturations (Broun et al., 1998), epoxide opening using non-natural nucleophiles as azide, nitrite, cianate and thiocyanate (Hasnaoui-Dijoux et al., 2008) have also been reported. Currently it is possible carry out glycosidation of dinitrophenyl-glycosides activated with deoxythiosugars (Jahn et al., 2003).

Resolutions and Asymmetric Synthesis with Enzymes

Two attributes that have contributed to the use of enzymes in organic synthesis are their efficiency and selectivity. Regarding efficiency, there is a wide spectrum of transformations that can be carried out with enzymes. Currently with the use of enzymes it is possible to bond a wide range of organic functions to generate any structure whose reactions yields are compared with classic non enzymatic methods. Regarding selectivity, with the proper enzyme it is possible to recognize specific structural features on a substrate and carry out transformations on a region-, chemo-, and stereoselective level. In the steric level, enzymes show selectivity and specification and can affect Kinetic Resolutions (KR) and symmetric transformations, respectively (Berglund, 2001).

From an enzymatic point of view, the first known method to obtain chiral compounds was the KR. Along with non-enzymatic KR, this is the most used method of resolution of racemic mixtures in industry (Li et al., 2016). KR takes advantage of the reaction rate differences of a pair of enantiomers in a racemic mixture respect to the same reaction with other chiral compound (an enzyme in this example) to carry out the total separation of enantiomers. In the ideal case, the yield of a KR achieves only 50% of transformation, namely total transformation of one of the enantiomers. The opposite process it is known as racemization and has also been useful in chemoenzymatic synthesis also (Huerta et al., 2001).

Examples of KR include almost any enzymatic transformation (Ghanem and Aboul-Enein, 2005), however the most studied methods are hydrolysis (Utczás et al., 2011), aminolysis (de Zoete et al., 1993; García-Urdiales et al., 1999; Aoyagi et al., 2004; Weiβ and Gröger, 2009), trans-esterification of esters and (Flores-

Sánchez et al., 2005; Li et al., 2012a) alcohols (Rotticci et al., 2001; Henke et al., 2002; Ottoson et al., 2002), resolution of amines (Gedey et al., 1999; Luna et al., 2002; Sigmund and DiCosimo, 2004; Goswami et al., 2005; Fitz et al., 2005; Shakeri et al., 2010), nitriles (Brady et al., 2004; Torre et al., 2004; Moeller et al., 2013), and glycosides (Shimod and Katsuragi, 2010). Enzymatic epoxidations (Choi, 2009) and azidolysis (Molinaro et al., 2010) at a resolutive level have been reported.

A special case of KR is the Dynamic Kinetic Resolution (DKR), which carries out an enantioselective transformation over one of the enantiomers of the racemic mixture and at the same time racemize the other enantiomer. Thus, an efficient DKR could transform one of the enantiomers with 100% of yield. Examples of DKR include both hydrolysis (Asano and Yamaguchi, 2005; Yasukawa and Asano, 2012) and aminolysis (Wegman et al., 1999; Choi et al., 2001), of racemic α-aminoacids to yield D-aminoacids; other examples are hydrolysis of thioesters (Um and Drueckhammer, 1998). Protocols where enzymes and metals (Martín-Matute and Bäckvall, 2007) are combined to carry out DKRs though the acylation of primary amines (Paetzold and Bäckvall, 2005), benzylic and allylic alcohols (Akai et al., 2006) have been reported.

Enzymes can distinguish stereotopic, both groups of meso and prochiral compounds to carry out stereospecific reactions. Enzymatic desymmetrization is an election method because of its high yields (Candy et al., 2009; Kołodziejska et al., 2010; Sapu et al., 2011; Palomo and Cabrera, 2012; González-García et al., 2018). Examples of asymmetric syntheses encompass carboxylases catalyzed aldol type condensations (Ward and Singh, 2000), reductive amination of α-aminoketoacids (Xue et al., 2018) to yield chiral aminoacids in both cases, as well as reductive amination of prochiral ketones (Koszelewski et al., 2008; Schätzle et al., 2011), asymmetric reduction of α-alkylated-α,β-unsaturated aldehydes (Stueckler et al., 2010), aldolic type condensations (Dean et al., 2007). Enzymes studied in this field are *Pseudomona cepacia* and *Candida antactica* type A or B lipases (Leuenberger et al., 1976; Johnson and Bis, 1992; Johnson and Braun, 1993; Gu et al., 1993; Wei et al., 1998; Kawai et al., 1998).

On the other hand, enzymatic asymmetric additions include the addition of enolates to benzylidenketone (Sano et al., 2013), Stetter reactions to form C–C bonds (Dresen et al., 2010), and additions to ketones to form chiral alcohols (Puertas et al., 1996; Sánchez et al., 1999; López-García et al., 2003, Müller, 2012).

Enzymatic Catalysis in Non-aqueous Media

The discovery of enzymatic activity in non-aqueous media opened a new panorama to the use of enzymes in synthesis (Klibanov et al., 1977; Zaks and Klibanov, 1984 and 1985). Despite the higher activity of the enzymes that is observed in aqueous media, many of them preserve it in organic solvents, ionic liquids and supercritical fluids. Replacement of water for any solvent allows the use of non-polar substrates or industrial scaling (Koskine and Klibanov, 1996; Schmid et al., 2001). However, the use of solvents is not an advantage at all times, and the effect of the same solvent with many enzymes or the same enzyme with many solvents it is not always equal.

Regarding enzyme activity in organic media, it is several orders lower than in aqueous solution (Klibanov, 1990), due to the fact that the solvent occupies some areas in the active site, which generate different recognition to those observed in the aqueous media. The enzyme stability in organic solvents depends of the interactions solvent-enzyme, some of them can inactivate the enzyme due to the loss of its tertiary or secondary structure, this occurs in water (when cofactor for coenzymes are required and are not present), DMSO, formamide, ethylene glycol and ionic liquids with hydrogen bond acceptors like acetate, lactate or nitrate (Singer, 1963; van Rantwijk et al., 2006). Other inactivating solvents are glycerol, 2,2,2-trifluroethanol, methanol, and phenol (Chin et al., 1994). In other solvents, enzymes are insoluble but preserve their activity. Due to these variants many enzymes are used in solid surfaces, which make them easy to isolate, recover and use them again.

Enzyme features as activity, selectivity and stability have been correlated with solvent physicochemical descriptors as $logP$, dielectric constant, dipolar momentum, Hildebrand solubility parameter and others, however, these correlations are not complete and change with each enzyme and substrate. In recent years, solvent engineering applied to reactions catalyzed by lipases have shown to be useful in the control of chemoselectivity. In this case, the control is achieved by rationalization of the thermodynamic activity coefficient of possible products and solvent polarity, both to displace the equilibrium to the desired product (Bellot et al., 2001; Rendón et al., 2001; Castillo et al., 2003; Priego et al., 2009; Rivera-Ramírez et al., 2015).

In Substrates and products, solvation is a common effect. Solvents that dissolve completely to the substrate, decrease its $\Delta G°$ value and increase its K_m and $\Delta G^{\ddagger}$ values, which means decrease in reaction rate in respect to the rate in aqueous solution (Ryu and Dordick, 1992). On the other hand, thermodynamic and stereoelectronic properties of solvents could allow newer selectivity routes and yield products that are not observed in aqueous medium.

The Role of Enzymes in Organic Synthesis

The use of enzymes in organic synthesis it is still growing, and, in many cases, the basic principles are still at an understanding stage. The rate orders of enzymes are $10^8–10^{10}$ times faster than non-catalyzed processes, however many of them still show difficulties and cannot as yet compete with pure organic synthesis.

However, the accepted idea about an enzyme is that it acts as catalyst in an organic transformation, it means, speeding up a slow but possible process.[1] Though currently it is known that enzymes promote processes with little chances of being achieved spontaneously, even in a slow manner. Thus, the concept of enzymatic catalysis could be extended.

Some examples of processes that are not spontaneous but can be induced by one enzyme are i) enantiomer resolution, ii) chemo-, and regioselective reactions, iii) asymmetric reactions, and iv) non-reactive positions activation. Strictly speaking, in these cases, enzymes do not work as catalysts because the transformations are neither spontaneous nor slow. Here, the enzymes are not catalysts, but reactants.[2]

Footnote: 1 and 2 at end of the Chapter on page 214.

Even these days, there are sectors in synthetic chemistry that do not believe in the use of enzymes arguing that in biocatalysis, chemists are unable to manipulate the result. Enzymatic machinery is the result of millions of years of selective evolution. One area where chemists could achieve this is enzymes.

REFERENCES

Abdulmalek, E., Arumugam, M., Mizan, H., Rahman, M., Basri, M., BaSalleh, A. Chemoenzymatic epoxidation of alkenes and reusability study of the phenylacetic acid. *The Scientific World Journal*, 2014, 2014, 1–7.

Abe, F., Yamauchi, T. Indole alkaloids from leaves and stems of Leuconotis eugenifolius. *Phytochemistry*, 1993, 35, 169–171.

Abe, I., Rohmer, M., Prestwich, G. Enzymatic cyclization of squalene and oxidosqualene to sterols and triterpenes. *Chemical Reviews*, 1993, 93, 2189–2206.

Abian, O., Grazú, V., Hermoso, J., González, R., García, J.L., Fernández, R., Guisán, J. Stabilization of penicillin G acylase from *Escherichia coli*: Site-directed mutagenesis of the protein surface to increase multipoint covalent attachment. *Applied and Environmental Microbiology*, 2004, 70, 1249–1251.

Aemissegger, A., Jaun, B., Hilvert, D. Investigation of the enzymatic and nonenzymatic Cope rearrangement of carbaprephenate to carbachorismate. *Journal of Organic Chemistry*, 2002, 67, 6725–6730.

Afagh, N., Yudin, A. Chemoselectivity and the curious reactivity preferences of functional groups. *Angewandte Chemie International Edition*, 2010, 49, 262–310.

Akai, S., Tanimoto, K., Kanao, Y., Egi, M., Yamamoto, T., Kita, Y. A dynamic kinetic resolution of allyl alcohols by the combined use of lipases and [VO(OSiPh₃)₃]. *Angewandte Chemie International Edition in English*, 2006, 45, 2592–2595.

Akhtar, M., Botting, N., Cohen, M., Gani, D. Enantiospecific synthesis of 3-substituted aspartic acids via enzymic amination of substituted fumaric acids. *Tetrahedron*, 1987, 43, 5899–5908.

Alcántara, A. Biotransformations in drug synthesis: A green and powerful tool for medicinal chemistry. *Journal of Medicinal Chemistry and Drug Design*, 2017, 1, 1–7.

Andrés, J.M., Maestro, A., Valle, M., Pedrosa, R. Chiral bifunctional thioureas and squaramides and their copolymers as recoverable organocatalysts. stereoselective synthesis of 2-substituted 4-amino-3-nitrobenzopyrans and 3-functionalized 3,4-diamino-4H-chromenes. *Journal of Organic Chemistry*, 2018, 83, 5546–5557.

Aoyagi, N., Kawauchi, S., T. Izumi, T. Different recognitions of (E)- and (Z)-1,1'-binaphthyl ketoximes using lipase-catalyzed reactions. *Tetrahedron Letters*, 2004, 45, 5189–5192.

Aresta, M., Quaranta, E., Liberio, R., Dileo, C., Tommasi, I. Enzymatic synthesis of 4-OH-benzoic acid from phenol and CO₂: the first example of a biotechnological application of a carboxylase enzyme. *Tetrahedron*, 1998, 54, 8841–8846.

Arnold, F. Engineering enzymes for non-aqueous solvents. *Trends in Biotechnology*, 1990, 8, 244–249.

Arnold, F. The nature of chemical innovation: New enzymes by evolution. *Quarterly Reviews of Biophysics*, 2015, 48, 404–410.

Asano, Y., Yamaguchi, S. Dynamic kinetic resolution of amino acida mide catalyzed by D-aminopeptidase and α-amino-ε-caprolactam racemase. *Journal of the American Chemical Society*, 2005, 127, 7696–7697.

Baldessari, A., Mangone, C. One-pot biocatalyzed preparation of substituted amides as intermediates of pharmaceuticals. *Journal of Molecular Catalysis B: Enzymatic*, 2001, 11, 335–341.

Banik, S.M., Levina, A., Hyde, A.M., Jacobsen, E.N. Lewis acid enhancement by hydrogen-bond donors for asymmetric catalysis. *Science*, 2017, 358, 761–764.

Bae, H.Y., Song, C.E. Unprecedented hydrophobic amplification in noncovalent organocatalysis "on water": Hydrophobic chiral squaramide catalyzed Michael addition of malonates to nitroalkenes. *ACS Catalysis*, 2015, 5, 3613–3619.

Bartlett, P., Cooper, J. A review of the immobilization of enzymes in electropolymerized films. *Journal of Electroanalytical Chemistry*, 1993, 362, 1–12.

Bashir, N., Phythian, S.J., Reason, A.J., Roberts, S.M. Enzymatic esterification and de-esterification of carbohydrates: Synthesis of a naturally occurring rhamnopyranoside of *p*-hydroxybenzaldehyde and a systematic investigation of lipase-catalysed acylation of selected arylpyranosides. *Journal of the Chemical Society*, 1995, 1, 2203–2222.

Baumann, L., Buvi, H., Guodong, D., Woo, L. Iron porphyrin catalyzed N-H insertion reactions with ethyl diazoacetate. *Organometallics*, 2007, 26, 3995–4002.

Belliotti, T.R., Capiris, T., Ekhato, I.V., Kinsora, J.J., Field, M.J., Heffner, T.G., Meltzer, L.T., Schwarz, J.B., Taylor, C.P., Thorpe, A.J., Vartanian, M.G., Wise, L.D., Zhi-Su, T., Weber, M.L., Wustrow, D.J. Structure–activity relationships of pregabalin and analogues that target the $\alpha 2$-δ protein. *Journal of Medicinal Chemistry*, 2005, 48, 2294–2307.

Bellot, J.C., Choisnard, L., Castillo, E., Marty, A. Combining solvent engineering and thermodynamic modeling to enhance selectivity during monoglyceride synthesis by lipase-catalyzed esterification. *Enzyme and Microbial Technology*, 2001, 28, 362–369.

Berglund, P. Controlling lipase enantioselectivity for organic synthesis. *Biomolecular Engineering*, 2001, 18, 13–22.

Bhawal, B.N., Morandi, B. Catalytic transfer functionalization through shuttle catalysis. *ACS Catalysis*, 2016, 6, 7528–7535.

Boland, W., Frößl, C., Lorenz, M. Esterolytic and lipolytic enzymes in organic synthesis. *Synthesis*, 1991, 1049–1072.

Bordeaux, M., Galarneau, A., Fajula, F., Drone, J. A regioselective biocatalyst for alkane activation under mild conditions. *Angewandte Chemie International Edition*, 2011, 50, 2075–2079.

Bordeaux, M., Galarneau, A., Drone, J. Catalytic, mild, and selective oxyfunctionalization of linear alkanes: Current challenges. *Angewandte Chemie International Edition*, 2012, 51, 10712–10723.

Bornscheuer, U., Pohl, M. Improved biocatalysts by directed evolution and rational protein design. *Current Opinion in Chemical Biology*, 2001, 5, 137–143.

Bornscheuer, U., Kazlauskas, R. Catalytic promiscuity in biocatalysis: Using old enzymes to form new bonds and follow new pathways. *Angewandte Chemie International Edition*, 2004, 43, 6032–6040.

Bornscheuer, U., Huisman, G., Kazlauzkas, R., Lutz, S., Moore, J., Robins, K. Engineering the third wave of biocatalysis. *Nature*, 2012, 485, 185–194.

Bornscheuer, U. The fourth wave of biocatalysis is approaching. *Philosophical Transactions of the Royal Society A*, 2017, 376, 1–7.

Bos, J., Fusetti, F., Driessen, A., Roelfes, G. Enantioselective artificial metalloenzymes by creation of a novel active site at the protein dimer interface. *Angewandte Chemie*, 2012, 124, 7590–7593.

Boyd, D., Dorrity, M., Hand, M., Malone, J., Sharma, N., Dalton, H., Gray, D., Sheldrake, G. Enantiomeric excess and absolute configuration determination of cis-dihydrodiols from bacterial metabolism of monocyclic arenes. *Journal of the American Chemical Society*, 1991, 113, 666–667.

Brady, D., Beeton, A., Zeevaart, J., Kgaje, C., van Rantwijk, F., Sheldon, R. Characterisation of nitrilase and nitrile hydratase biocatalytic systems. *Applied Microbiology and Biotechnology*, 2004, 64, 76–85.

Brahmachari, G., Nayek, N., Nurjamal, K., Karmakar, I., Begam, S. Triethylamine–A versatile organocatalyst in organic transformations: A decade update. *Synthesis*, 2018, 50, 4145–4164.

Braisted, A., Schultz, P. An antibody-catalyzed oxy-cope rearrangement. *Journal of the American Chemical Society*, 1994, 116, 2211–2212.

Branneby, C., Carlqvist, P., Hult, K., Brinck, T., Berglund, P. Aldol additions with mutant lipase: Analysis by experiments and theoretical calculations. *Journal of Molecular Catalysis B: Enzymatic*, 2004, 31, 123–128.

Broun, P., Shanklin, J., Whittle, E., Somerville, C. Catalytic plasticity of fatty acid modification enzymes underlying chemical diversity of plant lipids. *Science*, 1998, 282, 1315–1317.

Brown, J.M. Nobel Prizes in Chemistry 2001; Asymmetric hydrogenation recognised! *Advanced Synthesis & Catalysis*, 2001, 343(8), 755–756.

Brown, N.J., García-Trenco, A., Weiner, J., White, E.R., Allison, M., Chen, Y., Wells, P.P., Gibson, E.K., Hellgardt, K., Shaffer, M.S.P., Williams, C.K. From organometallic zinc and copper complexes to highly active catalysts for the conversion of CO_2 to methanol. *ACS Catalysis*, 2015, 5, 2895–2902.

Buist, P., Dimnik, G. Use of sulfur as a chemical connector. *Tetrahedron Letters*, 1986, 27, 1457–1460.

Bulfield, D., Huber, S.M. Halogen bonding in organic synthesis and organocatalysis. *Chemistry: A European Journal*, 2016, 22, 14434–14450.

Butler, L. Enzymes in non-aqueous solvents. *Enzyme and Microbial Technology*, 1979, 1, 253–259.

Candy, M., Audran, G., Bienayme, H., Bressy, C., Pons, J. Enantioselective enzymatic desymmetrization of highly functionalized *meso* tetrahydropyranyl diols. *Organic Letters*, 2009, 11, 4950–4953.

Candy, M., Durand, T., Galano, J.M., Oger, C. Total synthesis of the isoketal 5-D2-isoK natural product based on organocatalysis. *European Journal of Organic Chemistry*, 2016, 5813–5816.

Cao, W., Liu, X., Feng, X. Chiral organobases: Properties and applications in asymmetric catalysis. *Chinese Chemical Letters*, 2018, 29, 1201–1208.

Cao, Z.C., Shi, Z.J. Deoxygenation of ethers to form carbon-carbon bonds via nickel catalysis. *Journal of the American Chemical Society*, 2017, 139, 6546–6549.

Capobianco, A., Di Mola, A., Intintoli, V., Massa, A., Capaccio, V., Roiser, L., Waser, M., Palombi, L. Asymmetric tandem hemiaminal-heterocyclization-aza-Mannich reaction of 2-formylbenzonitriles and amines using chiral phase transfer catalysis: an experimental and theoretical study. *RSC Advances*, 2016, 6, 31861–31870.

Cariou, R., Shabaker, J.W. Iron-catalyzed chain growth of ethylene: *In situ* regeneration of $ZnEt_2$ by tandem catalysis. *ACS Catalysis*, 2015, 5, 4363–4367.

Carlqvist, P., Svedendahl, M., Branneby, C., Hult, K., Brinck, T., Berglund, P. Exploring the active-site of a rationally redesigned lipase for catalysis of Michael-type additions. *ChemBioChem*, 2005, 6, 331–336.

Castillo, E., Pezzotti, F., Navarro, A., López-Munguía, A. Lipase-catalyzed synthesis of xilitol monoesters: Solvent engineering approach. *Journal of Biotechnology*, 2003, 102, 251–259.

Cavallo, G., Metrangolo, P., Milani, R., Pilati, T., Priimagi, A., Resnati, G., Terraneo, G. The halogen bond. *Chemical Reviews*, 2016, 116, 2478–2601.

Chanda, T., Zhao, J.C. Recent progress in organocatalytic asymmetric domino transformations. *Advanced Synthesis & Catalysis*, 2018, 360, 2–79.

Chauhan, P., Mahajan, S., Enders, D. Achieving molecular complexity via stereoselective multiple domino reactions promoted by a secondary amine organocatalyst. *Accounts of Chemical Research*, 2017, 50, 2809–2821.

Chen, X.Y., Enders, D. Multisubstituted unnatural prolines for asymmetric catalytic domino reactions. *Chem*, 2018, 4, 21–23.

Cheng, D., Ishihara, Y., Tan, B., Barbas, C.F. Organocatalytic asymmetric assembly reactions: Synthesis of spirooxindoles via organocascade strategies. *ACS Catalysis*, 2014, 4, 743–762.

Chin, J., Wheeler, S., Klibanov, A. Commullication to the editor on protein solubility in organic solvents. *Biotechnology and Bioengineering*, 1994, 44, 140–145.

Choi, W. Biotechnological production of enantiopure epoxides by enzymatic kinetic resolution. *Applied Microbiology and Biotechnology*, 2009, 84, 239–247.

Choi, Y., Kim, M., Ahn, Y., Kim, M. Lipase/palladium-catalyzed asymmetric transformations of ketoximes to optically active amines. *Organic Letters*, 2001, 3, 4099–4101.

Christmann, M., Stefan, S. Asymmetric Synthesis the Essentials. Germany: WILEY-VCH, Verlag GmbH & Co.; 2007.

Coelho, P., Brustad, E., Kannan, A., Arnold, F. Olefin cyclopropanation via carbene transfer catalyzed by engineered cytochrome P450 enzymes. *Science*, 2013, 339, 307–310.

Coelho, P., Wang, Z., Ener, M., Baril, S., Kannan, A., Arnold, F., Brustad, E. Corrigendum: A serine-substituted P450 catalyzes highly efficient carbene transfer to olefins *in vivo*. *Nature Chemical Biology*, 2014, 9, 485–487.

Coker, J. Extremophiles and biotechnology: Current uses and prospects. *F1000Research*, 2016, 5, 1–7.

Collins, L.R., van Gastel, M., Neese, F., Fürstner, A. Enhanced electrophilicity of heterobimetallic Bi-Rh paddlewheel carbene complexes: A combined experimental, spectroscopic, and computational study. *Journal of the American Chemical Society*, 2018, 140, 13042–13055.

Conde, E., Bello, D., de Cózar, A., Sánchez, M., Vázquez, M.A., Cossío, F.P. Densely substituted unnatural l- and d-prolines as catalysts for highly enantioselective stereodivergent (3+2) cycloadditions and aldol reactions. *Chemical Science*, 2012, 3, 1486–1491.

Conde, S., López-Serrano, P., Martínez, A. Regioselective lipase-catalysed γ-monoamidation of D-glutamic acid diesters: Effect of the *N*-protecting group. *Tetrahedron: Asymmetry*, 2000, 11, 2537–2545.

Craig, R., Sorrentino, E., Connon, S.J. Enantioselective alkylation of 2-oxindoles catalyzed by a bifunctional phase-transfer catalyst: Synthesis of (−)-debromoflustramine B. *Chemistry: A European Journal*, 2018, 24, 4528–4531.

Dai, J., Zhu, L., Tang, D., Fu, X., Tang, J., Guo, X., Hu, C. Sulfonated polyaniline as a solid organocatalyst for dehydration of fructose into 5-hydroxymethylfurfural. *Green Chemistry*, 2017, 19, 1932–1939.

Das, R.K., Barnea, E., Andrea, T., Kapon, M., Fridman, N., Botoshansky, M., Eisen, M.S. Group 4 lanthanide and actinide organometallic inclusion complexes. *Organometallics*, 2015, 34, 742–752.

Daugulis, O., MacArthur, A.H.R., Rix, F.C., Templeton, J.L. A career in catalysis: Maurice Brookhart. *ACS Catalysis*, 2016, 6, 1518–1532.

de Gracia Retamosa, M., Ruiz-Olalla, A., Bello, T., de Cózar, A., Cossío, F.P. A three-component enantioselective cyclization reaction catalyzed by an unnatural amino acid derivative. *Angewandte Chemie International Edition*, 2018, 57, 668–672.

de Zoete, M.C., Kock-van Dalen, A.C., van Rantwijk, F., Sheldon, R.A. Ester ammoniolysis: A new enzymatic reaction. *Journal of the Chemical Society, Chemical Communications*, 1993, 1831–1832.

de Zoete, M.C., Kock-van Dalen, A.C., van Rantwijk, F., Sheldon, R.A. Lipase-catalysed ammoniolysis of lipids. A facile synthesis of fatty acid amides. *Journal of Molecular Catalysis B: Enzymatic*, 1996, 2, 141–145.

Dean, S., Greenberg, W., Wong, C. Recent advances in aldolase-catalyzed asymmetric synthesis. *Advanced Synthesis & Catalysis*, 2007, 349, 1308–1320.

Denard, C.A., Bartlett, M.J., Wang, L.L., Hartwig, J.F., Zhao, H. Development of a one-pot tandem reaction combining ruthenium-catalyzed alkene metathesis and enantioselective enzymatic oxidation to produce aryl epoxides. *ACS Catalysis*, 2015, 5, 3817–3822.

Dhake, K., Tambade, P., Singhal, R., Bhanage, B. ChemInform abstract: Promiscuous *Candida Antarctica* Lipase B-catalyzed synthesis of amino esters via aza-Michael addition of amines to acrylates. *Tetrahedron Letters*, 2010, 51, 4455–4458.

Dhake, K., Thakare, D., Bhanage, B. Lipase: A potential biocatalyst for the synthesis of valuable flavor and fragrance ester compounds. *Flavour and Fragrance Journal*, 2013, 28, 71–83.

Di Marco, L., Hans, M., Delaude, L., Monbaliu, J.C.M. Continuous-flow N-heterocyclic carbene generation and organocatalysis. *Chemistry: A European Journal*, 2016, 22, 4508–4514.

Dilek, Ö., Tezeren, M.A., Tilki, T., Ertürk, E. Chiral 2-(2-hydroxyaryl)alcohols (HAROLs) with a 1,4-diol scaffold as a new family of ligands and organocatalysts. *Tetrahedron*, 2018, 74, 268–286.

Dixon, R., Lamb, C., Masoud, S., Sewalta, V., Paiva, N. Metabolic engineering: Prospects for crop improvement through the genetic manipulation of phenylpropanoid biosynthesis and defense responses. *Gene*, 1996, 179, 61–71.

Dochain, S., Vetica, F., Puttreddy, R., Rissanen, K., Enders, D. Combining organocatalysis and lanthanide catalysis: A sequential One-pot quadruple reaction sequence/hetero-Diels-Alder asymmetric synthesis of functionalized tricycles. *Angewandte Chemie*, 2016, 128, 16387–16389.

Dordick, J.S. Designing enzymes for use in organic solvents. *Biotechnology Progress*, 1992, 8, 259–267.

Doukyu, N., Ogino, H. Organic solvent-tolerant enzymes. *Biochemical Engineering Journal*, 2010, 48, 270–282.

Drauz, K., Waldmann, H Enzyme Catalysis in Organic Synthesis. 2nd Ed. Germany: WILEY-VCH; 2002.

Dresen, C., Richter, M., Pohl, M., Lüdeke, S., Müller, M. The enzymatic asymmetric conjugate umpolung reaction. *Angewandte Chemie*, 2010, 122, 6750–6753.

Drienosvká, I., Rios-Martínez, A., Drakssharapu, A., Roelfes, G. Novel artificial metallo-enzymes by *in vivo* incorporation of metal-binding unnatural amino acids. *Chemical Science*, 2015, 6, 770–776.

Dumoulin, A., Masson, G. Lewis acids turn unreactive substrates into pure enantiomers. *Science*, 2016, 351, 918–919.

Durchschein, K., Ferreira-da Silva, B., Wallner, S., Macheroux, P., Kroutil, W., Glueck, S., Faber, K. The flavoprotein-catalyzed reduction of aliphatic nitro-compounds represents a biocatalytic equivalent to the Nef-reaction. *Green Chemistry*, 2010, 12, 616–619.

Effenberger, F., Ziegler, T., Förster, S. Enzyme-catalyzed cyanohydrin synthesis in organic solvents. *Angewandte Chemie International Edition*, 1987, 26, 458–460.

Enders, D., Niemeier, O., Henseler, A. Organocatalysis by N-heterocyclic carbenes. *Chemical Reviews*, 2007, 107, 5606–5655.

Erb, T., Jones, P., Bar-Even, A. Synthetic metabolism: Metabolic engineering meets enzyme design. *Current Opinion in Chemical Biology*, 2017, 37, 56–62.

Erkkilä, A., Majander, I., Pihko, P.M. Iminium catalysis. *Chemical Reviews*, 2007, 107, 5416–5470.

Estell, D., Graycar, T., Wells, J. Engineering an enzyme by site-directed mutagenesis to be resistant to chemical oxidation. *The Journal of Biological Chemistry*, 1985, 260, 6518–6521.

Faber, K. Biotransformations in Organic Chemistry. 6th Ed. Berlin: Springer-Verlag., 2011.

Fan, Y., Kass, S.R. Enantioselective friedel-crafts alkylation between nitroalkenes and indoles catalyzed by charge activated thiourea organocatalysts. *Journal of Organic Chemistry*, 2017, 82, 13288–13296.

Fang, X., Wang, C.J. Recent advances in asymmetric organocatalysis mediated by bifunctional amine-thioureas bearing multiple hydrogen-bonding donors. *Chemical Communications*, 2015, 51, 1185–1197.

Farwell, C., Zhang, R., McIntosh, J., Hyster, T., Arnold, F. Enantioselective enzyme-catalyzed aziridination enabled by active-site evolution of a cytochrome P450. *ACS Central Science*, 2015, 1, 89–93.

Findeis, M., Whitesides, G. Fumarase-catalyzed synthesis of L-threo-chloromalic acid and its conversion to 2-deoxy-D-ribose and D-erythro-sphingosine. *Journal of Organic Chemistry*, 1987, 52, 2838–2848.

Fitz, M., Lundell, K., Lindros, M., Fülöp, F., Kanerva, L. An effective approach to the enantiomers of alicyclic β-aminonitriles by using lipase catalysis. *Tetrahedron: Asymmetry*, 2005, 16, 3690–3697.

Flanigan, D.M., Romanov-Michailidis, F., White, N.A., Rovis, T. Organocatalytic reactions enabled by N-heterocyclic carbenes. *Chemical Reviews*, 2015, 115, 9307–9387.

Flores-Sánchez, P., Escalante, J., Castillo, E. Enzymatic resolution of N-protected-β^3-amino methyl esters, using lipase B from *Candida antarctica*. *Tetrahedron: Asymmetry*, 2005, 16, 629–634.

Fontaine, F.G., Rochette, E. Ambiphilic molecules: From organometallic curiosity to metal-free catalysis. *Accounts of Chemical Research*, 2018, 51, 454–464.

Formica, M., Sorin, G., Farley, A.J.M., Díaz, J., Paton, R.S., Dixon, D.J. Bifunctional iminophosphorane catalysed enantioselective sulfa-Michael addition of alkyl thiols to alkenyl benzimidazoles. *Chemical Science*, 2018, 9, 6969–6974.

Franssen, M.C.R., Alessandrini, L., Terraneo, G. Biocatalytic production of flavors and fragrances. *Pure and Applied Chemistry*, 2005, 77, 273–279.

Fuganti, C., Grasselli, P. Baker's yeast mediated synthesis of natural products. *In*: Whitaker J.R., Sonnet, P.E. (eds), Biocatalysis in Agricultural Biotechnology. ACS Symposium Series; 389; 1988; pp. 359–370.

Fujii, R., Nakagawa, Y., Hiratake, J., Sogabe, A., Sakata, K. Directed evolution of *Pseudomonas aeruginosa* lipase for improved amide-hydrolyzing activity. *Protein Engineering, Design and Selection*, 2005, 18, 93–101.

Galván, A., González-Pérez, A.B., Álvarez, R., de Lera, A.R., Fañanás, F.J., Rodríguez, F. Exploiting the multidentate nature of chiral disulfonimides in a multicomponent reaction for the asymmetric synthesis of pyrrolo[1,2-a]indoles: A remarkable case of enantioinversion. *Angewandte Chemie International Edition*, 2016, 55, 3428–3432.

Ganem, B. The mechanism of the Claisen rearrangement: Déja vu all over again. *Angewandte Chemie International Edition*, 1996, 35, 936–945.

García, M., Rebolledo, F., Gotor, V. Chemoenzymatic aminolysis and ammonolysis of β-ketoesters. *Tetrahedron Letters*, 1993, 34, 6141–6142.

García, M., Rebolledo, F., Gotor, V. Lipase-catalyzed aminolysis and ammonolysis of β-ketoesters. Synthesis of optically active β-ketoamides. *Tetrahedron*, 1994, 50, 6935–6940.

García-Junceda, E. Multi-step Enzyme Catalysis. Germany; Wiley-VCH, 2008.

García-Urdiales, E., Rebolledo, F., Gotor, V. Enzymatic ammonolysis of ethyl (±)-4-chloro-3-hydroxybutanoate. Chemoenzymatic syntheses of both enantiomers of pyrrolidine-3-ol and 5-(chloromethyl)-1,3-oxazolidin-2-one. *Tetrahedron: Asymmetry*, 1999, 10, 721–726.

Gatzenmeier, T., van Gemmeren, M., Xie, Y., Höfler, D., Leutzsch, M., List, B. Asymmetric Lewis acid organocatalysis of the Diels-Alder reaction by a silylated C–H acid. *Science*, 2016, 351, 949–952.

Gedey, S., Liljeblad, A., Fülöp, F., Kanerva, L. Sequential resolution of ethyl 3-aminobutyrate with carboxylic acid esters by *Candida antarctica* lipase B. *Tetrahedron: Asymmetry*, 1999, 10, 2573–2581.

Gershenson, A., Schauerte, J., Giver, L., Arnold, F. Tryptophan phosphorescence study of enzyme flexibility and unfolding in laboratory-evolved thermostable esterases. *Biochemistry*, 2000, 39, 4658–4665.

Ghanem, A., Aboul-Enein, H. Application of lipases in kinetic resolution of racemates. *Chirality*, 2005, 17, 1–15.

Gill, I., Patel, R. Biocatalytic ammonolysis of (5S)-4,5-dihydro-1H-pyrrole-1,5-dicarboxylic acid, 1-(1,1-dimethylethyl)-5-ethyl ester: preparation of an intermediate to the dipeptidyl peptidase IV inhibitor Saxagliptin. *Bioorganic & Medicinal Chemistry Letters*, 2006, 16, 705–709.

González-García, T., Verstuyf, A., Verlinden, L., Fernández, S., Ferrero, M. Enzymatic desymmetrization of 19-*nor*-vitamin D_3 A-ring synthon precursor: Synthesis, structure elucidation, and biological activity of 1α,25-dihydroxy-3-*epi*-19-*nor*-vitamin D_3 and 1β,25-dihydroxy-19-*nor*-vitamin D_3. *Advanced Synthesis & Catalysis*, 2018, 360, 2762–2772.

Goswami, A., Guo, A., Parker, W., Patel, R. Enzymatic resolution of sec-butylamine. *Tetrahedron: Asymmetry*, 2005, 16, 1715–1719.

Gu, J., Li, Z., Lin, G. Reductive biotransformation of carbonyl compounds-application of fungus, *Geotrichum* sp. G38 in organic synthesis. *Tetrahedron*, 1993, 49, 5805–5816.

Guo, J., Wong, M.W. Cinchona alkaloid-squaramide catalyzed sulfa-Michael addition reaction: Mode of bifunctional activation and origin of stereoinduction. *Journal of Organic Chemistry*, 2017, 82, 4362–4368.

Guo, W., Liu, X., Liu, Y., Li, C. Chiral catalysis at the water/oil interface. *ACS Catalysis*, 2018, 8, 328–341.

Gurung, N., Ray, S., Bose, S., Rai, V. A broader view: Microbial enzymes and their relevance in industries, medicine, and beyond. *BioMed Research International*, 2013, 2013, 1–18.

Gutman, A., Zuobi, K., Guibe-Jampel, E. Lipase catalyzed hydrolysis of γ-substituted α-aminobutyrolactones. *Tetrahedron Letters*, 1990, 31, 2037–2038.

Haghshenas, P., Langdon, S.M., Gravel, M. Thieme chemistry journals awardees-Where are they now? Carbene organocatalysis for stetter, benzoin, domino, and ring-expansion reactions. *Synlett*, 2017, 28, 542–559.

Han, X., Zhou, H.B., Dong, C. Applications of chiral squaramides: From asymmetric organocatalysis to biologically active compounds. *The Chemical Record*, 2016, 16, 897–906.

Hasnaoui-Dijoux, G., Elenkov, M., Spelberg, J., Hauer, B., Janssen, D. Catalytic promiscuity of halohydrin dehalogenase and its application in enantioselective epoxide ring opening. *ChemBioChem*, 2008, 9, 1048–1051.

He, X., Wang, X., Tse, Y.L(Steve)., Ke, Z., Yeung, Y.Y. Applications of selenonium cations as Lewis acids in organocatalytic reactions. *Angewandte Chemie International Edition*, 2018, 57, 12869–12873.

Heel, T., Mcintosh, J., Dodani, S., Meyerowitz, J., Arnold, F. Non-natural olefin cyclopropanation catalyzed by diverse cytochrome P450s and other hemoproteins. *ChemBioChem*, 2014, 15, 2556–2562.

Held, F.E., Tsogoeva, S.B. Asymmetric cycloaddition reactions catalyzed by bifunctional thiourea and squaramide organocatalysts: Recent advances. *Catalysis Science & Technology*, 2016, 6, 645–667.

Henke, E., Pleiss, J., Bornscheuer, U. Activity of lipases and esterases towards tertiary alcohols: Insights into structure-function relationships. *Angewandte Chemie International Edition*, 2002, 41, 3211–3213.

Higuchi, K., Suzuki, S., Ueda, R., Oshima, N., Kobayashi, E., Tayu, M., Kawasaki, T. Asymmetric total synthesis of (−)-leuconoxine via chiral phosphoric acid catalyzed desymmetrization of a prochiral diester. *Organic Letters*, 2015, 17, 154–157.

Hiroto, S., Miyake, Y., Shinokubo, H. Synthesis and functionalization of porphyrins through organometallic methodologies. *Chemical Reviews*, 2017, 117, 2910–3043.

Holland, P.L. Distinctive reaction pathways at base metals in high-spin organometallic catalysts. *Accounts of Chemical Research*, 2015, 48, 1696–1702.

Hu, Y., Zhou, B., Wang, C. Inert C–H bond transformations enabled by organometallic manganese catalysis. *Accounts of Chemical Research*, 2018, 51, 816–827.

Huerta, F., Minidis, A., Bäckvall, J. Racemization in asymmetric synthesis. Dynamic kinetic resolution and related processes in enzyme and metal catalysis. *Chemical Society Reviews*, 2001, 30, 321–331.

Hult, K., Berglund, P. Enzyme promiscuity: Mechanism and applications. *Trends in Biotechnology*, 2007, 25, 231–238.

Illanes, A. Enzyme Biocatalysis Principles and Applications. Netherlands: Springer; 2008.

Jahn, M., Marles, J., Warren, R., Withers, S. Thioglycoligases: Mutant glycosidases for thioglycoside synthesis. *Angewandte Chemie International Edition*, 2003, 45, 352–354.

Jeschek, M., Reuter, R., Heinisch, T., Trindler, C., Klehr, J., Panke, S., Ward, T. Directed evolution of artificial metalloenzymes for *in vivo* metathesis. *Nature*, 2016, 537, 661–665.

Jia, M., Bandini, M. Counterion effects in homogeneous gold catalysis. *ACS Catalysis*, 2015, 5, 1638–1652.

Johnson, C., Bis, S. Enzymatic asymmetrization of *meso*-2-cycloalken-1,4-diols and their diacetates in organic and aqueous media. *Tetrahedron Letters*, 1992, 33, 7287–7290.

Johnson, C., Braun, M. A two-step, three-component synthesis of PGE1: Utilization of .alpha.-iodo enones in Pd(0)-catalyzed cross-couplings of organoboranes. *Journal of the American Chemical Society*, 1993, 115, 11014–11015.

Kadokawa, J., Kobayashi, S. Polymer synthesis by enzymatic catalysis. *Current Opinion in Chemical Biology*, 2010, 14, 145–153.

Kaphan, D.M., Klet, R.C., Perras, F.A., Pruski, M., Yang, C., Kropf, A.J., Delferro, M. Surface organometallic chemistry of supported iridium(III) as a probe for organotransition metal-support interactions in C–H activation. *ACS Catalysis*, 2018, 8, 5363–5373.

Karan, R., Capes, M., Dassarma, S. Function and biotechnology of extremophilic enzymes in low water activity. *Aquatic Biosystems*, 2012, 8, 1–15.

Kaur, J., Chauhan, P., Singh, S., Chimni, S.S. Journey heading towards enantioselective synthesis assisted by organocatalysis. *The Chemical Record*, 2017, 18, 137–153.

Kawai, Y., Hayashi, M., Inaba, Y., Saitou, K., Ohno, A. Asymmetric reduction of α,β-unsaturated ketones with a carbon-carbon double–bond reductase from baker's yeast. *Tetrahedron Letters*, 1998, 39, 5225–5228.

Kenyon, P., Mecking, S. Pentafluorosulfanyl substituents in polymerization catalysis. *Journal of the American Chemical Society*, 2017, 139, 13786–13790.

Kikuchi, J., Aramaki, H., Okamoto, H., Terada, M. F10BINOL-derived chiral phosphoric acid- catalyzed enantioselective carbonyl-ene reaction: Theoretical elucidation of stereochemical outcomes. *Chemical Science*, 2019, 10, 1426–1433.

Kitazume, T., Ikeya, T., Murata, K. Synthesis of optically active trifluorinated compounds: Asymmetric Michael addition with hydrolytic enzymes. *Journal of the Chemical Society, Chemical Communications*, 1986, 0, 1331–1333.

Klibanov, A. Asymmetric transformations catalyzed by enzymes in organic solvents. *Accounts of Chemical Research*, 1990, 23, 114–120.

Klibanov, A., Samokhin, G., Martinek, K., Berezin, I. A new approach to preparative enzymatic synthesis. *Biotechnology and Bioengineering*, 1977, 19, 1351–1361.

Kołodziejska, R., Górecki, M., Frelek, J., Dramiński, M. Enantioselective enzymatic desymmetrization of the prochiral pyrimidine acyclonucleoside. *Tetrahedron: Asymmetry*, 2010, 23, 683–689.

Koskinen, A., Klibanov, A., Enzymatic Reactions in Organic Media. London: Blackie Academic Professional, 1996.

Koszelewski, D., Lavandera, I., Clay, D., Rozzell, D., Kroutil, W. Asymmetric synthesis of optically pure pharmacologically relevant amines employing ω-transaminases. *Advanced Synthesis & Catalysis*, 2008, 350, 2761–2766.

Koszelewski, D., Lavandera, I., Clay, D., Guebitz, G., Rozzell, D., Kroutil, W. Formal asymmetric biocatalytic reductive amination. *Angewandte Chemie International Edition*, 2018, 47, 9337–9340.

Lam, Y., Houk, K.N. Origins of stereoselectivity in intramolecular aldol reactions catalyzed by cinchona amines. *Journal of the American Chemical Society*, 2015, 137, 2116–2127.

Larios, A., García, H., Oliart, R., Valerio-Alfaro, G. Synthesis of flavor and fragrance esters using Candida Antarctica lipase. *Applied Microbiology and Biotechnology*, 2004, 65, 373–376.

Laszlo, J., Yu, Y., Lutz, S., Compton, D. Glycerol acyl-transfer kinetics of a circular permutated *Candida Antarctica* lipase B. *Journal of Molecular Catalysis B: Enzymatic*, 2011, 72, 175–180.

Lau, R., Sorgedrager, M., Carrea, G., van Rantwijk, F., Secundo, F., Sheldon, R. Dissolution of *Candida Antarctica* lipase B in ionic liquids: Effects on structure and activity. *Green Chemistry*, 2004, 6, 483–487.

Lavandera, I., Fernández, S., Ferrero, M., Gotor, V. First regioselective enzymatic alkoxy-carbonylation of primary amines. Synthesis of novel 5'- and 3'-carbamates of pyrimidine 3',5'-diaminonucleoside derivatives including BVDU analogues. *Journal of Organic Chemistry*, 2004, 69, 1748–1751.

Leak, D., Aikens, P., Seyed-Mahmoudian, M. The microbial production of epoxides. *Trends in Biotechnology*, 1992, 10, 256–261.

Lemière, G., Lepoivre, J., Alderweireldt, F. Hlad-catalyzed oxidations of alcohols with acetaldehyde as a coenzyme recycling substrate. *Tetrahedron Letters*, 1985, 26, 4527–4528.

Leuenberger, H.G.W., Boguch, W., Widmer, E., Zell, R. Synthese von optisch aktiven, natürlichen carotinoiden und strukturell verwandten naturprodukten. I. Synthese der chiralen schlüsselverbindung (4R, 6R)-4-hydroxy-2,2,6-trimethylcyclohexanon. *Helvetica Chimica Acta*, 1976, 59, 1832–1849.

Levin, M.D., Kaphan, D.M., Hong, C.M., Bergman, R.G., Raymond, K.N., Toste, F.D. Scope and mechanism of cooperativity at the intersection of organometallic and supramolecular catalysis. *Journal of the American Chemical Society*, 2016a, 138, 9682–9693.

Levin, M.D., Kim, S., Toste, F.D. Photoredox catalysis unlocks single-electron elementary steps in transition metal catalyzed cross-coupling. *ACS Central Science*, 2016b, 2, 293–301.

Li, N., Hu, S., Feng, G. Resolution of 2-nitroalcohols by Burkholderia cepacian lipase-catalyzed enantioselective acylation. *Biotechnology Letters*, 2012a, 34, 153–158.

Li, S., Yang, X., Yang, S., Zhu, M., Wang, X. Technology prospecting on enzymes: Application, marketing and engineering. *Computational and Structural Biotechnology Journal*, 2012b, 2, 1–11.

Li, P., Hu, X., Dong, X., Zhang, X. Recent advances in dynamic kinetic resolution by chiral bifunctional (thio)urea- and squaramide-based organocatalysts. *Molecules*, 2016, 21.

Liang, M., Zhang, S., Jia, J., Tung, C.H., Wang, J., Xu, Z. Synthesis of spiroketals by synergistic gold and scandium catalysis. *Organic Letters*, 2017, 19, 2526–2529.

Lillo, V.J., Mansilla, J., Saá, J.M. Organocatalysis by networks of cooperative hydrogen bonds: Enantioselective direct mannich addition to preformed arylideneureas. *Angewandte Chemie International Edition*, 2016, 55, 4312–4316.

List, B., Lerner, R.A., Barbas, C.F. Proline-catalyzed direct asymmetric aldol reactions. *Journal of the American Chemical Society*, 2000, 122, 2395–2396.

List, B. Introduction: Organocatalysis. *Chemical Reviews*, 2007, 107, 5413–5415.

Litjens, M., Straathof, J., Jongejan, J., Heijnen, J. Exploration of lipase-catalyzed direct amidation of free carboxylic acids with ammonia in organic solvents. *Tetrahedron*, 1999, 55, 12411–12418.

Liu, J. Catalysis by supported single metal atoms. *ACS Catalysis*, 2017, 7, 34–59.

Liu, L., Kaib, P.S.J., Tap, A., List, B. A general catalytic asymmetric prins cyclization. *Journal of the American Chemical Society*, 2016, 138, 10822–10825.

Liu, J., Wang, L. Recent advances in asymmetric reactions catalyzed by proline and its derivatives. *Synthesis (Stuttg)*, 2017, 49, 960–972.

Liu, Y., Lou, B., Shangguan, L., Cai, J., Zhu, H., Shi, B. Pillar[5]arene-based organometallic cross-linked polymer: Synthesis, structure characterization, and catalytic activity in the Suzuki-Miyaura coupling reaction. *Macromolecules*, 2018, 51, 1351–1356.

López-García, M., Alfonso, I., Gotor, V. Desymmetrization of dimethyl 3-substituted glutarates through enzymatic ammonolysis and aminolysis reactions. *Tetrahedron: Asymmetry*, 2003, 14, 603–609.

Lou, F., Liu, B., Wu, Q., Ly, D., Lin, X. *Candida antarctica* lipase B (CAL-B)-catalyzed carbon-sulfur bond addition and controllable selectivity in organic media. *Advanced Synthesis & Catalysis*, 2008, 350, 1959–1962.

Luk, L., Qian, Q., Tanner, M. A Cope rearrangement in the reaction catalyzed by dimethylallyltryptophan synthase? *Journal of the American Chemical Society*, 2011, 133, 12342–12345.

Luna, A., Alfonso, I., Gotor, V. Biocatalytic approaches toward the synthesis of both enantiomers of *trans*-cyclopentane-1,2-diamine. *Organic Letters*, 2002, 4, 3627–3629.

Madalińska, L., Kwiatkowska, M., Cierpiał, T., Kiełbasiński, P. Investigations on enzyme catalytic promiscuity: The first attempts at a hydrolytic enzyme-promoted conjugate addition of nucleophiles to a α,β-unsaturated sulfinyl acceptors. *Journal of Molecular Catalysis B: Enzymatic*, 2012, 81, 25–30.

Mansuy, D., Battoni, P. Activation and Functionalization of Alkanes. New York: Wiley, 1989.

Margolin, A. Enzymes in the synthesis of chiral drugs. *Enzyme and Microbial Technology*, 1993, 15, 266–280.

Martínez-Crespo, L., Escudero-Adán, E.C., Costa, A., Rotger, C. The role of N-methyl squaramides in a hydrogen-bonding strategy to fold peptidomimetic compounds. *Chemistry: A European Journal*, 2018, 24, 17802–17813.

Martínez-Cuesta, S., Rahman, S., Furnham, N., Thornton, J. The classification and evolution of enzyme function. *Biophysical Journal*, 2015, 109, 1082–1086.

Martín-Matute, B., Bäckvall, J. Dynamic kinetic resolution catalyzed by enzymes and metals. *Current Opinion in Chemical Biology*, 2007, 11, 226–232.

Meng, G., Shi, S., Szostak, M. Palladium-catalyzed Suzuki-Miyaura cross coupling of amides via site-selective N–C bond cleavage by cooperative catalysis. *ACS Catalysis*, 2016, 6, 7335–7339.

Min, C., Seidel, D. Asymmetric Brønsted acid catalysis with chiral carboxylic acids. *Chemical Society Reviews*, 2017, 46, 5889–5902.

Moczulski, M., Drelich, P., Albrecht, Ł. Bifunctional catalysis in the stereocontrolled synthesis of tetrahydro-1,2-oxazines. *Organic and Biomolecular Chemistry*, 2018, 16, 376–379.

Modrocká, V., Veverková, E., Mečiarová, M., Šebesta, R. Bifunctional amine-squaramides as organocatalysts in Michael/hemiketalization reactions of β,γ-unsaturated α-ketoesters and α,β-unsaturated ketones with 4-hydroxycoumarins. *Journal of Organic Chemistry*, 2018, 83, 13111–13120.

Moeller, K., Nguyen, G., Hollmann, F., Hanefeld, U. Expression and characterization of the nitrile reductase queF from *E. coli*. *Enzyme and Microbial Technology*, 2013, 52, 129–133.

Mohapatra, S., Bhakta, S., Baral, N., Nayak, S. Synthetic application of pipecolic acid. *Research on Chemical Intermediates*, 2015, 41, 4545–4553.

Molinaro, C., Guilbault, A., Kosjek, B. Resolution of 2,2-disubstituted epoxides via biocatalytic azidolysis. *Organic Letters*, 2010, 12, 3772–3775.

Momiyama, N., Okamoto, H., Shimizu, M., Terada, M. Synthetic method for 2,2′-disubstituted fluorinated binaphthyl derivatives and application as chiral source in design of chiral mono-phosphoric acid catalyst. *Chirality*, 2015, 27, 464–475.

Mordhorst, S., Siegrist, J., Müller, M., Richter, M., Andexer, J. Catalytic alkylation using a cyclic S-adenosylmethionine regeneration system. *Angewandte Chemie International Edition*, 2017, 56, 4037–4041.

Mugford, P., Wagner, U., Jiang, Y., Faber, K., Kazlauskas, R. Enantiocomplementary enzymes: Classification, molecular basis for their enantiopreference, and prospects for mirror-image biotransformations. *Angewandte Chemie International Edition in English*, 2008, 47, 8782–8793.

Mukherjee, S., Yang, J.W., Hoffmann, S., List, B. Asymmetric enamine catalysis. *Chemical Reviews*, 2007, 107, 5471–5569.

Müller, M. Recent developments in enzymatic asymmetric C–C bond formation. *Advanced Synthesis & Catalysis*, 2012, 354, 3161–3174.

Nagasawa, T., Yamada, H. Microbial transformations of nitriles. *Trends in Biotechnology*, 1989, 7, 153–158.

Neidleman, S., Geigert, J. Biohalogenation: Principles, Basic Roles and Applications. New York: Chichester: Ellis Horwood Limited, 1986.

Nesbit, M.A., Suess, D.L.M., Peters, J.C. E–H bond activations and hydrosilylation catalysis with iron and cobalt metalloboranes. *Organometallics*, 2015, 34, 4741–4752.

Newton, C.G., Kossler, D., Cramer, N. Asymmetric catalysis powered by chiral cyclopentadienyl ligands. *Journal of the American Chemical Society*, 2016, 138, 3935–3941.

Nguyen, N.T., Chen, P.A., Setthakarn, K., May, A.J. Chiral diol-based organocatalysts in enantioselective reactions. *Molecules*, 2018, 23, 2317.

Noyori, R., Ohkuma, T., Kitamura, M., Takaya, H., Sayo, N., Kumobayashi, H., Akutagawa, S. Asymmetric hydrogenation of .beta.-keto carboxylic esters. A practical, purely chemical access to .beta.-hydroxy esters in high enantiomeric purity. *Journal of the American Chemical Society*, 1987, 109, 5856–5858.

O'Brien, P., Herschlag, D. Catalytic promiscuity and the evolution of new enzymatic activities. *Chemistry & Biology*, 1999, 6, 91–105.

Ohta, H. Biocatalytic asymmetric decarboxylation. *Advances in Biochemical Engineering-Biotechnology*, 1999, 63, 1–30.

Oikawa, H., Katayama, K., Suzuki, Y., Ichihara, A. Enzymatic activity catalyzing *exo*-selective Diels-Alder reaction in solanapyrone biosynthesis. *Journal of the Chemical Society, Chemical Communications*, 1995, 13, 1321–1322.

Olivos-Suárez, A.I., Szécsényi, A., Hensen, E.J.M., Ruiz-Martínez, J., Pidko E.A., Gascon, E. Strategies for the direct catalytic valorization of methane using heterogeneous catalysis: Challenges and opportunities. *ACS Catalysis*, 2016, 6, 2965–2981.

Ottoson, J., Fransson, L., King, J., Hult, K. Size as a parameter for solvent effects on *Candida antarctica* lipase B enantioselectivity. *Biochimica et Biophysica Acta*, 2002, 1594, 325–334.

Paetzold, J., Bäckvall, J. Chemoenzymatic dynamic kinetic resolution of primary amines. *Journal of the American Chemical Society*, 2005, 127, 17620–17621.

Palomo, J., Cabrera, Z. Pyrimidine ring as building block for the synthesis of functionalized π-conjugated materials. *Current Organic Synthesis*, 2012, 9, 791–805.

Parmar, D., Sugiono, E., Raja, S., Rueping, M. Complete field guide to asymmetric BINOL-phosphate derived brønsted acid and metal catalysis: History and classification by mode of activation., brønsted acidity, hydrogen bonding, ion pairing, and metal phosphates. *Chemical Reviews*, 2014, 114, 9047–9153.

Parmar, D., Sugiono, E., Raja, S., Rueping, M. Addition and correction to complete field guide to asymmetric BINOL-phosphate derived brønsted acid and metal catalysis:

History and classification by mode of activation., brønsted acidity, hydrogen bonding, ion pairing, and metal phosphates. *Chemical Reviews*, 2017, 117, 10608–10620.

Patel, R. Biocatalytic synthesis of chiral alcohols and amino acids for development of pharmaceuticals. *Biomolecules*, 2013, 3, 741–777.

Pelletier, J.D.A., Basset, J.M. Catalysis by design: Well-defined single-site heterogeneous catalysis. *Accounts of Chemical Research*, 2016, 49, 664–677.

Pellissier, H. Use of TADDOLs and their derivatives in asymmetric synthesis. *Tetrahedron*, 2008, 64, 10279–10317.

Pellissier, H. Recent developments in the asymmetric organocatalytic Morita-Baylis-Hillman reaction. *Tetrahedron*, 2017, 73, 2831–2861.

Penning, T., Jez, J. Enzyme redesign. *Chemical Reviews*, 2001, 101, 3027–3046.

Phillips, R., May, S. Enzymatic sulphur oxygenation reactions. *Enzyme and Microbial Technology*, 1981, 3, 9–18.

Pleiss, J. Thermodynamic activity-based interpretation of enzyme kinetics. *Trends in Biotechnology*, 2017, 35, 379–382.

Pohl, M., Lingen, B., Müller, M. Thiamin-diphosphate-dependent enzymes: New aspects of asymmetric C–C bond formation. *Chemistry: A European Journal*, 2002, 8, 5288–5295.

Pohnert, G. Diels-alderases. *ChemBioChem*, 2001, 2, 873–875.

Polak, J., Jarosz-Wilkolazka, A. Fungal laccases as green catalysts for dye synthesis. *Process Biochemistry*, 2012, 47, 1295–1307.

Poornima, K., Preetha, R. Biosynthesis of food flavours and fragrances-a review. *Asian Journal of Chemistry*, 2017, 29, 2345–2352.

Powers, I.G., Uyeda, C. Metal-metal bonds in catalysis. *ACS Catalysis*, 2017, 7, 936–958.

Priego, J., Ortíz-Nava, C., Carrillo-Morales, M., López-Munguía, A., Escalante, J., Castillo, E. Solvent engineering: An effective tool to direct chemoselectivity in a lipase-catalyzed Michael addition. *Tetrahedron*, 2009, 65, 536–539.

Puertas, S., Rebolledo, F., Gotor, V. Enantioselective enzymatic aminolysis and ammonolysis of dimethyl 3-hydroxyglutarate. Synthesis of (R)-4-amino-3-hydroxybutanoic acid. *Journal of Organic Chemistry*, 1996, 61, 6024–6027.

Reisman, S.E., Doyle, A.G., Jacobsen, E.N. Enantioselective thiourea-catalyzed additions to oxocarbenium ions. *Journal of the American Chemical Society*, 2008, 130, 7198–7199.

Renata, H., Wang, Z., Arnold, F. Expanding the enzyme universe: Accessing non-natural reactions by mechanism-guided directed evolution. *Angewandte Chemie International Edition*, 2015, 54, 3351–3367.

Rendón, X., López-Munguía, A., Castillo, E. Solvent engineering applied to lipase-catalyzed glycerolysis of triolein. *Journal of the American Oil Chemists' Society*, 2001, 78, 1061–1066.

Ricca, E., Brucher, B., Schrittwieser, J. Multi-enzymatic cascade reactions: Overview and perspectives. *Advanced Synthesis & Catalysis*, 2011, 353, 2239–2262.

Ričko, S., Svete, J., Štefane, B., Perdih, A., Golobič, A., Meden, A., Grošelj, U. 1,3-Diamine-derived bifunctional organocatalyst prepared from camphor. *Advanced Synthesis & Catalysis*, 2016, 358, 3786–3796.

Rivera-Ramírez, J., Escalante, J., López-Munguía, A., Marty, A., Castillo, E. Thermo-dynamically controlled chemoselectivity in lipase-catalyzed *aza*-Michael additions. *Journal of Molecular Catalysis B: Enzymatic*, 2015, 112, 76–82.

Romero, P., Arnold, F. Exploring protein fitness landscapes by directed evolution. *Nature Reviews Molecular Cell Biology*, 2009, 10, 866–876.

Rosenthaler, L. Durch enzyme bewirkte asymmetrische synthesen. *Biochemische Zeitschrift*, 1908, 14, 238–253.

Rotticci, D., Ottosson, J., Norin, T., Hult, K. Candida antarctica lipase B a tool for preparation of optically alcohols. *Methods in Biotechnology*, 2001, 15, 261–276.

Ryu, K., Dordick, J. How do organic solvents affect peroxidase structure and function? *Biochemistry*, 1992, 31, 2588–2598.

Saito, K., Akiyama, T. Chiral phosphoric acid catalyzed kinetic resolution of indolines based on a self-redox reaction. *Angewandte Chemie International Edition*, 2016, 55, 3148–3152.

Sánchez, V., Rebolledo, F., Gotor, V. *Candida antarctica* lipase-catalyzed doubly enantio-selective aminolysis reactions. Chemoenzymatic synthesis of 3-hydroxypyrrolidines and 4-(silyloxy)-2-oxopyrrolidines with two stereogenic centers. *Journal of Organic Chemistry*, 1999, 64, 1464–1470.

Sano, K., Saito, S., Hirose, Y., Kohari, Y., Nakano, H., Seki, C., Tokiwa, M., Takeshita M., Uwai, K. Development of a novel method for warfarin synthesis via lipase-catalyzed stereoselective Michael reaction. *Heterocycles*, 2013, 87, 1269–1278.

Sapu, M., Bäckvall, J., Deska, J. Enantioselective enzymatic desymmetrization of prochiral allenic diols. *Angewandte Chemie International Edition*, 2011, 50, 9731–9734.

Schafer, A.G., Wieting, J.M., Fisher, T.J., Mattson, A.E. Chiral silanediols in anion-binding catalysis. *Angewandte Chemie International Edition*, 2013, 52, 11321–11324.

Schätzle, S., Steffen-Munsberg, F., Thontowi, A., Höhne, M., Robins, K., Bornscheuer, U. Enzymatic asymmetric synthesis of enantiomerically pure aliphatic, aromatic and arylaliphatic amines with (R)-selective amine transaminases. *Advanced Synthesis & Catalysis*, 2011, 353, 2439–2445.

Schmid, A., Dordick, J., Hauer, B., Kiener, A., Wubbolts, M., Witholt, B. Industrial biocatalysis today and tomorrow. *Nature*, 2001, 409, 258–268.

Schmidt-Kastner, G., Egerer, P. Amino acids and peptides. *In*: Rehm, H.J., Reed, G. (eds), Biotechnology, Vol. 6a, Kieslich, K. (Vol.-ed.). Weinheim, Verlag Chemie; 1984; pp. 387–419.

Schwab, J., Henderson, B. Enzyme-catalyzed allylic rearrangements. *Chemical Reviews*, 1990, 90, 1203–1245.

Seffernick, J., Wackett, L. Rapid evolution of bacterial catabolic enzymes: A case study with atrazine chlorohydrolase. *Biochemistry*, 2001, 40, 12747–12753.

Shaabani, A., Tabatabaei, A.T., Hajishaabanha, F., Shaabani, S., Seyyedhamzeh, M., Keramati nejad, M. KMnO$_4$/guanidinium-based sulfonic acid: As an efficient Brønsted acid organocatalyst for the selective oxidation of organic compounds. *Journal of Sulfur Chemistry*, 2018, 39, 367–379.

Shakeri, M., Engström, K., Sandström, A., Bäckvall, J. Highly enantioselective resolution of β-amino esters by *Candida antarctica* lipase A immobilized in mesocellular foam: Application to dynamic kinetic resolution. *ChemCatChem*, 2010, 2, 534–538.

Shaw, M.H., Twilton, J., MacMillan, D.W.C. Photoredox catalysis in organic chemistry. *Journal of Organic Chemistry*, 2016, 81, 6898–6926.

Shimod, K., Katsuragi, H. Enzymatic resolution of (RS)-1-phenylalkyl β-D-glucosides to (R)-1-phenylalkyl β-primeverosides and (S)-1-phenylalkyl β-D-glucosides via plant xylosyltransferase. *Tetrahedron: Asymmetry*, 2010, 21, 2060–2065.

Shirakawa, S., Maruoka, K. Recent developments in asymmetric phase-transfer reactions. *Angewandte Chemie International Edition*, 2013, 52, 4312–4348.

Shu, T. Ni, Q. Song, X. Zhao, K. Wu, T. Puttreddy, R. Rissanen, K. Enders, D. Asymmetric synthesis of cyclopentanes bearing four contiguous stereocenters via an NHC-catalyzed Michael/Michael/esterification domino reaction. *Chemical Communications*, 2016, 52, 2609–2611.

Sigmund, A., DiCosimo, R. Enzymatic resolution of (RS)-2-(1-aminoethyl)-3-chloro-5-(substituted)pyridines. *Tetrahedron: Asymmetry*, 2004, 15, 2797–2799.

Sih, C., Abushanab, E., Jones, J. Chater 30. Biochemial procedures in organic synthesis. *Annual Reports in Medicinal Chemistry*, 1977, 12, 298–308.

Singer, S. The properties of proteins in nonaqueous solvents. *Advances in Protein Chemistry*, 1963, 17, 1–68.

Singh, R., Kumar, M., Mittal, A., Mehta, P. Microbial enzymes: Industrial progress in 21st century. *3 Biotech*, 2016, 6, 1–15.

Smajlagic, I., Durán, R., Pilkington, M., Dudding, T. Cyclopropenium enhanced thiourea catalysis. *Journal of Organic Chemistry*, 2018, 83, 13973–13980.

Sonsona, G.I., Marqués-López, E., Häring, M., Díaz, D.D., Herrera, P.R. Urea activation by an external Brønsted acid: Breaking self-association and tuning catalytic performance. *Catalysts*, 2018, 8, 305.

Sperger, T., Sanhueza, I.A., Kalvet, I., Schoenebeck, F. Computational studies of synthetically relevant homogeneous organometallic catalysis involving Ni, Pd, Ir, and Rh: An overview of commonly employed DFT methods and mechanistic insights. *Chemical Reviews*, 2015, 115, 9532–9586.

Sperl, J., Sieber, V. Multienzyme cascade reactions-status and recent advances. *ACS Catalysis*, 2018, 8, 2385–2396.

Stecher, H., Twengg, M., Ueberbacher, B., Remler, P., Schwab, H., Griengl, H., Gruber-Khadjawi, M. Biocatalytic Friedel-Crafts alkylation using non-natural cofactors. *Angewandte Chemie*, 2009, 48, 9546–9548.

Stueckler, C., Mueller, N., Winkler, C., Glueck, S., Gruber, K., Steinkellner, G., Faber, K. Bioreduction of α-methylcinnamaldehyde derivatives: Chemo-enzymatic asymmetric synthesis of Lilial™ and Helional™. *Dalton Transactions*, 2010, 39, 8472–8476.

Svedendahl, M., Carlqvist, P., Branneby, C., Allnér, O., Frise, A., Hult, K., Berglund, P., Brink, T. Direct epoxidation in *Candida antarctica* lipase B studies by experiment and theory. *ChemBioChem*, 2008, 9, 2443–2451.

Svedendalhl, M., Hult, K., Berglund, P. Fast carbon-carbon bond formation by a promiscuous lipase. *Journal of the American Chemical Society*, 2005, 127, 17988–17989.

Sweers, H., Wong, C. Enzyme-catalyzed regioselective diacylation of protected sugars in carbohydrate synthesis. *Journal of the American Chemical Society,* 1986, 108, 6421–6422.

Taylor, M.S., Jacobsen, E.N., Asymmetric catalysis by chiral hydrogen-bond donors. *Angewandte Chemie International Edition*, 2006, 45, 1520–1543.

Taylor, S., Sutherland, A., Lee, C., Wisdom, R., Thomas, S., Roberts, S., Evans, C. Chemo enzymatic synthesis of (–)-carbovir utilizing a whole cell catalyzed resolution of 2-azabicyclo[2.2.1]hept-5-en-3-one. *Journal of the Chemical Society, Chemical Communications*, 1990, 0, 1120–1121.

Toone, E., Simon, E., Bednarski, M., Whitesides, G. Enzyme-catalyzed synthesis of carbohydrates. *Tetrahedron*, 1989, 45, 5365–5422.

Torre, O., Alfonso, I., Gotor, V. Lipase catalysed Michael addition of secondary amines to acrylonitrile. *Chemical Communications*, 2004, 15, 1724–1725.

Um, P., Drueckhammer, D. Dynamic enzymatic resolution of thioesters. *Journal of the American Chemical Society*, 1998, 120, 5605–5610.

Utczás, M., Székely, E., Tasnádi, G., Monek, E., Vida, L., Forró, E., Fülöp, F., Simándi, B. Kinetic resolution of 4-phenyl-2-azetidinone in supercritical carbon dioxide. *The Journal of Supercritical Fluids*, 2011, 55, 1019–1022.

Uritsky, N., Shokhen, M., Albeck, A. Stepwise versus concerted mechanism in general-base catalysis by serine proteases. *Angewandte Chemie International Edition*, 2016, 55, 1680–1684.

van de Waterbeemd, H., Gifford, E. ADMET in silico modelling: Towards prediction paradise? *Nature Reviews Drug Discovery*, 2003, 2, 192–204.

van Rantwijk, F., Secundo, F., Sheldon, R. Structure and activity of *Candida antarctica* lipase B in ionic liquids. *Green Chemistry*, 2006, 8, 282–286.

Vanbésien, T., Monflier, E., Hapiot, F. Supramolecular emulsifiers in biphasic catalysis: The substrate drives its own transformation. *ACS Catalysis*, 2015, 5, 4288–4292.

Velonia, K., Flomenbom, O., Loos, D., Masuo, S., Cotlet, M., Engelburghs, Y., Hofkens, J., Rowan, A.E., Klafter, J., Naite, R.J., Shryver, F.C. Single-enzyme kinetics of CALB-catalyzed hydrolysis. *Angewandte Chemie International Edition*, 2005, 43, 560–564.

Vidossich, P., Lledós, A., Ujaque, G. First-principles molecular dynamics studies of organometallic complexes and homogeneous catalytic processes. *Accounts of Chemical Research*, 2016, 49, 1271–1278.

Walsh, C., Chen, Y. Enzymic Baeyer-Villiger oxidations by flavin-dependent monooxygenases. *Angewandte Chemie International Edition*, 1988, 27, 333.

Walsh, C. Correction: Enabling the chemistry of life. *Nature*, 2001, 409, 226–231.

Wang, G., Xu, L., Li, P. Double N,B-type bidentate boryl ligands enabling a highly active iridium catalyst for C–H borylation. *Journal of the American Chemical Society*, 2015, 137, 8058–8061.

Wang, D., Zhu, N., Chen, P., Lin, Z., Liu, G. Enantioselective decarboxylative cyanation employing cooperative photoredox catalysis and copper catalysis. *Journal of the American Chemical Society*, 2017, 139, 15632–15635.

Wang, F., Gong, J., Zhang, X., Ren, Y., Zhang, J. PLMA-b-POEGMA amphiphilic block copolymers as nanocarriers for the encapsulation of magnetic nanoparticles and indomethacin. *Polymers*, 2018a, 10, 1–14.

Wang, W., Cui, L., Sun, P., Shi, L., Yue, C., Li, F. Reusable N-heterocyclic carbene complex catalysts and beyond: A perspective on recycling strategies. *Chemical Reviews*, 2018b, 118, 9843–9929.

Wang, X., Wang, L., Dong, C.L., Menendez Rodríguez, G., Huang, Y.C., Macchioni, A., Shen S. Activating Kläui-type organometallic precursors at the metal oxide surfaces for enhanced solar water oxidation. *ACS Energy Letters*, 2018c, 3, 1613–1619.

Ward, O., Singh, A. Enzymatic asymmetric synthesis by decarboxylases. *Current Opinion in Biotechnology*, 2000, 11, 520–526.

Wardencki, W., Curylo, J., Namiesnik, J. Green chemistry principles. *Polish Journal of Environmental Studies*, 2005, 14, 389–395.

Wegman, M., Hacking, M., Rops, J., Pereira, P., van Rantwijk, F., Sheldon, R. Dynamic kinetic resolution of phenylglycine esters via lipase-catalysed ammonolysis. *Tetrahedron: Asymmetry*, 1999, 10, 1739–1750.

Wei, Z., Li, Z., Lin, G. *anti*-Prelog microbial reduction of aryl α-halomethyl or α-hydroxymethyl ketones with *Geotrichum* sp. 38. *Tetrahedron*, 1998, 54, 13059–13072.

Weiß, M., Gröger, H. Practical, highly enantioselective chemoenzymatic one-pot synthesis of short-chain aliphatic β-amino acid esters. *Synlett*, 2009, 8, 1251–1254.

Welin, E.R., Warkentin, A.A., Conrad, J.C., MacMillan, D.W.C. Enantioselective α-alkylation of aldehydes by photoredox organocatalysis: Rapid access to pharmacophore fragments from β-cyanoaldehydes. *Angewandte Chemie International Edition*, 2015, 54, 9668–9672.

Wilkinson, A., Fersht, A., Blow, D., Winter, G. Site-directed mutagenesis as a probe of enzyme structure and catalysis: tyrosyl-tRNA synthetase cysteine-35 to glycine-35 mutation. *Biochemistry*, 1983, 22, 3581–3586.

Williams, R. Total synthesis and biosynthesis of the paraherquamides: An intriguing story of the biological Diels-Alder construction. *Chemical and Pharmaceutical Bulletin*, 2002, 50, 711–740.

Winter, G., Fersht, A., Wilkinson, A., Zoller, M., Smith, M. Redesigning enzyme structure by site-directed mutagenesis: tyrosyl tRNA synthetase and ATP binding. *Nature*, 1982, 299, 756–758.

Wöhler, F., von Liebig, J. Untersuchungen über das radikal der benzoesäure. *Annalen der Pharmacie*, 1832, 3, 249–282.

Wolczanski, P.T., Chirik, P.J. A career in catalysis: John E. Bercaw. *ACS Catalysis*, 2015, 5, 1747–1757.

Xiao Zhu, X., Yang, Y., Yu, H. In silico enzyme modelling. *Australian Biochemist*, 2014, 45, 12–15.

Xie, J., Xing, X.Y., Sha, F., Wu, Z.Y., Wu, X.Y. Enantioselective synthesis of spiro[indoline-3,4'-pyrano[2,3-c]pyrazole] derivatives via an organocatalytic asymmetric Michael/cyclization cascade reaction. *Organic and Biomolecular Chemistry*, 2016, 14, 8346–8355.

Xu, F., Li, Y.J., Huang, C., Xu, H.C. Ruthenium-catalyzed electrochemical dehydrogenative alkyne annulation. *ACS Catalysis*, 2018, 8, 3820–3824.

Xue, Y., Cao, C., Zheng, Y. Enzymatic asymmetric synthesis of chiral amino acids. *Chemical Society Reviews*, 2018, 47, 1516–1561.

Yamamoto, Y., Yamamoto, K., Nishioka, T., Oda, J. Asymmetric synthesis of optically active lactones from cyclic acid anhydrides using lipase in organic solvents. *Agricultural and Biological Chemistry*, 1988, 52, 3087–3092.

Yao, W., Zhu, J., Zhou, X., Jiang, R., Wang, P., Chen, W. Ferrocenophane–based bifunctional organocatalyst for highly enantioselective Michael reactions. *Tetrahedron*, 2018, 74, 4205–4210.

Yasukawa, K., Asano, Y. Enzymatic synthesis of chiral phenylalanine derivatives by a dynamic kinetic resolution of corresponding amide and nitrile substrates with a multi-enzyme system. *Advanced Synthesis & Catalysis*, 2012, 354, 3327–3332.

Young, D., Nichols, J., Kelly, R., Deiters. A. Microwave activation of enzymatic catalysis. *Journal of the American Chemical Society*, 2008, 130, 10048–10049.

Young, C., Britton, Z., Robinson, A. Recombinant protein expression and purification: A comprehensive review of affinity tags and microbial applications. *Biotechnology Journal*, 2012, 7, 620–634.

Yuan, J., Li, M., Ji, N., He W., Liu, Y. Recent progress in chiral guanidine-catalyzed Michael reactions. *Current Organic Chemistry*, 2017, 21, 1205–1226.

Zaks, A., Klibanov, A. Enzymatic catalysis in organic media at 100°C. *Science*, 1984, 224, 1249–1251.

Zaks, A., Klibanov, A. Enzyme-catalyzed processes in organic solvents. *Proceedings of the National Academy of Sciences, USA*, 1985, 82, 3192–3196.

Zhang, D., Poulter, C. Biosynthesis of archaebacterial ether lipids. Formation of ether linkages by prenyltransferases. *Journal of the American Chemical Society*, 1993, 115, 1270–1277.

Zhang, J.-Q. Li, N.-K. Yin, S.-J. Sun, B.-B. Fan, W.-T. Wang, X.-W. Chiral N-heterocyclic carbene-catalyzed asymmetric Michael-intramolecular aldol-lactonization cascade for enantioselective construction of β-propiolactone-fused spiro[cyclopentane-oxindoles. *Advanced Synthesis & Catalysis*, 2017, 359, 1541–1551.

Zhao, K., Zhi, Y., Shu, T., Valkonen, A., Rissanen, K., Enders, D. Organocatalytic domino oxa-Michael/1,6-addition reactions: asymmetric synthesis of chromans bearing oxindole scaffolds. *Angewandte Chemie International Edition*, 2016, 55, 12104–12108.

Zhou, J., Wakchaure, V., Kraft, P., List, B. Primary-amine-catalyzed enantioselective intramolecular aldolizations. *Angewandte Chemie International Edition*, 2008, 47, 7656–7658.

Zhou, Q.L. Privileged Chiral Ligands and Catalysts. Weinheim: Wiley-VCH, 2011.

Zhu, J., Mix, E., Winblad, B. The antidepressant and antiinflammatory effects of rolipram in the central nervous system. *CNS Drug Reviews*, 2001, 7, 387–398.

Zlotin, S.G., Kochetkov, S.V. C_2-Symmetric diamines and their derivatives as promising organocatalysts for asymmetric synthesis. *Russian Chemical Reviews*, 2015, 84, 1077–1099.

---

[1]The IUPAC definition for catalyst is: A substance that increases the rate of a reaction without modifying the overall standard Gibbs energy change in the reaction; the process is called catalysis. (…) The term catalysis is also often used when the substance is consumed in the reaction (for example: base-catalysed hydrolysis of esters). Strictly, such a substance should be called an activator. PAC, 1996, 68, 155 *(Glossary of terms used in chemical kinetics, including reaction dynamics). Notice that the definition does not establish amount of catalyst.*

[2]The IUPAC definition for reactant is: A substance that is consumed in the course of a chemical reaction. It is sometimes known, especially in the older literature, as a reagent, but this term is better used in a more specialized sense as a test substance that is added to a system in order to bring about a reaction or to see whether a reaction occurs (e.g., an analytical reagent). PAC, 1996, *68*, 149 *(A glossary of terms used in chemical kinetics, including reaction dynamics (IUPAC Recommendations 1996)).*

Chapter 12

Biodegradation of Pesticides

Nimisha P. Vijayan[1], Haridas Madathilkovilakathu[2]
and Sabu Abdulhameed[1,3]*

[1]Department of Biotechnology and Microbiology, Kannur University, Thalassery
Campus, Kannur- 670661, Kerala, India.
[2]Inter University Centre for Biosciences, Kannur University, Thalassery Campus,
Kannur- 670661, Kerala, India.
[3]Centre for Bio-innovation and Product Development, Department
of Biotechnology and Microbiology, Kannur University,
Thalassery Campus, Kannur- 670661, Kerala, India.

INTRODUCTION

Pesticides are chemicals which are used to eliminate or control the pests causing diseases, to improve crop productivity (Damalas, 2011). Use of such chemicals has enhanced crop productivity and better control of pathogens. Pesticides are used all over the world for the protection of agricultural crops. But, the wide spread use of such chemicals has resulted in contaminating the environment with harmful chemicals. Pesticides will remain in the soil with little degradation and will eventually reach into nearby water bodies and make their entry into the food chain. Use of these harmful chemicals has lead to the study of new technologies to overcome the harmful effect of pesticides Biodegradation is a natural process which involves the degradation of pesticides by microorganisms (Singh, 2008).

The degradation rates of pesticides greatly vary among themselves. Some pesticides may be recalcitrant remaining for a very long time in the environment making their entry into the food chain decades after their application. Whereas,

*For Correspondence: Department of Biotechnology and Microbiology, Kannur University,
Thalassery Campus, Kannur-670661, Kerala, India. Email: drsabu@gmail.com

some pesticides like organophosphates readily undergo biodegradation making them more preferable than recalcitrant organochlorine compounds. Considering the harmful effects of pesticides, the removal of these chemical pollutants from the environment in a proper way is a mandatory involvement to be accomplished.

PESTICIDES

Current agricultural practices mainly aim at maximum production of crops and require control of weeds, destructive insects, pests and pathogenic fungi (Mancy and El-Bestawy, 2002). Pesticide is a broad term covering a large number of compounds including fungicides, insecticides, herbicides and bactericides. A wide variety of pests attack agricultural crops causing depletion in their yield. These chemicals are capable of killing a wide variety of pests or weeds. They prevent the spreads of insect borne diseases. Through the application of these chemicals, crop yield can be increased. Some chemicals target specific pests or pathogens. Many of these pesticides are characterized by high lipophilicity, bioaccumulation, long half-life and potential of long range transport (Jayaraj et al., 2016). They are artificially synthesized, toxic and non-biodegradable compounds. Pesticides mainly cause disturbance in the physiological activities of the target organism which lead to dysfunction and reduced vitality. They are subjected to some process of chemical degradation, hydrolysis and oxidation (Ormad et al., 1997; Arias-Estevez et al., 2008). Based on their degrading nature, pesticides can be classified into persistent and non-persistent pesticides. Persistent pesticides will have long half-lives. They accumulate in the environment and they get transferred through the food chain, thereby finally reaching into humans (Kodaka et al., 2003). Based on their chemical nature they are classified into organochlorines, organophosphates or carbamates etc. (Jayaraj et al., 2016).

HARMFUL EFFECTS OF PESTICIDES

Increased demand for agro products has resulted in increase of pesticide applications (Shetty et al., 2008). Application of pesticides without taking care of the environment will cause adverse impacts on the environment. A pesticide may be expected to be lethal to target pests only, but in most of the cases this is not the scenario. Many of them target other organisms, including humans. By gross estimation it has been found that only 10% of the applied pesticides reached its target area. The presence of these persistent chemicals in soil and water sediments would pollute the environment (Parte et al., 2017). Because of this problem, many pesticides are being banned. Treatment of soil with pesticides also affects and often alters the soil environment unsuitable for beneficial and essential microflora. Continuous use of these pesticides may cause an inhibitory or stimulatory action on the growth and other activities of the soil microflora. Pesticide residues in the soil also affect organisms like earthworms, bees, spiders, etc. (Singh et al., 2014). In some cases, the applied pesticides may also affect non-target vegetation. It also poses a risk to other organisms like birds, fishes and other beneficial insects. Surface water contamination is a major problem associated with the use of pesticides. Applied pesticides may reach into nearby water bodies by runoff and leach from soil (Aktar et al., 2009). There is evidence which proves the potential risk of these chemicals on human health (Kanekar et al., 2004; Jokanovic and Prostran, 2009; Verma et al., 2014). Their persistent nature and continuous use make them a major threat to the environment. Some pesticides undergo biomagnifications, hence causing more problems in the environment. So, proper management of such hazardous pesticides must be done by removing them from the environment.

BIODEGRADATION OF PESTICIDES

Due to the problems associated with the extensive use of pesticides, their removal in a safe, efficient and economical way has become critical to a healthy environment. There could be two ways for the degradation of pesticides. One is chemical degradation involving photolysis, and oxidation-reduction reactions. Another way of degradation is by microorganisms, known as microbial degradation or biodegradation (Sassman et al., 2004). Various chemical treatments of pesticides are not getting enough importance due to the emission of potentially toxic by-products and the high cost of treatment. These methods also are inefficient for complete removal of pesticides from the environment (Nyakundi et al., 2011). An ideal treatment process should not involve the generation of any toxic intermediates or end products. So, degradation of pesticides using microorganisms will provide a promising method for their efficient removal.

Biodegradation of pesticides is a cost effective, eco-friendly, minimally hazardous and efficient method (Finley et al., 2010). It involves the complete mineralization of pesticides without the generation of any toxic intermediates. Biodegradation involves the breakdown of a toxic chemical into nontoxic compounds or breakdown of a complex organic compound into simple inorganic chemicals or metabolic utilization of a toxic chemical by microorganisms (Finley et al., 2010; You and Liu, 2004; Wood, 2008). Biodegradation studies mainly involve the search for pesticide degrading organisms, study of their genetics and biochemistry for developing methods for their application in fields (Mcgharaj et al., 2011). Many of the microorganisms in the soil can use the pesticides for their growth. They mineralize the compounds, thereby detoxifying them (Kanekar et al., 2004). Microorganisms have the capacity to interact with substances both physically and chemically which may lead to structural changes or complete degradation. They can use pesticides as a source of energy and nutrients. But in some cases the applied pesticides may be toxic to microorganisms. Many microorganisms possess the ability to degrade pesticides and can be used to transform pesticides (Parte et al., 2017). The transformation of such compounds not only depends on the presence of the organism in the soil but also on a wide range of environmental factors (Aislabie and Lloyd-Jones, 1995; Alves et al., 2010). Hence, it is possible to exploit the ability of microorganisms to degrade exotic chemicals for the removal of pesticides from the environment. Due to the continuous exposure of soil microorganisms to the pesticides, they will develop a genetically modified system against pesticides (Parsek et al., 1995).

The rate of biodegradation in soil depends on different factors like, availability of pesticides to the microorganism, physiological status of the microorganism, survival or proliferation of the pesticide degrading microorganism, sustainable population of these microorganisms. In addition to that certain other factors like temperature, pH, water potential and nutrients also affect the biodegradation rate. In some cases, where the pesticide degrading microorganism is absent, addition of pesticide degrading microorganism is recommended (Singh, 2008). Fungi and bacteria are the main organisms involved in the biodegradation of pesticides (Briceño et al., 2007). The role of fungi and bacteria in the mineralization of the pesticides atrazine, alachlor, malathion and carbofuran have been studied. It has been found that the mineralization of alkyl side chains of alachlor and alkylamio side chains of atrazine was mainly due to fungal activity, whereas the mineralization of malathion and carbofuran was mainly due to bacterial activity (Levanon, 1993).

Biodegradation of Organophosphorous Pesticides

Among the pesticides used, organophosphorous (OP) compound accounts for more than 36% of the total world market (Kanekar et al., 2004). They are broad spectrum, and used on a wide variety of crops including vegetables, grains, fruits and ornamentals. Many of these pesticides act by inhibiting acetyl choline esterase enzyme, animal nervous system acting enzyme (Rani and Dhania, 2014). They are esters of phosphoric acid. They include aliphatic, phenyl and heterocyclic derivatives. They are biodegradable, but their residues are found in the environment (Kanekar et al., 2004). In most cases, OP compounds get mineralized by microorganisms. The primary step in the degradation of organophosphorous compounds is the hydrolytic cleavage of the organophosphate bond, making them inactive (Rani and Lalithakumari, 1994). They are readily soluble in water which contributes to the health hazards associated with these pesticides (Kumar et al., 2018). Most organophosphorous compounds are used by microorganisms as a source of carbon and phosphorous. Degradation of OP compounds, mainly involves oxidation, hydrolysis, alkylation and dealkylation (Singh and Walker, 2006). Biodegradation of OP pesticides mainly takes place through the hydrolysis of P-O alkyl or aryl bonds with the help of various enzymes. Phosphotriesterase, phosphatases and carboxylesterases are involved in the detoxification process (Kumar et al., 2018). Bacteria, fungi, algae and cyanobacteria are found to be efficient degraders of OP compounds. The first microorganism reported to degrade OP pesticide was *Flavobacterium* (Singh and Walker, 2006). Several other microorganisms are also reported to degrade OP compounds (Table 12.1).

Table 12.1 List of organophosphorous pesticides and the microorganisms involved in their degradation

Pesticides	Degrading microorganisms	References
Parathion	*Flavobacterium* ATCC 27551 *Leuconostoc mesenteroides, L. brevis,* *L. plantarum, L. sakei*	(Sethunathan and Yoshida, 1973) (Cho et al., 2009)
Diazinon	*Flavobacterium* ATCC 27551 *Lactobacillus brevis* *Leuconostoc mesenteroides, L. brevis,* *L. Plantarum, L. sakei*	(Sethunathan and Yoshida, 1973) (Zhang et al., 2006) (Cho et al., 2009)
Fenthion	*Bacillus* sp.	(Patel and Gopinathan, 1986)
Chlorpyrifos	*Synechosystis* sp. *Shingomonas* *Bacillus pumilus* *Enterobacter* *Cupidavirus*	(Singh et al., 2011) (Li et al., 2007) (Anwar et al., 2009) (Singh et al., 2004) (Lu et al., 2013)
Methyl parathion	*Ochrobactrum* sp.	(Qiu et al., 2007)
Fenamiphos	*Pseudomonas putida,* *Acinetobacter rhizosphaerae*	(Chanika et al., 2011)
Malathion	*Lactobacillus brevis*	(Zhang et al., 2006)
Dichlorovos	*Flavobacterium* sp *Proteus vulgaris, Vibrio* sp., *Serratia* sp., *and Acinetobacter* sp	(Ning et al., 2012) (Agarry et al., 2013)

Biodegradation of Organochlorine Pesticides

Organochlorine pesticides (OCP) are compounds that contain at least one chlorine atom. They are mainly insecticides containing carbon, hydrogen and chlorine (Parte et al., 2017). Due to their harmful effects many countries have banned their use, but some developing countries still continue their use. The important negative features of these compounds are high toxicity, slow degradation rate and bioaccumulation. They belong to the class of persistent organic pollutants. *In situ* biodegradation of organochlorine compounds using pure cultures has proved to be successful. Bacteria comprise a major group among the OCP degrading microorganisms (Langlois et al., 1970). *Bacillus, Pseudomonas, Arthrobacter,* and *Micrococcus* are the important bacterial genera involved in the biodegradation of organochlorine compounds. They undergo degradation by two major pathways, reductive dechlorination which is an anaerobic process and dehydrochlorination, an aerobic process. In the anaerobic process, OCPs are used as electron acceptors (Dennie et al., 1998; Futagami et al., 2008; Bisaillon et al., 2010). Many fungi were reported to degrade organochlorine compounds using oxidative enzymes. Peroxidases like laccase, lignin peroxidase, manganese peroxidase and cytochrome P450 are the main oxidative enzymes involved in organochlorine pesticide degradation (Ning and Wang, 2012; Itoh et al., 2013). Aerobic OCP degrading bacteria use these chemicals as a source of carbon and energy. There are several reports on the degradation of organochlorine pesticides by microorganisms (Table 12.2).

Table 12.2   List of organochlorine pesticides and the microorganisms involved in degradation

Pesticides	Degrading microorganisms	References
DDT	*Pseudomonas* sp *Micrococcus* sp *Arthrobacter* sp *Bacillus* sp *Sphingobacterium* sp *Trichoderma viridae*	(Patil et al., 1970) (Wedemeyer, 1967)
Endosulfan	*Aspergillus niger* *Pseudomonas aeruginosa, Arthrobacter* sp., *Burkholderia cepaeia*	(Bhalerao and Puranik, 2009)
Dieldrin	*Pseudomonas* sp	(Matsumura et al., 1968)
Endrin	*Micrococcus* sp, *Arthrobacter* sp, *Bacillus* sp	(Patil et al., 1970)
Aldrin	*Pseudomonas* sp, *Micrococcus* sp, *Bacillus* sp, *Trichoderma viridae*	(Patil et al., 1970)
Lindane	*Streptomyces* sp, *Pleurotus ostreatus* *Basea thiooxidans,* *Sphingomonas paucimobilis*	(Benimeli et al., 2008) (Pesce and Wunderlin, 2004)
Heptachlor	*Phanerochaete chrysosoporium*	(Arisoy and Kolankaya, 1998)

Biodegradation by Bacteria

Major bacterial genera involved in biodegradation of pesticides are *Bacillus, Pseudomonas, Flavobacterium, Moraxella, Acinetobacter, Arthrobacter, Paracoccus, Aerobacter, Alkaligens, Burkholderia and Spingomonas* (Parte et al., 2017). Rani and Lalithakumari, 1994, reported the utilization of methyl parathion by *Pseudomonas putida*, where the organism utilized the pesticide as a source of carbon and phosphorous (Fig. 12.1). The enzyme involved was organophosphorous acid anhydrase which hydrolyzed methyl parathion to *p*-nitrophenol. It was further degraded into hydroquinone and 1,2,4-benzenetriol which was then cleaved by benzenetriol oxygenase to maleyl acetate.

Flavobacterium and *Pseudomonas diminuta* strain isolated from the soil have been studied for their ability to degrade organophosphorous pesticide diazinon and parathion (McDaniel et al., 1988; Harper et al., 1988). *Enterobacter* strain B-14 was found to hydrolyze chlorpyrifos into TCP and DETP, and they utilized DETP for their growth and energy (Singh et al., 2004). Methyl parathion, an important pesticide of organophosphorous class, has been studied for its degradation by bacteria. Bacteria can use this pesticide as a source of carbon (Liu et al., 2003; Hong et al., 2005). *In situ* bioremediation of pesticides usually involves a consortium of microorganisms. But pure culture studies allow the study of mechanism by which the pesticide is metabolized or degraded by the bacteria. The degradation pathway can also be studied (Aislabie and Lloyd-Jones, 1995). The metabolism of pesticides is a three phase process. The initial properties of the parent compound were transformed in the first phase through oxidation, reduction or hydrolysis. It results in the production of a more water soluble and less toxic compound. The second phase involves conjugation of the pesticide to a sugar or amino acid making the pesticide water soluble and less toxic than the parent compound. The third phase involves the conversion of the second phase metabolites into secondary conjugates which are non-toxic (Van Herwijnen et al., 2003). Pino and Peñuela, 2011 studied the simultaneous degradation of methyl parathion and chlorpyrifos by an isolated bacterial consortium from a contaminated site. Identified microorganisms in the consortium were, *Acinetobacter* sp., *Pseudomonas putida*, *Bacillus* sp., *Pseudomonas aeruginosa*, *Proteus vulgaris*, *Flavobacterium* sp., *Pseudomonas* sp., *Citrobacter freundii*, *Stenotrophomonas* sp., *Acinetobacter* sp., *Klebsiella* sp. and *Acinetobacter* sp., Microbial consortium is more useful in the degradation of pesticides, because it reduces the accumulation of toxic substances. Use of enriched cultures from contaminated sites helps to isolate microorganisms capable of degrading pesticides since they are exposed to extreme conditions which make them more potent in the breakdown of toxic chemicals (Pino and Peñuela, 2011).

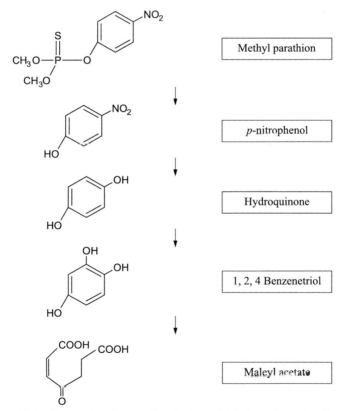

Figure 12.1   Proposed pathway of methyl parathion degradation by *P. putida* (Rani and Lalithakumari, 1994).

Pesticide degrading enzymes are found in bacteria (Table 12.3) Phosphotriesterases are a group of enzymes that can degrade OP compounds.

Organophosphorous hydrolase, methyl parathion hydrolase, organophosphorous acid anhydrolase are the three well characterized bacterial phosphotriesterases (Singh, 2009). The presence of atrazine dechlorinating enzyme has been reported in bacteria and it was involved in the hydrolytic dechlorination of atrazine to hydroxyatrazine (Wackett et al., 2002).

Table 12.3   List of pesticide degrading enzymes and their microbial sources

Pesticide	Degrading enzyme	Microbial source	References
Methyl parathion	Organophosphoros acid anhydrase	*Pseudomonas putida*	(Rani and Lalithakumari, 1994)
Monocrotophos	Phosphatase and Esterase	*Arthrobacter atrocyaneus, Bacillus megaterium*	(Bhadbhade et al., 2002)
Coumaphos	Parathion hydrolase	*Flavobacterium*	(Karns et al., 1987)
Paraoxon, Parathion, Diazinon, Methyl parathion, Cyanophos, Fensulfothion, Coumaphos	Phosphotriesterase	*Pseudomonas diminuta*	(Dumas et al., 1989)
Glyphosate	Oxidoreductases	*Pseudomonas* sp. *LBr Agrobacterium* strain T 10	(Scott et al., 2008)
Endosulfan and Endosulfate	Monooxygenases-ESd, ESe	*Mycobacterium* sp, *Arthrobacter* sp	(Sutherland et al., 2002; Weir et al., 2006)
Hexachloro-cyclohexane	Haloalkane dehalogenase	*Sphingobium* sp	(Ito et al., 2007)

Genetic studies of pesticide degrading bacteria involve cloning and characterization of enzymes involved in pesticide degradation. In bacteria, genes involved in pesticide degradation are found in plasmids which encode enzymes responsible for pesticide degradation. Sayler et al., 1990 reported these catabolic plasmids from *Pseudomonas, Alcaligenes*, *Acinetobacter*, *Cytophaga*, *Moraxella*, *Klebsiella*, and *Arthrobacter*. Serdar et al., 1982 reported the involvement of plasmid in the degradation of parathion by *Peudomanas diminiuta*. The gene opd which encode for organophosphate phosphotriesterase was cloned into M13 mp10 and found to express parathion hydrolase under control of lac promoter in *E. coli*. Chaudhary and Huang, 1988 reported a *Pseudomonas* sp. capable of degrading parathion and methyl parathion to *p*-nitrophenol. DNA from this organism showed homology with the opd gene reported from parathion hydrolyzing *Flavobacterium*. The nucleotide sequence of the gene was determined and could be expressed in *E. coli*. Somara and Siddavattam, 1995 reported the role of an indigenous plasmid (86kb) in the degradation of methyl parathion by *Flavobacterium balustinum*. The enzyme parathion hydrolaze was found to be encoded by this plasmid. Guha et al., 1997 reported the involvement of plasmid in the degradation of chlorpyrifos and malathion. *Sphingomonads indicum* contains two lin A genes which encode HCH dehydrochlorinase. It mediates the first two steps of dechlorination of γHCH (Lal et al., 2006). Chen and Mulchandani, 1998 studied the use of live biocatalysts for the degradation of pesticides. They discussed the use of genetically engineered *E. coli* with surface expressed organophosphorous hydrolase in the degradation of organophosphorous pesticides. *Plesiomonas* sp. which has the capacity of degrading methyl parathion into *p*-nitrophenol possesses an organophosphorous hydrolase gene (mpd). It was selected from its genomic library prepared by shotgun cloning. The nucleotide sequence of the gene was determined and could be effectively expressed in *E. coli* (Zhongli et al., 2001). *Moraxella* sp was genetically engineered for the degradation of organophosphorous and *p*-nitrophenol (PNP). The truncated ice nuclear protein (INPNC) anchor was used to target the organophosphorous hydrolase onto the surface of *Moraxella* sp. A shuttle vector (pPNCO33) coding for INPNCO-OPH was constructed. The translocation, surface display and functionality of OPH were studied in both *E. coli* and *Moraxella* sp. Whole cell activity was found to be higher in *Moraxella* sp. than in *E. coli*. The *Moraxella* sp. efficiently degraded both organophosphates and PNP (Shimazu et al., 2001). The monocrotophos degradation capacity of *Pseudomonas mendocina* MCM B-24 is plasmid born and it can be transferred and expressed in *E. coli* Nova Blue (Bhadbhade et al., 2002). By understanding the gene of interest and the specific enzyme involved, superbugs can be created for fast degradation of pesticides within a short time period.

Biodegradation by Fungi

Fungi act on pesticides by causing minor structural changes and converting them into nontoxic compounds and releasing them into the environment where further degradation could be done by bacteria (Diez, 2010). The ability of fungi to form extended mycelia and low specificity of their catabolic enzymes makes fungi a suitable candidate for the biodegradation process (Harms et al., 2011).

The main genera involved in the biodegradation of pesticides are *Fusarium*, *Aspergillus niger*, *Penicillium*, *Lentinula edodes*, *Lecanicillium*, *Oxysporum* (Parte et al., 2017). *Trichoderma harzianum* was found to degrade DDT, dieldrin, endosulfan, pentachloronitrobenzene and pentachlorophenol. The fungus degraded endosulfan at various nutritional conditions. Major fungal metabolites of endosulfan were endosulfan sulfate and endosulfandiol (Katayama and Matsumura, 1993). Harish et al., 2013, reported the biodegradation of organophosphorous pesticide, chlorpyrifos and ethion by soil fungal isolates, *T. harzianum* and *Rhizopus nodosus*. The organisms were shown capable of utilizing the insecticide as a sole source of carbon. Some recent reports have shown that the filamentous fungi can act in an interdependent manner with bacteria to enhance the degradation of the contaminant by the bacteria. Fungi can transform the pesticides into a more accessible form for bacteria to degrade them. In addition to that, the hyphal growth of fungi may help the bacteria to reach inaccessible contaminants (Ellegaard-Jensen et al., 2014; Lade et al., 2012). Fungi were reported to be more tolerant to pollutants than bacteria (Evans and Hedger, 2001). *Cladosporium cladosporioides* was found to degrade chlorpyrifos. Chlorpyrifos was first hydrolyzed to 3, 5, 6 trichloro 2 pyridonol

(TCP) and diethylthiophosphoric acid (DETP). TCP was then subjected to ring breakage resulting in complete detoxification (Fig. 12.2) (Chen et al., 2012).

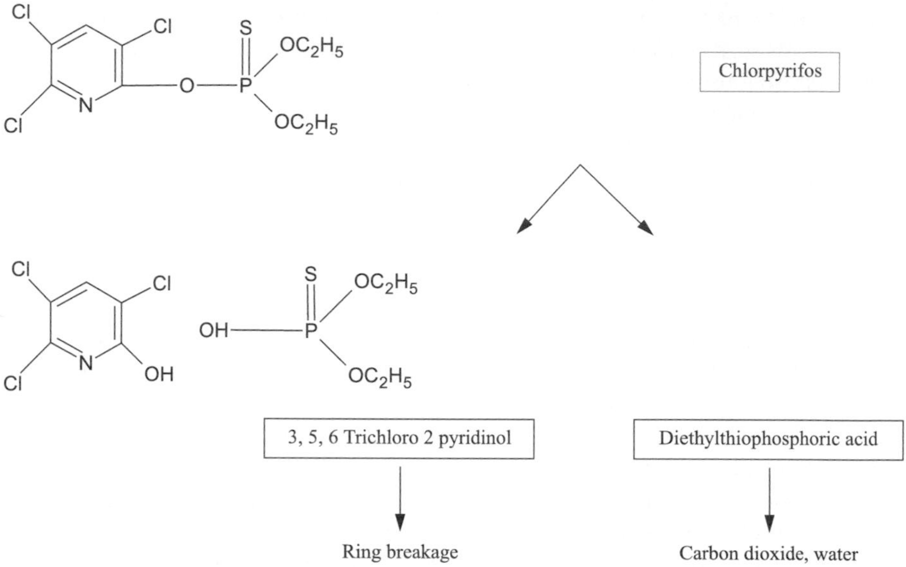

Figure 12.2　Chlorpyrifos degradation by *Cladosporium cladosporioides* (Chen et al., 2012).

Biodegradation of organochlorine pesticide endosulfan has been studied by Bhalerao and Puranik 2009. They reported the complete mineralization of endosulfan by *Aspergillus niger*. The main fungal genera involved in the biodegradation of pesticides are *Fusarium*, *Aspergillus niger*, *Penicillium*, *Lentinula edodes*, *Lecanicillium*, *Oxysporum* (Parte et al., 2017).

Biodegradation by Cyanobacteria

Cyanobacteria are photoautotrophic organisms and can be found in a wide variety of environments. They can be seen in various ecological habitats especially in rice fields (Kumar et al., 2013). Nitrogen fixing cyanobacteria are common in paddy fields where they contribute to soil fertility. They also play an important role in the biodegradation of pesticides. They can form mats called blooms and some of them are capable of fixing atmospheric nitrogen (Holt et al., 1994).

There are many reports showing their ability to degrade different classes of pesticides. Singh, 1973 first reported the tolerance of paddy field isolates of *Cylindrospermum sp, Aulosira fertilissima* and *Plectonema boryanum* to commercial preparations of lindane. Benzet and Knowles, 1981 reported the degradation of chlordimeform by *Oscillatoria* sp. The study of using these organisms as a tool for biodegradation was started when the cyanobacterial bloom formation occurred around the oil spill caused by the 1991 war in the Persian Gulf (Sorkokh et al., 1992). *Anabaena* sp and *N. ellipsosporum* required nitrate in the medium for the degradation of lindane (Kuritz and Wolk, 1995). Singh et al., 2011 reported the degradation of chlorpyrifos by the unicellular cyanobacterium *Synechocystis* sp. strain PUPCC 64. One of the degradation products was 3,5,6-trichloro 2 pyridinol. Degradation of fluometuron by six selected cyanobacterial species has been reported. Biodegradation of fluometuron was species dependent. All the species showed great ability to degrade the compound (Mancy and El-Bestawy, 2002). Degradation of glyphosate, an organophophorous herbicide, is mainly carried out by soil microorganisms. But, once they make their entry into the aquatic system, cyanobacterial strains were found to be involved in the biodegradation process. A high tolerance level was reported in the cyanobacterial strains towards glyphosate. The presence of a resistant form of target enzyme 5-enolpyruvylshikimate-3-phosphate and metabolism of glyphosate by the organism are the reported reasons for their high tolerance level. *Spirulina* sp. showed the ability to degrade glyphosate (Arunakumara et al., 2013). Biodegradation of methyl parathion by soil isolates of micro algae and cyanobacteria have been reported (Megharaj et al., 1994). Three filamentous cyanobacterial strains, *A. oryzae, N. muscorum,* and *S. platensis* were studied for their ability to degrade and utilize malathion as a source of phosphorous. Among such strains, *N. muscorum* dominated by removing 90% of malathion (Ibrahim et al., 2014). Biodegradation of fenamiphos, an organophosphorous pesticide, by cyanobacteria has been reported. All the species tested were able to transform the pesticide into its primary oxidation product, fenamiphos sulfoxide (FSO) and most of the cultures were able to hydrolyze FSO to fenamiphos sulfoxide phenol (Cáceres et al., 2008).

Since cyanobacteria are photoautotrophic organisms, their use in bioremediation of pollutants would overcome the need to supply biodegradative heterotrophs with organic nutrients. Their widespread appearance in the polluted areas also makes them a better candidate for the biodegradation of pesticides.

Conclusion

Uncontrolled use of pesticides is causing a great threat to the environment and living organisms. It is a major cause of environmental pollution. There are several physical and chemical methods for the degradation of pesticides. Since they are inefficient and costly, degradation through the use of microorganisms is the widely accepted method. It ensures the complete removal of pesticides from the environment. Biodegradation involves complete breakdown of these toxic chemicals into nontoxic compounds. Bacteria, fungi and blue green algae have proved to be efficient in the degradation of pesticides. Many microorganisms can be genetically improved to increase the efficiency of degradation. Many research groups are active in this area and the prime focus is to identify and develop new strains of microorganisms for the efficient degradation of pesticides.

REFERENCES

Agarry, S.E., Olu-arotiowa, O.A., Aremu, MO., Jimoda, LA. Biodegradation of dichlorovos (Organophosphate Pesticide) in soil by bacterial isolates. *Journal of Natural Sciences Research*, 2013 3, 12–16.

Aislabie, J., Lloyd-Jones, G. A review of bacterial-degradation of pesticides. *Australian Journal of Soil Research*, 1995, 33, 925–942.

Aktar, W., Sengupta, D., Chowdhury, A. Impact of pesticides use in agriculture: Their benefits and hazards. *Interdisciplinary Toxicology*, 2009, 2, 1–12.

Alves, S.A., Albergaria, J.T., Fernandes, D.V., Alvim-Ferraz, M.C., Delerue-Matos, C. Remediation of soils combining soil vapor extraction and bioremediation: Benzene. *Chemosphere*, 2010, 80, 823–828.

Anwar, S., Liaquat, F., Khan, Q.M., Khalid, Z.M., Iqbal, S. Biodegradation of chlorpyrifos and its hydrolysis product 3,5,6trichloro-2-pyridinol by *Bacillus pumilus* strain C2A1. *Journal of Hazardous Materials*, 2009, 168, 400–405.

Arias-Estevez, M., Lopez-Periago, E., Martinez-Carballo, M., Simal-Gandara, J., Mejuto, J.C., Garcia-Rio, L. The mobility and degradation of pesticides in soils and the pollution of groundwater resources. *Agriculture, Ecosystems & Environment*, 2008, 123, 247–260.

Arisoy, M., Kolankaya, N. Biodegradation of heptachlor by *Phanerochaete chrysosporium* ME 446: The toxic effects of heptachlor and its metabolites on mice. *Turkish Journal of Biology*, 1998, 22, 427–434.

Arunakumara, K.K.I.U., Walpola, B.C., Yoon, M. Metabolism and degradation of glyphosate in aquatic cyanobacteria: A review. *African Journal of Microbiology Research*, 2013, 7, 4084–4090.

Benimeli, C.S., Fuentes, M.S., Abate, C.M., Amoroso, M.J. Bioremediation of lindane-contaminated soil by *Streptomyces* sp. M7 and its effects on Zea mays growth. *International Biodeterioration & Biodegradation*, 2008, 61, 233–239.

Benzet, H.J., Knowles, C.O. Degradation of chlor- dimeform by algae. *Chemosphere*, 1981, 10, 909–917.

Bhadbhade, B.J., Sarnaik, S.S., Kanekar, P.P. Biomineralization of an organophosphorous pesticide Monocrotophos by soil bacteria. *Journal of Applied Microbiology*, 2002, 93, 224–34.

Bhalerao, T.S., Puranik, P.R. Microbial degradation of monocrotophos by *Aspergillus oryzae*. *International Biodeterioration and Biodegradation*, 2009, 63, 503–508.

Bisaillon, A., Beaudet, R., Lepine, F., Deziel, E., Villemur, R. Identification and characterization of a novel CprA reductive dehalogenase specific to highly chlorinated phenols from *Desulfitobacterium hafniense* strain PCP-1. *Applied and Environmental Microbiology*, 2010, 76, 7536–7540.

Briceño, G., Palma, G., Duran, N. Influence of organic amendment on the biodegradation and movement of pesticides. *Critical Reviews in Environmental Science and Technology*, 2007, 37, 233–271.

Cáceres, T.P., Megharaj, M., Naidu, R. Biodegradation of fenamiphos by ten different species of green algae and cyanobacteria. *Current Microbiology*, 2008, 57, 643–646.

Chanika, E., Georgiadou, D., Soueref, E., Karas, P., Karanasios, E., Tsiropoulos, N.G., Tzortzakakis, E.A., Karpouzas, D.G. Isolation of soil bacteria able to hydrolyze both organophosphate and carbamate pesticides. *Bioresource Technology*, 2011, 102, 3184–3192.

Chaudhary, G.R., Huang, G.H. Isolation and characterization of a new plasmid from *Flavobacterium* sp., which carried the genes for degradation of 2, 4, dichlorophenoxy acetate. *Journal of Bacteriology*, 1988, 170, 3897.

Chen, S., Liu, C., Peng, C., Liu, H., Hu, M., Zhong, G. Biodegradation of chlorpyrifos and its hydrolysis product 3,5,6-trichloro-2-pyridinol by a new fungal strain *Cladosporium cladosporioides* Hu-01. *PLoS One*, 2012, 7, 1–12

Chen, W., Mulchandani, A. The use of live biocatalysts for pesticide detoxification. *Trends in Biotechnology,* 1998, 16, 71–76.

Cho, K.M., Math, R.K., Islam, S.M.A., Lim, W.J., Hong, S.Y., Kim, J.M., Yun, M.G., Cho, J., Yun, H.D. Biodegradation of chlorpyrifos by lactic acid bacteria during kimchi fermentation. *Journal of Agricultural and Food Chemistry,* 2009, 57, 1882–1889.

Damalas, C.A., Eleftherohorinos, IG. Pesticide exposure, safety issues, and risk assessment indicators. *International Journal of Environmental Research and Public Health,* 2011, 8, 1402–1419.

Dennie, D., Gladu, I.I., Lepine, F., Villemur, R., Bisaillon, J., Beaudet, R. Spectrum of the reductive dehalogenation activity of *Desulfitobacteriumfrappieri* PCP-1. *Applied and Environmental Microbiology,* 1998, 64, 4603–4606.

Diez, MC. Biological aspects involved in the degradation of organic pollutants. *Journal of Soil Science and Plant Nutrition,* 2010, 244–267.

Dumas, D.P., Caldwell, S.R., Wild, J.R., Raushel, F.M. Purification and properties of the phosphotriesterase from *Pseudomonas diminuta. Journal of Biological Chemistry,* 1989, 264, 19659–19665.

Ellegaard-Jensen, L., Knudsen, B.E., Johansen, A., Albers, C.N., Aamand, J., Rosendahl, S. Fungal–bacterial consortia increase diuron degradation in water-unsaturated systems. *The Science of the Total Environment,* 2014, 466–467, 699–705.

Evans C.S, Hedger J.N. Degradation of plant cell wall polymers. *In*: Gadd, G.M. (ed.), Fungi in Bioremediation. Cambridge; Cambridge University Press; 2001; p. 26.

Finley, S.D., Broadbelt, L.J., Hatzimanikatis, V. In silico feasibility of novel biodegradation pathways for 1, 2, 4-trichlorobenzene. *BMC Systems Biology,* 2010, 4, 4–14.

Futagami, T., Goto, M., Furukawa, K. Biochemical and genetic bases of dehalorespiration. *Chemical Record,* 2008, 8, 1–12.

Guha , A., Kumari, B., Bora, T.C., Roy, M.K. Posible involvement of plasmids in degradation of malathion and chlorpyrifos by *Micrococcus* sp. *Folia Microbiologica,* 1997, 42, 574–576.

Harish, R., Supreeth, M., Chauhan, J.B. Biodegradation of organophosphate pesticide by soil fungi. *Advanced BioTech,* 2013, 12, 2319–6750.

Harms, H., Schlosser, D., Wick, L.Y. Untapped potential: Exploiting fungi in bioremediation of hazardous chemicals. *Nature Reviews. Microbiology,* 2011, 9, 177–192.

Harper, L.L., McDaniel, C.S., Miller, C.E., Wild, J. Dissimilar plasmids isolated from *Pseudomonas diminuta* MG and a *Flavobacterium* sp. (ATCC 27551) contain identical opd genes. *Applied and Environmental Microbiology* 1988, 54, 2586–2589.

Holt, J.G., Krieg, N.R., Sneath, P.H.A. Group 11. Oxygenic phototrophic bacteria. *In*: Hensyl, W.R. (ed.), Bergey's Manual of Determinative Bacteriology. Baltimore, USA: Williams & Wilkins; 1994; pp. 377–425.

Hong, L., Zhang, J.J., Wang, S.J., Zhang, X.E., Zhou, N.Y. Plasmid-borne catabolism of methyl parathion and p-nitrophenol in *Pseudomonas* sp. strain WBC-3. *Biochemical and Biophysical Research Communications,* 2005, 334, 1107–1114.

Ibrahim, W.M., Karam, M.A., El-Shahat, R.M., Adway, A.A.. Biodegradation and utilization of organophosphorus pesticide malathion by cyanobacteria. *BioMed Research International,* 2014, Article ID 392682.

Ito, M., Prokop, Z., Klavana, M., Ostubo, Y., Tsuda, M., Damborsky, J., Nagata, Y. Degradation of β-hexachlorocyclohexane by haloalkane dehalogenase LinB from γ-hexachlorocyclohexane utilizing bacterium *Sphingobium* sp MI1205. *Archives of Microbiology,* 2007, 188, 313–25.

Itoh, K., Kinoshita, M., Morishita, S., Chida, M., Suyama, K. Characterization of 2,4-dichlorophenoxyacetic acid and 2,4,5-trichlorophenoxyacetic acid-degrading fungi in Vietnamese soils. *FEMS Microbiology Ecology*, 2013, 84, 124–132.

Jayaraj, R., Megha, P., Sreedev, P. Organochlorine pesticides, their toxic effects on living organisms and their fate in the environment. *Interdisciplinary Toxicology*, 2016, 9, 90–100.

Jokanovic, M., Prostran, M. Pyridinium oximes as cholinesterase reactivators structure-activity relationship and efficacy in the treatment of poisoning with organophosphorus compounds. *Current Medicinal Chemistry*, 2009, 16, 2177–2188.

Kanekar, P.P., Bhadbhade, B., Deshpande, N.M., Sarnaik, S.S. Biodegradation of organo-phosphorous pesticides. *Proceedings-Indian National Science Academy*, 2004, B70, 57–70.

Karns, J.S., Muldoon, M.T., Nlulbry, W.W., Derbyshire, M.K., Kearney, P.C. Use of micro-organisms and microbial systems in the degradation of pesticides. *In*: LeBaron, H.M., Mumma, R.O., Honeycutt, R.C., Duesing, J.H. (eds), Biotechnology in Agricultural Chemistry. Washington, DC: American Chemical Society; 1987; pp. 156–170.

Katayama, A., Matsumura, F. Degradation of organochlorine pesticides, particularly endosulfan by *Trichoderma harzianum*. *Environmental Toxicology and Chemistry*, 1993, 12, 1059–1065.

Kodaka, R., Sugano, T., Katagi, T., Takimoto, Y. Clay-catalyzed nitration of a carbamate fungicide diethofencarb. *Journal of Agricultural and Food Chemistry*, 2003, 51, 7730–7737.

Kumar, J.I.N., Bora, A., Rita, N.K., Amb, M.K., Khan, S. Toxicity analysis of pesticides on cyanobacterial species by 16S rDNA molecular characterization. *Proceedings of the International Academy of Ecology and environmental Sciences*, 2013, 3, 101–132.

Kumar, S., Kaushik, G., Dar, M.A., Nimesh, S., L´opez-Chuken, U.J., Villarreal-Chiu, J.F. Microbial degradation of organophosphate pesticides: A review. *Pedosphere*, 2018, 28, 190–208.

Kuritz T., Wolk C.P. Use of filamentous cyanobacteria for biodegradation of organic pollutants. *Applied and Environmental Microbiology*, 1995, 61, 234–238.

Lade, H.S., Waghmode, T.R., Kadam, A.A., Govindwar, S.P. Enhanced biodegradation and detoxification of disperse azo dye Rubine GFL and textile industry effluent by defined fungal-bacterial consortium. *International Biodeterioration & Biodegradation*, 2012, 72, 94–107.

Lal, R., Dogra, C., Malhotra, S., Sharma, P., Pal, R. Diversity, Distribution and Divergence of lin genes in hexachlorocyclohexane degrading *Sphingomonads*. *Trends in Biotechnology*, 2006, 24, 121–129.

Langlois, B.E., Collins, J.A., Sides, K.G. Some factors affecting degradation of organochlorine pesticide by bacteria. *Journal of Dairy Science*, 1970, 53, 1671–1675.

Levanon, D. Roles of fungi and bacteria in the mineralization of the pesticides atrazine, alachlor, malathion and carbofuran in soil. *Soil Biology and Biochemistry*, 1993, 25, 1097–1105.

Li, X., He, J., Li, S. Isolation of a chlorpyrifos-degrading bacterium, *Sphingomonas* sp. strain Dsp-2, and cloning of the mpd gene. *Research in Microbiology*, 2007, 158, 143–149.

Liu, Z., Hong, Q., Xu, J.-H., Wu, J., Zhang, X.-Z., Zhang, X.-H., Ma, A.-Z., Zhu, J., Li, S.-P. Cloning, analysis and fusion expression of methyl parathion hydrolase. *Acta Genetica Sinica*, 2003, 30, 1020–1026.

Lu, P., Li, Q., Liu, H., Feng, Z., Yan, X., Hong, Q., Li, S. Biodegradation of chlorpyrifos and 3,5,6-trichloro-2-pyridinol by *Cupriavidus* sp. DT-1. *Bioresource Technology*, 2013, 127, 337–342.

Mancy, A., El-Bestawy, E. Toxicity and biodegradation of fluometuron by selected cyanobacterial species. *World Journal of Microbiology and Biotechnology*, 2002, 18, 125–131.

Matsumura, F., Boush, G.M., Tai, A. Breakdown of dieldrin in the soil by a microorganism. *Nature*, 1968, 219, 965–967.

McDaniel, C.S., Harper, L.L., Wild, J.R. Cloning and sequencing of a plasmid-borne gene (opd) encoding a phosphotriesterase. *Journal of Bacteriology*, 1988, 170, 2306–2311.

Megharaj, M., Madhavi, D.R., Sreenivasulu, C., Umamaheswari, A., Venkateswarlu, K. Bio-degradation of methyl parathion by soil isolates of microalgae and cyanobacteria. *Bulletin of Environmental Contamination and Toxicology*, 1994, 53, 292–297.

Megharaj, M., Ramakrishnan, B., Venkateswarlu, K., Sethunathan, N., Naidu, R. Bioremediation approaches for organic pollutants: A critical perspective. *Environment International*, 2011, 37, 1362–1375.

Ning, D., Wang, H. Involvement of cytochrome P450 in pentachlorophenol transformation in a white rot fungus *Phanerochaete chrysosporium*. *PLoS One*, 2012, 7, e45887.

Ning, J., Gang, G., Bai, Z., Qing, Hu., Hongyan, Qi., Anzhou, Ma., Zhuan, X., Zhuang, G. In situ enhanced bioremediation of dichlorvos by a phyllosphere *Flavobacterium* strain. *Frontiers of Environmental Science & Engineering*, 2012, 6, 231–237.

Nyakundi, W.O., Magoma, G., Ochora, J., Nyende, A.B. Biodegradation of diazinon and methomyl pesticides by white rot fungi from selected horticultural farms in rift valley and central Kenya. *Journal of Applied Technology in Environmental Sanitation*, 2011, 1, 107–124.

Ormad, P., Cortes, S., Puig, A., Ovelleiro, J.L. Degradation of organochloride compounds by O_3 and O_3/H_2O_2. *Water Research*, 1997, 31, 2387–2391.

Parsek, M.R., Mc Fall, S.M., Chakrabarty, A.M. Microbial degradation of toxic environment pollution: Ecological and evolutionary consideration. *International Biodeterioration & Biodegradation*, 1995, 35, 175–188.

Parte, G.S., Mohekar, D.A., Kharat, A.S. Microbial degradation of pesticide: A review. *African Journal of Microbiology Research*, 2017, 11, 992–1012.

Patel, M.N., Gopinathan, K.P. Lysozyme sensitive bioemulsifier for immiscible organophosphorus pesticides. *Applied and Environmental Microbiology*, 1986, 52, 1224–6.

Patil, K.C., Matsumura, F., Boush, G.M. Degradation of endrin, aldrin, and DDT by Soil Microorganisms. *Applied Microbiology*, 1970, 19, 879–881.

Pesce, S.F., Wunderlin, D.A. Biodegradation of lindane by a native bacterial consortium isolated from contaminated river sediment. *International Biodeterioration & Biodegradation*, 2004, 54, 255–260.

Pino, N., Peñuela, G. Simultaneous degradation of the pesticides methyl parathion and chlorpyrifos by an isolated bacterial consortium from a contaminated site. *International Biodeterioration & Biodegradation*, 2011, 65, 827–831.

Qiu, X., Zhong, Q., Li, M., Bai, W., Li, B. Biodegradation of pnitrophenol by methyl parathion-degrading *Ochrobactrum* sp. B2. *International Biodeterioration & Biodegradation*, 2007, 59, 297–301.

Rani, K., Dhania, G. Bioremediation and biodegradation of pesticide from contaminated soil and water—A noval approach. *International Journal of Current microbiology and Applied Sciences*, 2014, 10, 23–33.

Rani, N.L., Lalithakumari, D. Degradation of methyl parathion by *Pseudomonas putida*. *Canadian Journal of Microbiology*, 1994, 40, 1000–1006.

Sassman, S.A., Lee, L.S., Bischoff, M., Turco, R.F. Assessing N, N'-Dibutylurea (DBU) formation in soils after application of n butyl isocyanate and benlate fungicides. *Journal of Agricultural and Food Chemistry*, 2004, 52, 747–754.

Sayler, G.S., Hooper, S.W., Layton, A.C., King, J.M.H. Catabolic plasmids of environmental and ecological significance. *Microbial Ecology*, 1990, 19, 1–20.

Scott, C., Pandey, G., Hartley, C.J., Cheesman, M.J., Taylor, M.C., Pandey, R., Khurana, J.L., Teese, M., Coppin, C.W., Weir, K.M., Jain, R.K., Lal, R., Russell, R.J., Oakeshott, J.G. The enzymatic basis for pesticide bioremediation. *Indian Journal of Microbiology*, 2008, 48, 65–79.

Serdar, C.M., Gibson, D.T., Munnecke, D.M., Lancaster, J.H. Plasmid involvement in parathion hydrolysis by *Pseudomonas diminuta*. *Applied and Environmental Microbiology*, 1982, 44, 246–249.

Sethunathan, N., Yoshida, Y. A *Flavobacterium* that degrades diazinon and parathion. *Canadian Journal of Microbiology*, 1973 19, 873–875.

Shetty, P.K., Murugan, M., Sreeja, K.G. Crop protection stewardship in India: Wanted or unwanted. *Current Science*, 2008 95, 457–464.

Shimazu, M., Mulchandani, A., Chen, W. Simultaneous degradation of organophosphorous pesticides and *p*-nitrophenol by a genetically engineered *Moraxella* sp. with surface expressed organophosphorous hydrolase. *Biotechnology and Bioengineering*, 2001, 76, 318–324.

Singh, B.K., Allan, S., Walker, A., Alun, J., Morgan, W., Denis, J. Biodegradation of chlorpyrifos by *Enterobacter strain* B-14 and its use in bioremediation of contaminated soils. *Applied and Environmental Microbiology*, 2004, 70, 4855–4863.

Singh, B.K., Walker, A. Microbial degradation of organophosphorus compounds. *FEMS Microbiology Reviews*, 2006, 30, 428–471.

Singh, B.K. Organophosphorus-degrading bacteria: Ecology and industrial applications. *Nature Reviews Microbiology*, 2009, 7, 156–163.

Singh, D.P., Khattar, J.I.S., Nadda, J., Singh, Y., Garg, A., Kaur, N., Gulati, A. Chlorpyrifos degradation by the cyanobacterium *Synechocystis* sp. strain PUPCCC 64. *Environmental Science and Pollution Research*, 2011, 18, 1351–1359.

Singh, K.D. Biodegradation and bioremediation of pesticide in soil: Concept, method and recent developments. *Indian Journal of Microbiology*, 2008, 48, 35–40.

Singh, P.K. Effect of pesticides on blue-green algae. *Archives of Microbiology*, 1973, 89, 317–320.

Singh, R., Singh, P., Sharma, R. Microorganism as a tool of bioremediation technology for cleaning environment: A review. *Proceedings of the International Academy of Ecology and Environmental Sciences*, 2014, 4, 1–6.

Somara, S., Siddavattam, D. Plasmid mediated organophosphate pesticide degradation by *Flavobacterium balustinum*. *Biochemistry and Molecular Biology International*, 1995, 36, 627–631.

Sorkokh, N., Al-Hasan, R., Radwan, S., Hoppner, T. Self-cleaning of the gulf. *Nature*, 1992, 359, 109.

Sutherland, T.D., Horne, I., Harcourt, R.L., Russel, R.J., Oakeshott, J.G. Isolation and characterization of a mycobacterium strain that metabolizes the insecticide endosulfan. *Journal of Applied Microbiology*, 2002, 93, 380–389.

Van Herwijnen, R., Van de Sande, B.F., Van der Wielen, F.W.M., Springael, D., Govers, H.A.J., Parsons, J.R. Influence of phenanthrene and fluoranthene on the degradation of fluorine and glucose by *Sphingomonas* sp. strain LB126 in chemostat cultures. *FEMS Microbiology Ecology*, 2003, 46, 105–111.

Verma, J.P., Jaiswal, D.K., Sagar, R. Pesticide relevance and their microbial degradation: A-state-of-art. *Reviews in Environmental Science and Bio/Technology*, 2014, 13, 429–466.

Wackett, L.P., Sadowsky, M., Martinez, B., Shapir, N. Biodegradation of atrazine and related s-triazine compounds: From enzymes to field studies. *Applied Microbiology and Biotechnology*, 2002, 58, 39–45.

Wedemeyer, G. Dechlorination of 1, 1, 1-Trichloro-2, 2-bis(pchlorophenyl)ethane by *Aerobacter aerogene. Applied Microbiology*, 1967, 15, 569–574.

Weir, K.M., Sutherland, T.D., Horne, I., Russell, R.J., Oakeshott, J.G. A Single mono-oxygenase, ese, is involved in the metabolism of the organochlorides endosulfan and endosulfate in an *Arthrobacter* sp. *Applied and Environmental Microbiology*, 2006, 72, 3524–3530

Wood, T.K. Molecular approaches in bioremediation. *Curr Opin Biotechnol*, 2008, 19, 572–578.

You, M., Liu, X. Biodegradation and bioremediation of pesticide pollution. *Chinese Journal of Ecology*, 2004, 23, 73–77.

Zhang, Z.H., Hong, Q., Xu, J.H., Zhang, X.Z., Li, S.P. Isolation of fenitrothion-degrading strain *Burkholderia* sp. FDS-1 and cloning of mpd gene. *Biodegradation*, 2006, 17, 275–283.

Zhongli, C., Shunpeng L., Guoping, F. Isolation of methyl parathion degrading strain M6 and cloning of the methyl parathion hydrolase gene. *Applied and Environmental Microbiology*, 2001, 67, 4922–4925.

Chapter 13

Real-time Analysis for Pollution Prevention

Maria Isabel Martinez Espinoza*

Department of Chemistry, Materials, and Chemical Engineering "Giulio Natta",
Polytechnic University of Milan, Milan, Italy.

INTRODUCTION

"Analytical methodologies need to be further developed to allow for real-time, in-process monitoring and control prior to the formation of hazardous substances"

To better understand this principle, imagine a match with a blindfolded boxer, without being able to see what happens in the surroundings, the shots that come or where the opponent is. It would be hard not to take a punch in the face. Now removing the blindfold, it is much easier for the boxer to move, avoid the blows and counterattack. In this case, the sense of sight provides the boxer with means to monitor the surrounding environment in real time and allows it to adapt and react to achieve the goal. In the chemical sectors, the analytical techniques can be considered the eyes to see what happens inside of an experiment.

Today around of the world, high quantity of many types of experiments are carried out every day to synthesize a novel molecule, optimize a prodecure or study the chemical properties of new material of interest. In many cases, the most sophisticated or the simplest analytical techniques are used to monitor the experiment. The analytical techniques allow -among other performances- to control in real time the progress of a reaction or following the decomposition process of a lot of types of bio- or organic molecules; also they play a crucial role in

*For Correspondence: Via Luigi Mancinelli, 7, 20131, Milan Italy; Email: q.isabel09@gmail.com

monitoring environmental organic/inorganic contaminants (Turner, 2013; de Marco et al., 2019). For this reason, the development of new analytical methodologies is one of the most active areas of green chemistry. It is important to note that this principle is largely linked to the first principle of prevention and both work closely together to improve a chemical process and at the same time decrease and prevent the generation of waste.

REAL-TIME ANALYSIS FOR POLLUTION PREVENTION

Seen at the atomic level and in real time what happens inside an experiment is one of the goals of health. Unfortunately, technology has not yet reached these levels, but it is true that it is very close. However, to date, analytical techniques have allowed identifying phenomena within different chemical processes, and the role of "eyes" in an experiment has been well done.

Green chemistry, in its 12 principles, has proposed an excellent solution to control and monitor the formation of dangerous substances in real time thanks to these analytical "eyes".

The real time analysis for the pollution prevention principle shows how scientific activity in general, and in particular the analytical method, must satisfy the needs of society avoiding errors in a chemical process that can be translated into actions that can cause damage to human health and the ecosystem. In this context, research and development in chemical disciplines must consider the different aspects related to the conservation of health and the environment. On the other hand, the analytical techniques applied to make a chemical reaction "greener" must also operate within the 12 principles and consequently, most of the traditional analytical methods need to be modified to contribute to the ultimate goal of prevention.

However, the principle real-time analysis for pollution prevention requires additional efforts that involve an efficient collaboration of differents areas with the aim of preventing the formation of dangerous substances in each phase of the production process and constant monitoring to avoid errors that could lead to the appearance of dangerous substances harmful to the environment and human health and for this reason green analytical methods are in continous development.

ANALYTICAL TECHNIQUES FOR REAL-TIME ANALYSIS

In most scientific fields from biology to materials chemistry, chemical reactions are done every day, and an analytical technique is applied to monitoring the reaction. Today, there exists a large number of analytical instruments, and their choice depends on the sample preparation, analyte and the type of information that the analysis can bring. The tendency of the scientific community is developing instrumental methods to avoid the sample preparation and waste treatment, increasing the analysis reliability, precision and is time-saving. In some cases, there is a choice of direct techniques of analysis or solventless processes of analysis, which are green processes (Korany et al., 2017).

Chromatographic techniques

1) **High-Performance Liquid Chromatography (HPLC).** HPLC technique is different from GC because in this case, the mobile phase can be a source of pollution, The principal disadvantage is a high solvent consumption that produce about 500 L/year of eluent (columns 4.6x150 mm) (Yabré et al., 2018) of which about 50% consists of acetonitrile (ACN) which it is classified as a toxic and harmful solvent, as well as being relatively expensive. From the instrumental point of view, different strategies have been proposed to reduce the amount of solvent by reducing the dimensions of the column or using only water as a solvent. In the last case, for organic molecules, it is limited due to the solubity and polarity of these molecules. On the other hand, analytical scientists continue to develop new chromatografic tecniques that can be used with greener solvents such as ethanol or ionic liquids as eluent.

2) **Gaschromatography (GC)** is the separation technique most used in chemical analysis. The GC, using helium as transport gas and not using organic solvents, is more sustainable than HPLC (Nuttall et al., 2012). However, efforts to improve the analysis are mainly focused on the implementation of the sample preparation steps and derivatization of the analyte.

3) **High Temperature Liquid Chromatography (HTLC).** This technique is considered a good candidate to decrease the amount of eluent required in the separation and analysis of the sample. The temperature is an essential factor that favors the differential migration phenomenon, and with increasing temperature there is a decrease in viscosity and an increase in diffusivity that allow operating at higher flow rates without compromising the resolution and accuracy of the analysis (Teutenberg, 2010).

4) **Ultra-High Performance Liquid Chromatography (UHPLC).** UHPLC systems are gradually replacing conventional HPLC instruments. They have been introduced to overcome the high back pressures generated by small particle size. Decreasing column length will decrease the runtime which decreases the solvent consumption and waste generation. In addition, using a short column with a small internal diameter will lead to a shorter run time, resulting in more solvent reduction (Dong and Zhang, 2014).

5) **Supercritical Fluid Chromatography (SFC).** SFC is often used to analyze reduced concentrations of high molecular weight components and molecules. It is expected in the analysis of drugs, as well as in the analysis of foods, explosives, petroleum, polymers and propellants. The SFC is similar to gas and liquid chromatography but the most important differences are the kind of eluent and the chromatography conditions. The carbon dioxide (CO_2), that at room condition is a gas, is used as the mobile phase in the SFC systems and for this reason, this technique requires the extreme conditions of temperature and pressure in order to maintain the fluidic state of CO_2. In other words, the highly pressurized flux of CO_2 plays a crucial role in the separation of complex mixtures (a.e.chiral and natural compounds). In addition, this chromatography technique is faster, uses much less solvent, and overall is a less expensive and greener method that exceeds HPLC in performance for the separation of several types of compounds (Taylor, 2009).

Spectroscopic methods

Spectroscopy detection systems are generally sustainable as they do not require complex sample treatments and operate with low energy consumption.

1) **Mass Spectroscopy**. Mass spectroscopy thanks to high sensitivity allows using a small quantity of the sample, unfortunately this technique has high energy consumption and often requires an adequate treatment of the sample, but thanks to several improvements such as nanoelectrospray techniques, it is possible to perform direct analysis on the sample and, in some cases, it is also possible to obtain a derivatization/detection system without the use of reagents (Venter et al., 2008).

2) **Infrared/Near-infrared Spectroscopy (IR/NIR)**. Infrared spectroscopy is a non-destructive technique that offers several advantages such as the use of a small sample quantity, high sensitivity and can be used directly on the sample without any treatment. These features allow real-time monitoring without generating any residue, low cost, fast and easy to use and for this reason are considered a green analytical technique. To date there are several devices including portable ones, which eliminates the need for sampling and allows measuring in the field or directly in the process through different probes. Both IR and NIR are techniques used in industry for continuous monitoring of the emission of gas and particles, in the biomedical and pharmaceutical area for the monitoring of biomolecules and natural compounds (Vanarase et al., 2010).

3) **Raman Spectroscopy**. Raman spectroscopy is a rapid, non-destructive, non-invasive method which does not require sample preparation and measurements can be done in aqueous environments and provides a molecular footprint of each sample. This technique is used as a technique of analysis applied inside and outside the laboratories, and is considered an analytical research tool very efficient in the analysis of different samples of different natures such as liquids, cells, materials, pharmaceuticals and bioprocesses. Current pharmaceutical applications include identifying polymorphs, monitoring real-time processes, detection of counterfeit and adulterated pharmaceutical products (Davis et al., 2007).

4) **Fluorescence Spectroscopy**. It is an analytical technique that offers a rapid and non-invasive analysis. It is considered one of the most sensitive spectroscopic approaches employed in the identification of organic compounds and can be used for qualitative and quantitative measurments. In the industry, fluorescence spectroscopy helps to obtain useful information from spectral data utilized in the characterization of food, drug, wastewater and natural samples. This technique can be applied to solid samples, powders, films and solutions of materials. On the other hand, several samples can be analyzed without any preparation, by placing samples on silica quartz disks or using a fiber-optic sampler or from a solution (Ahmad et al., 2017).

Electrochemical methods

The major goal of green chemistry is the omission of hazardous materials from the analytical method. In electrochemistry, for many years, mercury was the choice of electrode material due to its very attractive behavior but it is well known that mercury is one of the most dangerous heavy metals. In recent years, the electrochemical measurements using alternative work-electrodes have been studied such as bismuth-film electrodes that offer high-quality trace-metal measurements, carbon electrodes that present a wide potential window, chemical inertness, low cost and suitability for various sensing and detection applications, surface-modified electrodes that can be used to detect selective contaminants and exclusion of unwanted materials (Wang, 2002).

THE ROLE OF GREEN ANALYTICAL CHEMISTRY

According to the publication of the green chemist pioneer P. Anastas: "New methods and techniques that reduce and eliminate the use and generation of hazardous substances through all aspects of the life cycle of chemical analysis are the goal of Green Analytical Chemistry (GAC)" (Anastas, 1999). In other words, it is convenient to apply green chemistry principles to generate new methods and analytical techniques to real-time monitoring and prevention of pollution. Following this trend, Galuszka et al. suggested the principles of GAC (Table 13.1) to address the analytical chemistry of the elimination or reduction in the use of dangerous substances and preventing the generation of additional polluting chemicals, these principles also aim to reduce the energy consumption in the analysis and protection of the integrity and safety of the operator/worker (Gałuszka et al., 2013).

Table 13.1 12 Principles of Green Analytical Chemistry proposed by Galuszka, Migaszewski and Namiénski

1. Direct analytical techniques should be applied to avoid the sample treatment step
2. Minimal sample size and minimal number of samples are goals
3. *In situ* measurements should be performed.
4. Integration of analytical processes and operations saves energy and reduces the use of reagents
5. Automated and miniaturized methods should be selected
6. Derivatization should be avoided
7. Generation of a large volume of analytical waste should be avoided and proper management of analytical waste should be provided
8. Multi-analyte or multi-parameter methods are preferred versus methods using one analyte at a time
9. The use of energy should be minimized
10. Reagents obtained from a renewable source should be preferred
11. Toxic reagents should be eliminated or replaced
12. The safety of the operator should be increased

GAC involves the choice of the instrument, method, reagents, solvents, accessories and sample treatment in order to minimize the generation of residues at each step of the analysis and at the same time to be an instrument for monitoring other processes giving rise to a continuous cycle of prevention.

In fact, it is possible to observe that the conventional analytical methods present different disadvantages such as the use of high amounts of samples for analysis, abundant solvents and a high consumption of energy while the recent advances in analytical chemistry have allowed to improve and develop new precision, sensitivity and compact instruments that allow you to work at lower concentrations and with much less solvent, reducing the use of solvents in sample preparation and miniaturization of columns (Gallo and Ferranti, 2016).

After the optimization of analytical tools, the scientific community has shown great interest in the development of new greener simple, economic, rapid, easy, selective, accurate, reproducible, accurate, robust and indicative of stability for the rapid methods that allow monitoring a process with the same or better efficiency but reducing the environmental and economic impact. In sample preparation until the analysis, different approches have been explored. The new methods exclude or reduce the use of organic volatile solvents and in many cases they are replaced with water, ionic liquids or other ecological friendly solvents. Direct analysis methods, selective columns and acqueous eluents have been developed to minimize the derivatization reactions and when it is not possible greener derivatization reactions have been used (Eldin et al., 2016; Tobiszewski et al., 2009; Płotka-Wasylka et al., 2017; Korany et al., 2017; Galyan and Reilly, 2018; Lavilla et al., 2014; Armenta et al., 2015; Tobiszewski and Namiesnik, 2017; Koel and Kaljurand, 2006; de la Guardia and Garrigues, 2011b; Garrigues et al., 2010).

Sample preparation. Generally, the analysis begins with the treatment of the sample and its preparation for the further separation stage in the constituent components and which are subsequently appropriately detected so as to allow identification and quantification. At these classical stages it is also necessary to include the phases concerning sampling, handling and transport of the sample (Koel and Kaljurand, 2006). The latter can have a marked impact on the economics of the analytical process. In fact, sampling can contribute both to the energy cost and to the consumption of non-reusable materials (eg filters) (de la Guardia and Garrigues, 2011b). Many samples also need to be maintained under specific chemical-physical conditions by adding reagents or by thermostating them at a given temperature to preserve their characteristics. In these cases, energy expenditure is required as well as the use of chemical reagents. Finally, the transport of samples affects consumption of the analytical procedure.

Many samples must be maintained under specific chemical-physical conditions by adding reagents or by thermostating them at a given temperature to preserve their characteristics. In these cases it is necessary as an energy expenditure and the use of chemical reagents. However as regards the analysis of organic pollutants in environmental matrices, the number of direct analyzes is currently limited for which they use indirect methods carried out in the laboratory on aliquots of sample. Here it is possible to minimize the environmental impact of the analytical process.

Greener Sample Preparation. In a final product, the analyte under study is very complicated mixtures with interfering matrices not allowing the use of wasteless methods. A smart apporoach in the sample preparation is a decrease in sample volume needed for analysis that substantially decreases the generation of waste. Another way, is the use of alternative solvents and greener techniques to prepare the sample.

1) Greener solvents extraction. Water is the best known and most used solvent. It is at the same time a unique substance, non-toxic, easily available, economical, with a high boiling temperature and polarity. Extraction with water or superheated water as a solvent may be a potential alternative to conventional organic solvents (Płotka-Wasylka et al., 2017; Tobiszewski and Namiesnik, 2017). Despite the limited solubility of many organic compounds in water, the extraction of superheated water provides a selective, rapid and ecological method for the extraction of polar and moderately polar analytes from solid samples. Other alternatives as eco-friendly solvents are Ionic Liquids (ILs). In recent years they have attracted increasing interest due to their unique physicochemical properties. In addition, ILs have low vapor pressure, high thermal stability and a high degree of solubility for a wide range of organic and inorganic substances and miscibility in different solvents (Tobiszewski and Namiesnik, 2017). These properties make them excellent solvents for sample preparation. Also the properties of ILs have attracted the interest of analytical chemists not only in sample preparation but also as eluent in liquid chromatography. Surfactant agents also should be included in the green solvents group. Surfactants have been used for extraction for many years and a broad spectrum of sample preparation techniques has been developed. These techniques have advantages, such as low cost, nontoxic extractant, simplicity and high capacity to concentrate a wide range of analytes. A supercritical fluid such as carbon dioxide is commonly used in extraction processes for a wide range of compounds. Supercritical carbon dioxide posseses several advantageous such as low viscosity and high diffusivity, high extraction yields with smaller volume, not producing toxic residues, and is economical and environmentally friendly (Koel and Kaljurand, 2006).

2) Ultrasound extraction. For this type of extraction, the solvent allows the ultrasonic waves to propagate in the medium, consequently these sound waves collapse among themselves and generate the energy necessary to cause variations in the pressure and temperature of the system, furthermore it favors the contact of the solvent with the sample at a microscopic level and in a very short time. These factors favor the extraction of different types of compounds and preserve the integrity of all the molecules, whether thermolabile, thermostable, water-soluble or liposoluble (Tobiszewski et al., 2009). The extraction process assisted by ultrasound (UAE) reduces the operating time and increases the purity of the final product, simplifying the plant thanks to the reduction of solvent used and eliminating the post-treatment of wastewater. These goals are achieved by consuming only part of the energy needed for a conventional process.

3) Microwave extraction. The MAE (Microwave Assisted Extraction) technique is already widely used in the laboratory for the extraction of organic pollutants from different matrices and for the isolation of natural products (Tobiszewski et al., 2009; Lavilla et al., 2014). It is a reduction in the process times and solvent volumes used compared to the classic extraction conducted through other techniques. Also some years ago, the solvent-free microwave technology, SFME (Solvent-Free Microwave Extraction) was created, which allows extracting volatile substances from a matrix, without adding organic solvents or water.

Waste. Another often overlooked step is the treatment of generated waste from the analytical process itself and the possibility of including treatments of reclamation or recovery of waste products (Garrigues et al., 2010). Pretreatment of the samples is generally the stage that requires greater operator exposure to chemical reagents. In fact in this phase the raw sample is treated to remove potential interferents, reduce the complexity of the matrix and increase concentration analyte; these procedures generally require the use of organic solvents. It is therefore evident that an ideal protocol of analysis avoid both the sampling and the treatment phases and are therefore oriented to online and in-line direct survey methods, preferably to be performed in the field. The technology to operate them according to these criteria is available. In some cases direct methods of analysis, generally based on spectroscopic or electrochemical techniques can be used (de la Guardia and Garrigues, 2011a).

APPLICATION OF GREEN ANALYTICAL METHODS REAL-TIME ANALYSIS

Pharmaceutical industries. For the pharmaceutical industry, the synthesis of Active Pharmaceutical Ingredients (API) and the analytical investigation of bulk drug materials, intermediates, drug products, drug formulations, impurities and degradation products and biological samples containing the drugs and their metabolites is very important, and it allows obtaining crutial infomation about the safety and therapeutic efficacy of drugs and the production process (Raza Siddiqui et al., 2017). With the emergence of green chemistry and then the concepts of green analytical chemistry, conventional analytical tools and methods have been changed and optimized in such a way as to obtain reliable results but with a lower environmental and economic impact. Recently several works have been dedicated to the design and validation of analytical methods that follow the "green" trend. Yang and Doctor (Doctor and Yang, 2018) have reported the separation of aspirin and metformin HCl using a subcritical water chromatography on an XBridge C18 column at high temperature (95–125°C) and as an eluent a 100% phosphate buffer was used. As regards the traditional HPLC methods, the mobile phase is not toxic and is cheap. Results obtained by this method have shown a recovery rate of 99% for aspirin and metformin HCl demonstrating that this green separation technique for the separation and analysis of aspirin and metformin HCl is efficient and precise.

N. Haq et al. (Haq et al., 2017) proposed a robust, simple, selective, rapid, precise and greener RP-HPLC method applied to analyze olmesartan medoxomil in several commercially available formulations. The method was shown as very simple, rapid, sensitive, environmentally benign, without the need of sample preparation and an internal standard. In addition, the conventional solvents were replaced with new, innocuous and less toxic ones providing environmentally benign alternatives in the field of drug/pharmaceutical analysis.

Ferey et al. (Ferey et al., 2018) in their work, showed the utility of an analytical quality-by-design method to develop a specific and robust UHPLC method for the simultaneous analysis of 16 APIs with very different physicochemical properties using ethanol in mobile phases. The method reduces the environmental impact and analyst exposure due to the low toxicity of ethanol and is sourced from renewable sources and it can be considered a potential tool for the chromatographic analysis of complex mixtures in pharmaceutical sciences.

Two years ago, Leme de Figueiredo and Nunes Salgado (Leme de Figueiredo and Nunes Salgado, 2017) published a new analytical method to quantify aztreonam, a synthetic antimicrobial, in powder form using fourier-transform infrared (FT-IR) spectroscopy. The biggest advantage is that this technique is solvent-free, which contributes to minimize the generation of organic solvent waste and reduces the impact on the environment. The method based on absorbance measurements of the band corresponding to several functional groups presents accuracy and precision. This validated method can be used as an environmentally friendly alternative for the routine analysis in quality control.

In the same year, Kotadiya and Khristi (Kotadiya and Khristi, 2017) published a similar approach for quantitative analysis of an oral antidiabetic agent in which an active pharmaceutical ingredient is the teneligliptine hydrobromide hydrate by using FTIR spectroscopy for routine quality control testing. Also, in this case, several characteristic bands, which correspond to the antidiabetic agent, were selected for optimizationand validation of the method. The method includes the potassium bromide (KBr) pellets preparation of standards and samples, which means that it is solvent-free and due to the low toxicity of KBr, there is no environmental impact.

Foodomics. In the food industry , many types of analytical techniques are used, to mention a few which are spectroscopic such as Mass Spectrometry (MS), infrared (IR), High-Performance Liquid Chromatography (HPLC), Gas Chromatography (GC), Capillary Electrophoresis (CE), Supercritical Fluid Chromatography (SFC), and others (Pallone et al., 2018). But sample preparation in this field is the critical point because the analytes is inside a complex matrix and a sample treatment in most cases is inevitable. Nevertheless, several papers have been dedicated to minimizing the generation waste in this step using alternative methods such as Solid Phase Extraction (SPE); Supercritical Fluid Extraction (SFE) and microwave-assisted (MAE) and Ultrasound-Assisted Extraction (UAE) (García-Cañas et al., 2012).

F.-F. Chen et al. (Chen et al., 2018) developed a green two-dimensional HPLC-DAD/ESI-MS method for analyzing anthocyanins (type of pigment) from natural sources and improving their stability in energy drinks by the addition of phenolic acids. This method uses ethanol and tartaric acid solutions as mobile phases for quantitative analysis by HPLC-DAD and their identification using two-dimensional HPLC-MS.The results of this work showed that this method is promising for the identification of pigments in the food industry.

Some years ago, B.G. Botelho et al. (Botelho et al., 2015) in their publication proposed a new screening method using Attenuated Total Reflectance (ATR) mid-infrared spectroscopy. This method is characterized by being fast and simple, requires a small amount of the sample and does not require any pre-treatment before the analysis. Furthermore, the innovative method allows the detection of water, starch, sodium citrate, formaldehyde and sucrose in the raw milk of cow and it was considered suitable for quality control both for the production phase and for the inspection of the milk already on the market.

Cosmetics. In trials of green cosmetics analysis, researchers have performed and proposed green HPLC methods to determine several UV filters (that are authorized worldwide in sunscreen formulations), using environment-friendly solvents (Chisvert et al., 2001, 2013; Salvador and Chisvert, 2005; De Orsi et al., 2006). Also several works have been dedicated to developing an environment-friendly determination of ascorbic acid and its derivatives for skin-whitening cosmetics without using highly toxic organic solvents or hazardous chemicals (Balaguer et al., 2008).

Conclusions

In recent years the scientific community has presented numerous scientific publications concerning green chemistry and how they have positively contributed to drawing the attention of chemists to the well-being of the society in which they operate.

Analytical chemistry has contributed and can make further efforts to increase the dissemination of these concepts and research in this area. Significant progress has been made to make many of the analytical methodlogies more sustainable and scientific research shows a marked interest in the development of alternative and innovative approaches that combine the principles of green chemistry with the increase in the performance of the analysis method. This objective is achieved through the development of methods capable of reaching even lower quantification limits and with a concomitant reduction in the consumption of organic solvents/ reactants, analysis times and energy consumption of the entire analytical process.

However, it must be remembered that in addition to innovative developments, the analytical community also needs more routine operations, such as validation, without which the proposed methods cannot find a wide application and therefore an appreciable effect on the society at large.

REFERENCES

Ahmad, M.H., Sahar, A., Hitzmann, B. Fluorescence spectroscopy for the monitoring of food processes. *Advances in Biochemical Engineering/Biotechnology*, 2017, 161, 121–151.

Anastas, P.T., Green chemistry and the role of analytical methodology development. *Critical Reviews in Analytical Chemistry*, 1999, 29, 167–175.

Armenta, S., Garrigues, S., de la Guardia, M. The role of green extraction techniques in green analytical chemistry. *Trends in Analytical Chemistry*, 2015, 71, 2–8.

Balaguer, A., Chisvert, A., Salvador, A. Environmentally friendly LC for the simultaneous determination of ascorbic acid and its derivatives in skin-whitening cosmetics. *Journal of Separation Science*, 2008, 31, 229–236,

Botelho, B.G., Reis, N., Oliveira, L.S., Sena, M.M. Development and analytical validation of a screening method for simultaneous detection of five adulterants in raw milk using mid-infrared spectroscopy and PLS-DA. *Food Chemistry*, 2015, 181, 31–37.

Chen, F.-F., Sang, J., Zhang, Y., Sang, J. Development of a green two-dimensional HPLC-DAD/ESI-MS method for the determination of anthocyanins from Prunus cerasifera var. atropurpurea leaf and improvement of their stability in energy drinks. *International Journal of Food Science & Technology*, 2018, 53, 1494–1502.

Chisvert, A., Pascual-Martí, M.C., Salvador, A. Determination of the UV filters worldwide authorised in sunscreens by high-performance liquid chromatography: Use of cyclo-dextrins as mobile phase modifier. *Journal of Chromatography A*, 2001, 921 207–215.

Chisvert, A., Tarazona, I., Salvador, A. A reliable and environmentally-friendly liquid-chromatographic method for multi-class determination of fat-soluble UV filters in cosmetic products. *Analytica Chimica Acta*, 2013, 790, 61–67.

Davis, K.L., Kemper, M.S., Lewis, I.R. Raman spectroscopy for monitoring real-time processes in the pharmaceutical industry. *In*: Šašić, S. (ed.), Pharmaceutical Applications of Raman Spectroscopy. New Jersey, United States: John Wiley & Sons, Inc.; 2007; pp. 117–162. DOI:10.1002/9780470225882

de la Guardia, M., Garrigues, S. (eds), Challenges in Green Analytical Chemistry. The Royal Society of Chemistry, UK, 2011a; pp. 13–43.

de la Guardia, M., Garrigues, S. (eds), Challenges in Green Analytical Chemistry, The Royal Society of Chemistry, UK, 2011b; pp. 286–301.

de Marco, B.A., Rechelo, B.S., Tótoli, E.G., Kogawa, A.C., Nunes Salgado, H.R. Evolution of green chemistry and its multidimensional impacts: A review. *Saudi Pharmaceutical Journal*, 2019, 27, 1–8.

De Orsi, D., Gagliardi, L., Porra, R., Berri, S., Chimenti, P., Granese, A., Carpani, I., Tonelli, D. A environmentally friendly reversed-phase liquid chromatography method for phthalates determination in nail cosmetics. *Analytica Chimica Acta*, 2006, 555, 238–241.

Doctor, N., Yang, Y. Separation and analysis of aspirin and metformin HCl using green subcritical water chromatography. *Molecules*, 2018, 23, 2258–2266.

Dong, M.W., Zhang K. Ultra-high-pressure liquid chromatography (UHPLC) in method development. *Trends in Analytical Chemistry TrAC*, 2014, 63, 21–30.

Eldin, A.B., Ismaiel, O.A., Hassan, W.E., Shalaby, A.A. Green analytical chemistry: Opportunities for pharmaceutical quality control. *Journal of Analytical Chemistry*, 2016, 71, 861–871.

Ferey, L., Raimbault, A., Rivals, I., Gaudin, K. UHPLC method for multiproduct pharma-ceutical analysis by Quality-by-Design. *Journal of Pharmaceutical and Biomedical Analysis*, 2018, 148, 361–368.

Gałuszka, A., Migaszewski, Z., Namiesnik, J. The 12 principles of green analytical chemistry and the SIGNIFICANCE mnemonic of green analytical practices. *Trends in Analytical Chemistry TrAC*, 2013, 50, 78–84.

Gallo, M., Ferranti, P. The evolution of analytical chemistry methods in foodomics. *Journal of Chromatography A*, 2016, 1428, 3–15.

Galyan, K., Reilly, J. Green chemistry approaches for the purification of pharmaceuticals. *Current Opinion in Green and Sustainable Chemistry*, 2018, 11, 76–80.

García-Cañas, V., Simó, C., Herrero, M., Ibáñez, E., Cifuentes, A. Present and future challenges in food analysis: Foodomics. *Analytical Chemistry*, 2012, 84, 10150–10159.

Garrigues S., Armenta, S., de la Guardia, M. Green strategies for decontamination of analytical wastes. *Trends in Analytical Chemistry TrAC*, 2010, 29, 592–601.

Haq, N., Iqbal, M., Alanazi, F.K., Alsarra, I.A., Shakeel, F. Applying green analytical chemistry for rapid analysis of drugs: Adding health to pharmaceutical industry. *Arabian Journal of Chemistry*, 2017, 10, S777–S785.

Koel, M., Kaljurand, M. Application of the principles of green chemistry in analytical chemistry. *Pure and Applied Chemistry*, 2006, 78, 1993–2002.

Korany, M.A., Mahgoub, H., Haggag, R.S., Ragab, M.A.A., Elmallah, O.A. Green chemistry: Analytical and chromatography. *Journal of Liquid Chromatography & Related Technologies*, 2017, 40, 839–852.

Kotadiya, M., Khristi, A. Quantitative determination and validation of teneligliptine hydrobromide hydrate using FTIR spectroscopy. *Journal of Chemical and Pharmaceutical Research*, 2017, 9, 109–114.

Lavilla, I., Romero, V., Costas, I., Bendicho, C. Greener derivatization in analytical chemistry. *Trends in Analytical Chemistry TrAC*, 2014, 61, 1–10.

Leme de Figueiredo, A., Nunes Salgado, H.R. Validation of a green analytical method for the quantitative analysis of antimicrobial aztreonam in lyophilized powder for injection by fourier-transform infrared spectroscopy (FT-IR). *E-Cronicon Microbiology*, 2017, 8, 254–265.

Nuttall W.J., Clarke, R.H., Glowacki, B.A. Stop squandering helium, *Nature*, 2012, 485, 573–575

Pallone, J.A.L., Carames, E.T.S., Alamar, P.D. Green analytical chemistry applied in food analysis: Alternative techniques. *Current Opinion in Food Science*, 2018, 22, 115–121.

Płotka-Wasylka, J., Rutkowska, M., Owczarek, K., Tobiszewski, M., Namiesnik, J. Extraction with environmentally friendly solvents. *Trends in Analytical Chemistry TrAC*, 2017, 91, 12–25.

Raza Siddiqui, M., AlOthman, Z.A., Rahman, N. Analytical techniques in pharmaceutical analysis: A review. *Arabian Journal of Chemistry*, 2017 10, S1409–S1421.

Salvador, A., Chisvert, A. An environmentally friendly ("green") reversed-phase liquid chromatography method for UV filters determination in cosmetics. Chisvert. *Analytica Chimica Acta*, 2005, 537, 15–24.

Taylor, L.T., Supercritical fluid chromatography for the 21st century. *The Journal of Supercritical Fluids,* 2009, 47, 566–573.

Teutenberg, T. High-Temperature Liquid Chromatography A User's Guide for Method Development., RSC Chromatography Monographs. Published by the Royal Society of Chemistry; 2010. www.rsc.org

Tobiszewski, M., Mechlinska, A., Zygmunt, B., Namiesnik, J. Green Chemistry in sample preparation for determination of trace organic pollutants, *Trends in Analytical Chemistry TrAC*, 2009, 28, 943–951.

Tobiszewski, M., Namieśnik, J. Direct chromatographic methods in the context of green analytical chemistry. *Trends in Analytical Chemistry TrAC*, 2012, 35, 67–73.

Tobiszewski, M., Namiesnik, J. Greener organic solvents in analytical chemistry. *Current Opinion in Green and Sustainable Chemistry,* 2017, 5, 1–4.

Turner, C., Sustainable analytical chemistry—more than just being green. *Pure and Applied Chemistry*, 2013, 85, 2217–2229.

Vanarase, A.U., Alcalà, M., Jerez Rozo, J.I., Muzzio, F.J., Romañach, R.J. Real-time monitoring of drug concentration in a continuous powder mixing process using NIR spectroscopy. *Chemical Engineering Science*, 2010, 65, 5728–5733.

Venter, A., Nefliu, M., Cooks, R.G. Ambient desorption ionization mass spectrometry. *Trends in Analytical Chemistry TrAC*, 2008, 27, 284–290.

Wang, J., Real-time electrochemical monitoring: Toward green analytical chemistry. *Accounts of Chemical Research*, 2002, 811–816.

Yabré, M., Ferey, L., Somé, I.T., Gaudin, K., Greening Reversed-Phase Liquid Chromatography Methods Using Alternative Solvents for Pharmaceutical Analysis. *Molecules*, 2018, 23, 1065(1-25).

Chapter 14

Inherent Safer Chemistry for Accident Prevention

**M. Andrade-Guel[1], C. Cabello-Alvarado[2]*,
Carolina Caicedo[2], Leticia Melo[2] and C. Ávila-Orta[2]**

[1]Departamento de Materiales Avanzados,
Centro de Investigación en Química Aplicada. Blvd.
Enrique Reyna Hermosillo No.140, San Jose de los Cerritos,
CP. 25315, Saltillo, México.
[2]CONACYT-Consorcio de Investigación y de Innovación del Estado de Tlaxcala.
Calle 1 de mayo No. 22, Colonia Centro.CP.90000, Tlaxcala, Mexico.

INTRODUCTION

The most significant step to minimize the risk of accidents in the industry, laboratory and/or company is the design of a sustainable eco-friendly process. These accidents include risks of fire, explosions and emissions to the environment caused by chemical reagents, liquids, gases and halogenated substances.

To achieve balance between accident prevention and pollution control the procedures for obtaining a product must be modified or discarded to use different types of materials, with the purpose of decreasing a wide range of risks on chemical substances, processes and to consider avoiding contamination in its diverse ways. This chapter shows strategies for prevention of accidents as well to apply the approach in 12 different areas such as the health, food, automotive and textile industries.

During the technological development of industrial processes, acquisition of devices, equipment and implementation of more efficient safety systems have been

*For Correspondence: christian.cabello@ciqa.edu.mx

achieved with the aim to reduce and prevent accidents in the industry. However, in the last years significant accidents have occurred and industries like chemical, oil, metal processing, production and explosive storage sector are among the ones with the highest number of accidents (Abidin et al., 2018).

Due to the reasons mentioned above there is an increasing need of processes, synthesis and technological developments to be oriented towards eco-friendly chemistry, by proper selection of reagents, raw materials, use of solvents and catalysts, as well as the process design itself, thus preventive actions must be found within the 12 principles of eco-friendly chemistry. This is considered an effective strategy to obtain greater yield, to minimize byproducts and to avoid damage to the environment, as well as to prevent accidents (García et al., 2007; Kirchhoff, 2003).

Figure 14.1 shows the parameters to be taken into account for the design of a safe chemical process to avoid accidents. First, a chemical inventory must be performed, this means to make a list of reagents and materials to be used in the process, considering their hazard for classificaiton according to toxicity, pressure, reactivity, temperature and flammability (Edwards and Lawrence, 1993).

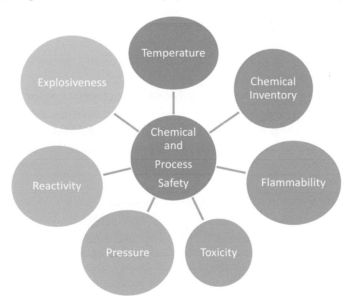

Figure 14.1   Scheme of parameters to consider in developing a chemical inventory.

Another significant strategy is to calculate the index of an ambient risk, this study will avoid failures or accidents in the chemical industry. Environmental risk index is used to assess risks for using reagents, raw materials and during their handling, besides inspecting in general the process route. The method for defining such an index is simple, fast and efficient and takes into account the amount of accidents and damage to the environment by exposure to chemical substances (Cave and Edwards, 1997 Warnasooriya and Gunasekera, 2017).

The safety goal is to prevent accidents, damages and ambient pollution. Primary prevention encourages changes in production technology, replacing consumables for eco-friendly options, redesign and reengineering processes, as well as reformulation of products for eliminating toxic compounds (Ashford, 2013). Control systems can be used in the industrial process to reduce accidents, for example relief valves in process, alarm systems, measurements of human and organization control like defining work instructions, safety culture and use of protection equipment. Besides feedback of accidents where accident reports are researched, data is collected and processed, defining lessons learned, information is distributed, a solution and decision are given on prevention actions and finally proceeding to their implementation (Kidam et al., 2014).

On the other hand if dangerous materials are handled, some safe solutions can be the following: to minimize dangerous material using only a small portion of it in order to avoid significant risk, replacing the harmful product by a non-harmful or attenuating or moderating the product or element; this means to use dangerous material in a less risky way with storage at a low temperature and low pressure, and also to corroborate refrigeration equipments having no leakages (Kletz, 2003). Considering some of these solutions and the organization of chemical industry, inherent safety must be followed when handling dangerous materials. Four catergories have been classified to reduce accident frequency, in descending order of feasibility and strength:

- **Inherent**. Danger elimination by use of non dangerous materials and processes.
- **Passive**. Minimize risk by a characteristic process and equipment design.
- **Active**. Use of controls, emergency shut off systems, mitigation devices like sprinklers, active systems to detect danger and take corrective actions.
- **Procedure**. Using operative and administrative procedure like response controls to face emergencies and management to prevent accidents (Hendershot, 1997).

Another way to prevent accidents is to replace explosive and flammable organic solvents by new solvents, called ionic liquids that have been developed like alternative solvents for organic synthesis. These compounds are advantageous as they have very low pressure value and therefore are non volatile (Warner et al., 2004).

In last few years inherent safety and chemical engineering have been present in the process design, while the conventional process design dealt with problems to the extent of the limitations and opportunities detected, safety had a passive role such as verification at the end of the process; in contrast, in actual design from the beginning inherent safety is considered for elimination or reduction of dangerous materials and has several levels of protection throughout the design process (Palaniappan et al., 2004).

An eco-friendly technology approach has become a need for the industry of the future, since design of safer methods for staff personnel and the need to manufacture materials for different applications like health, pharmaceutical, treatment and water purification, biosensors and food production.

HEALTH SAFETY

Pharmaceutical industry deals with development, manufacturing and marketing of medicine for human health, but is also considered one of the most pollutant industries worldwide and with more accidents when working with toxic substances (Catensson, 2008). Solvents used in this industry cause a big pollution problem and can cause accidents since most of organic compunds are flammable.

In organic synthesis of pharmaceutical products solvents play an essential role since they are used for dissolving, extracting, washing, separating, dispersing and cleaning products. In consequence, green chemistry proposes more clean solvents like ionic liquids and water. And it has to be natural, cheap, easily available and non toxic in its handling (Li and Trost, 2008).

Medicine manufacturing in a continuous flow within pharmaceutical chemistry, is a way to minimize the risk of creating exothermal and unstable energy. It can accelerate production from gram level to kilogram level, and besides can be performed in small reactors at low pressure, therefore volatile gases of a toxic and flammable reagent like hydrogen, cyanide, hydrazoic acid, can be kept in solution during the reaction. Rate increase does not affect reagents and avoids gases from reaching a high concentration level, conditions under which explosion can occur.

Conditions for continuous flow process can modify dangerous chemical conditions and offer quality in intermediate and final products, which are significant for patient safety (Martin et al., 2018).

Green chemistry can be an important partner in pharmaceutics since it offers significant benefits to improve the environment and products at a lower cost with excellent quality in production line, with the regulations and other legislative approachs.

Some examples of green chemistry are: small molecules development, synthesis with green solvents in the engineering field like continuous flow chemistry (Koenig and Dillon 2017). One if its principles is accident prevention eliminating toxic and flammable products, which can lead to the design of small molecules with less toxicity, using continuos flow chemistry, and that are biodegradable and do not pollute the environment.

Theoretical chemistry is a useful tool for molecules design, mainly for schemes of pharmaceutical products like Quantitative Structure-Activity-Relationship (QSAR), help in the design of new molecules with some biological activity (Leder et al., 2015). This philosophy must also be implemented in the analysis of pharmaceutical products, in methods of analysys by HPLC the proposed methods with no toxic and flammable products, as well as more eco-friendly solvents. An example of this is the analysis of clonazepam pills, where a simple and effective method was proposed, where some of the advantages are simple extractions, low retention times, an internal standard is not required, green solvents that help in personnel safety and accidents prevention (Eldin et al., 2014). Specifically some studies in the health sector like chromatography, need solvents with high purity, one way to eliminate or minimize the use of toxic solvents in a moving phase is to achieve a method that reduces retention times and the amount of the solvent, in the case of toxic non-polar solvents like diclometane can be replaced by isopropyl alcohol in heptane or a mixture of ethyl acetate with ethanol (Tobiszewski and Namieśnik, 2017).

Flammability and explosion risk of solvents used in analysis of pharmaceutical products is high, due to this use of solvents like water, ethanol, isopropyl alcohol and ionic liquids are preferred, and have several advantages like flammability, at ambient temperature, vapor low pressure and can be reusable (Mohamed, 2015).

Rivaroxoban is a drug used in thromboembolic diseases, the synthesis method is complex and in general has a low yield with impurities. For producing it, a more efficient method has been developed with 99.85% yield, designing a process that avoids use of dangerous chemical products, critical operations and complex tasks. For impurity control, reaction conditions are optimized, and crystallization parameters are adjusted in order to obtain the final product at less cost and without affecting the environment (De Marco et al., 2018).

FOOD SAFETY

Food safety these days is a priority, due to the market and its numerous consumers, this area has several risks such as: microorganism contamination, waste contamination, chemical products, incorrect labeling and physical deterioration (Wieczerzak et al., 2016). The responsibility for food being safe is shared between the government and food companies, making it fundamental to define safety policies, put in practice controls in the process, fulfill the defined standards for accidents prevention.

Some defects found in food industry are:

- Lack of knowledge in operation rules.
- Low levels of basic food hygiene.
- Low alphabetization levels and little training.
- Excessive use of toxic chemical products mainly in the farms.
- Personnel protection equipment (Mali et al., 2015).

A recent study corroborated using a Bayesian network to identify and how to quantify relationships and interactions of risks for food safety, using a notification in a fast alert system, for fruits and vegetables. Weather, soil contamination, economical data and agronomic studies are factors to consider. Building this network allows gathering data from product category, country, daily production, environmental risks, and also allows assessing risks with precision of 95% (Rastogi et al., 2014).

Contaminated food is a risk for consumer health, in general food is contaminated with heavy metals coming from irrigation water or from the soil where it is harvested, one of the techniques used for extracting arsenic from rice is microextraction with hydrolytic enzymes and ultrasound radiation, this eliminates toxic solvents, and is considered an innovative, eco-friendly and simple method and has no risks to health. After performing extraction from samples, these undergo coupled mass spectrometer analysis for quantitatively defining the amount of arsenic in rice (Simijonović et al., 2018). A way to avoid food contamination is packaging, which is a food protection and quality system, where several materials have been used for food packaging like glass, metal, paper, carton and plastics. Ruling agencies worldwide assess food packaging safety, like FDA, Canda Health and European Commision, but they differ in the application of methods for measurement of migration of packaging materials towards food (Spiric et al., 2015). Actually, there are required standards for packing manufacturing, food harvest and food treatment, as well as for analysis methods with an eco-friendly approach, since the food industry represents health risks for workers, who must follow strict handling in food safety, as well as for human and animal consumption. Since 2001 the food industry has seen an increase in leave of absence by disease on a long-term basis and permanent incapacity caused by exposure to chemicals, use of sharp tools and close contact to equipment. A study about safety in three different fields of food industry has been taken: red meat, bakery and dairy products. These include manual tasks like production, packaging, assembly line, maintenance and reparation. The automation level varies according to the final product, but it was noted in general that the safety approach was from a technical perspective, using a passive system control forming barriers or protection equipment and that companies usually lack safety systems knowledge (Khalid, 2016).

To solve this problem a more efficient safety system must be implemented, replaced or to minimize the risks, decreased exposure to toxic solvents and reagents, as well as to pesticides and fertilizers used in intensive agriculture. In 2012 worlwide consumption of nitrogenated fertilizers was 100 million tons, although industrial agriculture has successfully eliminated millions of these practices, the intensified use of these kinds of fertilizers causes great damage to the environment, some approachs to reduce pollution and guarantee food safety are:

- Improve independent productivity increasing yields can reduce the use of soil and mitigate emissions.
- Use of agricultural wastes like raw material for biofuel production.
- Change in the diet, consuming more vegetables and less processed products.
- Improve efficiency in consumption of agrochemicals and replace them by natural extracts (Bouzembrak and Marvin, 2019; Yilmaz, 2018).

SAFE CHEMISTRY IN NANOMATERIALS

Nanotechnology has been considered a multidisciplinary field, in which knowledge and methodologies of physics, chemistry, biology, engineering and materials science get together. Nanotechnology allows understanding and manipulating matter at the nanometric level (Karmaus et al., 2018; Stave and Törner, 2007).

Chemical synthesis and functionalization of nanoparticles has become a great challenge since it involves expensive techniques and toxic substances, therefore, for nanoparticles synthesis use of renewable and eco-friendly reagents have been proposed, as well as using non-harmful solvents, alongwith alternative energies like microwave and ultrasound to avoid byproducts generation (Yang et al., 2012).

Gold, silver and copper nanoparticles have diverse applications that are actually promising, which is why the method based on green chemistry principles has been developed to eliminate use of reagents like hydrazine, formaldehyde, sodium boroydrate and aniline, which are toxic, flammable and noxious to health substances, to replace them by natural source products (Sun et al., 2017). Bacteria, fungi, plant extracts, yeast and algae have been used for the nanostructure synthesis process which include renewable materials, low temperature and pressures, as well as safe conditions in the procedure for accidents prevention (Olawoyin, 2018). Chenopodium album is an extract used in synthesis of spherical gold and silver nanoparticles, obtaining sizes from 10–30 nm, in this study temperature, pH, amount of extract and metallic ions concentrations were assessed. It was defined that when the metallic ions concentrations increase, the particle size also increased (Asmatulu et al., 2013).

Another extract used in silver nanoparticles is Raphanus Sativus, the size of synthesized nanoparticles was from 3 to 6 nm. With an increase of ions concentrations, the extract increases particle size (Villaseñor and Ríos, 2018).

Studies using *Capsicum annuum* L. extract report the formation mechanism of silver nanoparticles, by reduction, limited nucleation and growth in solution (Banach and Pulit-Prociak, 2017), that can be simplified in three important steps, the first is called induction phase and involves a fast reduction of ions and nucleation, then the unstable small crystals are added and spontaneously transform into big agglomerates (growth phase), when the size and shape of the agglomerates turn into favorable energy, some biomolecules act like trapping agents and stabilize nanoparticles (termination phase), the plants metabolism and how they interact in synthesis is still unknown (Kadziński et al., 2018).

Metallic nanoparticles have been synthesized with plants extracts like biological materials, due to their simplicity and ecology, the obtained reaction yields are high and reaction conditions are not aggressive, using solvents that are not considered toxic to avoid dangerous compounds. Examples of this kind of synthesis are silver nanoparticles synthesized from Gongronema latifolium extract, belonging to Asclepiadaceae family, a food plant of medicinal origin (Mageswari et al., 2016).

Besides plants extracts, fruits from a plant have been used for ZnO nanoparticles synthesis, like Averrhoa bilimbi, a tropical tree that produces mature fruits that contain diverse phytochemicals, in these studies the reaction safety outstands and use of non flammable reagents allow to implement the green chemistry approach (Khan et al., 2015).

Conventional methods for nanoparticles synthesis use toxic and flammable reagents like sodium borohydride, hydrazine and dimethylformamide, the solvents are organic and flammable. The use of non toxic reagents, biodegradable material, green solvents like water, are fundamental for nanoparticles synthesis and processing to provide safe conditions and eco-friendly processes (Marchiol et al., 2014).

There are three steps to implement eco-friendly technologies to nanomaterial synthesis using biological systems:

- Selection of a green solvent,
- Selection of an ecological product,
- Finding an environmentally benign reducing agent.

It also needs to taken into account the selection of non toxic material like a coating reagent that allows to stabilize nanoparticles. This technology has advantages compared to direct methods, such as biological components that are available for nanoparticles formation, are ecologically benign and can be used in diverse medical applications (Issaabadi et al., 2017).

Chemical functionalization of carbon based nanomaterials in general is made with toxic and corrosive substances like inorganic acids, mixtures of nitric and sulfuric acids, defects in graphite network occur during long reaction times, therefore for this kind of procedure, care must be taken with experimental conditions to prevent accidents (Ramanarayanan et al., 2018).

In a recent study nanopellets of graphene with gallic acid were functionalized, in place of inorganic acids. Gallic acid is a natural product found in polyphenol extracts, in green tea, berries, grapes, wine and some wood plants, thanks to its structure and green properties this acid is an alternative for graphene nanopellets functionalization, it is worth mentioning that this processing is of low cost and easily scalable to the industrial level (Shameli et al., 2012).

Other kinds of nanometric structures are carbon nanotubes that can have two types of functions: covalent and non-covalent. These modifications are performed by processes involving high temperatures and can last 48 hours, as well as toxic products that can cause accidents, in order to prevent these energy sources like ultrasound and microwaves have been used (Cabello et al., 2014).

More friendly chemical modification methods have been promoted instead of conventional methods, with the approach to reduce toxic and flammable reagents that can cause accidents, and cleaner energy sources like ultrasound, treatment with organic acids like citric and acetic for modification of multiple wall carbon nanotubes has been promising for giving a yield of 19.7% in carboxyl groups on the surface when acetic acid is used and 8.9% of carboxylic groups using citric acid (Price et al., 2018). Modification involving chemical reduction and/or oxidation allows important nanoparticles dispersion for diverse applications. Graphene particles reducing reagents have been used to eliminate these kinds of compounds harmful for human health, it was proposed to use graphene oxide found in a green tea solution, instead of using aromatic compounds like polyphenols, thus guaranteeing proper dispersion of graphene oxide in organic and aqueous solvents (Andrade et al., 2012).

On the other hand, development of energy materials for space and defense applications has focused on producing eco-friendly materials, affecting their processing and production. Some strategies for adopting a green chemistry approach in materials for energy production are:

- To assess environmental impacts and problems related to production of materials for energy and define opportunity areas.
- To change explosive bonding by thermoplastic elastomers that have good mechanical properties and recycling.
- To use technologies that use CO_2 and supercritical enzymes for synthesis of materials in energy.
- To promote biodegradation of materials to relieve problems of water and soil contamination.
- To use less solvents in preparation of weapon propellants (Sadri et al., 2017).

There are other strategies to follow when nanometric size materials are available, these prevention measurements were published in 2007 Nanomaterials Forum for environment, health and safety, and the following are some among them:

- Written procedures must be documented, among them are the required engenineering controls, personnel care, safe storage methods, nanoparticles handling and exposure, high efficiency particles filter.
- Clothes for personnel protection must be used every time to minimize skin contact.
- Dry nanoparticles, or in liquid form, must be treated like special waste (Rajasekhar and Kanchi, 2018).

The effects of nanoparticles in the environment and health are still not conclusive, since they can enter the human body through the lungs, intestines and to a lesser extent through the skin, however they have some advantages in nanotechnology development in the health area, resulting in properties like high surface area and the possibility to cross the hemotoencephalic barrier, some nanoparticles have antimicrobian properties, helping in applications like medication administration, disinfection, tissue imaging and reparation (Osman, 2008).

Green chemistry is an alternative to develop synthesis methods safer for the personnel working with nanoparticles, minimize byproducts, use renewable or biological origin reducing reagents, green solvents like water, replacement of toxic reactives by eco-friendly reactives, safe disposal of wastes, engineering design of continuous flow and processes more friendly with the environment (Talawar et al., 2009).

Safety in the Automotive Industry

There are many industries in the manufacturing sector that demand the use of harmful chemicals in their production processes and the automotive industry is one of them, the operation requires employees to work with chemical substances that are dangerous and with a high risk for health in case of exposure. These chemical products have several forms; they can be solid, dust, liquid or gases, pure chemical products or mixtures. Depending on their physical version, they can enter the body in diverse ways, depending on handling or exposure levels. In many cases, when people perform their activities the kind of exposure is partial for noxious chemical products and the effects can appear after some years, and finally affecting health. Common risks from work that involves handling dangerous chemicals include poisoning, irritation, chemical burns, cancer and genetic disorders. In this way, the main adsorption paths are ingestion, dermal, parenteral and inhalation, this last is the most common and important since substance penetration begins in the lungs and then to the blood, and could affect other organs such as the brain, liver, kidney or cross the placenta.

Industries constantly make sustainability studies, technological innovation, economical dynamics, environmental issues and financial elements for which tools like Life Cycle Assessment (LCA), Multiple Criteria Decision Analysis (MCDA) are used as well as the method of sustainable development division for accountability of environmental management (EMA), Data Envolvent Analysis (DEA), Process Analysis Method (PAM), Industrial Ccology (EI), among others (Stoycheva et al., 2018; Manley et al., 2008). The results of these studies are consider technical, economical, social and ambiental factors where the use of clean technologies and the alternatives of materials are to a great extent suggest fulfilling normatives, increasing competitivity and productivity (Ljungberg and Edwards, 2003). Despite these efforts the earlier criterion did not completely involve minimization of risks of chemical accidents to prevent adverse occupational events.

The automotive industry is complex since it includes a series of interconnected companies that collaborate in the chain supply, which starts with extraction of raw materials up to the finishing of the final product , including users and operators for maintenance or reparation (Slušná and Balog, 2015). The automotive industry has suffered constant changes in the last two decades due the exigence of global economy, as well as improvements in science and technology mainly in autoparts development. The following are some examples related to safety aplications for this sector:

Friction Elements

Brake pads are used in the automobile and other vehicles brake systems to control speed converting kinetic energy in heat that dissipates to the atmosphere. Brake pads are steel support plates with friction material bonded to a disk surface (Idris et al., 2013). This system needs to keep the friction coefficient stable and be high enough in the disk (Efendy et al., 2010).

Brake pads in general consist of asbestos fiber embedded in polymeric matrix among other substances, which are manufactured molding in a plate at a high temperature and high pressure conditions. Most automotive brake linings bonded to steel brake pad are made of shorter fibers (<5 μm) by means of extrusion process. Brakes contain 30 to 70% of asbestos according to application (Pye, 1979; Rosato, 1959). Linings of drum brakes contain around 60% of asbestos, and disk brake pads, from 25 to 30%. It is worth noting that formulas vary according to the vehicle type.

Actually there are strict measurements for monitoring mounting, inspection, disassembly and the repairs of friction systems in cars due the asbestos and other fibers content, since they are considered as a threat to health (Gómez et al., 2012; Sulivan and Krieger, 2001,) for being the cause of asbestosis, mesothelioma and lung cancer in humans according to Occupational Safety and Health Administration (OSHA) (Gaggero and Ferretti, 2018; OSHA, 1995). The values for threshold limit for breathable asbestos and synthetic vitreous fibers (wool, glass and slag) are 0.1 and 1 f/mL, respectively (Limón et al., 2013). The standard method to define occupational exposure to asbestos in most of countries is the PCM method (Marioryad et al., 2011). Although this method is a proper exposure index in the labor area, its counting scheme can only determine total concentration (NIOSH, 1998; OSHA,1997). To surpass these limitations characterization techniques have been implemented that define the composition and morphology of asbestos subtypes, from assessment of raw materials and samples in areas with the purpose of understanding the effects that different microstructures in asbestos cause to health.

Asbestos covers a group of metamorphic mineal fibers (silicates) well known by their high mechanical resistance, electrical, chemical and high fusion point. The epidemiological importance of this substance lies in the wide range of exposure sources from their near of 1500 industrial applications such as building, shipbuilding, railway and automotive industry, plastics, chemical and food products (Virta, 2005). Metallurgy, and all types of flame-retardant fabrics (Gilson, 1989). There are two kinds of exposures: occupational and ambiental, the first is strictly occupational when the mineral is directly handled by an employee, or para-occupational by proximity with other people in the same workplace. This last can be domestic, by inhalation of asbestos fibers in the home after being transported by workers in their working clothes, or by aging of building materials. Ambient exposure can also be geographical, urban or industrial, and can occur by ingestion of liquid or solid products that could contain the mineral. Besides implementing protection elements and redesigning old products, this problem can be solved by the design of new products (Rao and Babji, 2015). Within new design products several asbestos substitutes have been assessed like aramid fibers, cellulose fibers, ceramic fibers, fibrous glass, graphite flakes and fibers, polypropylene fibers, mica, polytetrafluorethylene fibers, steel fibers and wollastonite. Examples of alternative products that cover other sectors include aluminum, vinile and wood coating, aluminum tube and sheet, asphalt coatings, ductile iron tubes, fiberglass sheet, polyvinyl chloride tube, prestressed and reinforced concrete tubes, semimetallic brakes, urethane coatings and vinyl flooring (Neira, 2015). However, neither substitute has proven to be as versatile as asbestos.

These alternative materials have few regulations for occupational exposure, and the potential effects to health in the long term are not fully documented. In general, some studies conclude that significant risks are caused by those fibers that are breathable, with sizes from $1 > 5$ μm, $d < 3$ μm, $l/d \geq 3$ (Virta, 2005). In other applications, some alternative materials of high value are: amorphous silica fabrics, having high thermal resistance, widely used in the aerospace industry. Cellulose fibers composed by a ~85% of recycled material (Ikpambese et al., 2016). Polyurethane foam widely used as isolation in the automotive, aeronautical and building sectors produces low heat conduction and offers an excellent thermal resistance by trapping gas bubbles within a foam structure. The non-interconnecting cells of polystyrene also provide this property. The thermostable plastics of flour and refilled with flour are produced from natural materials (based on rice, nuts, wheat, and others and their corresponding ashes) are useful as thermal and/or electrical isolation to repair cracks in buildings or applied in autoparts.

There are several patents for asbestos free organic materials with good properties for friction (Dagwa, 2005). The changes in formulas of brake pads were also driven by release of average corporative requirements for fuel efficiency at the end of the decade of 1970 and beginning of the decade of 1980. These requirements lead automotive industry to change from rear wheel drive to front wheel drives. This modification required more frontal brake, which lead to higher temperatures and preference for semimetallic brakes (Gudmand-Høyer et al., 1999).

Paints, Adhesives and Aealants

The paint process in automotive assembly companies is the neural area for finished product quality, but is also the area that requires more caution on safety and health, especially when working processes are not automated. In particular, micro-companies that offer bodyshop services are among these critical cases due to low budget involved in their formation, and associated to lack of organization, of interest and information about risks. There is evidence to believe health and ambient impact of dangerous substances is greater than is known (Matchaba-Hove, 1996).

Exposure of workers to dangerous particles or vapors from the grinding process, cut, welding, automotive preparation and spray painting in automotive bodyshops is thoroughly identified (Jayjock and Levin, 1984; Rongo, 2004) Jobs in which high exposure to organic solvents has beed described are automotive spray painting (Kalliokoski, 1986) and fumigation in furniture manufacturing (Hooiveld et al., 2006) Recent analysis in an automated painting facility (Daniell et al., 1992) identified up to 100 different organic compunds, 90% of which are acetone, xylene, toluene, methyl ethyl ketone, methyl isobutyl ketone and hexane. Several combinations of these solvents are used together with other additives (retardants, accelerators and levellers) to reduce material viscosity to a sprayable level and are adaptable to weather conditions.

These solvents are particularly dangerous and inhalation is the main exposure medium for paint workers. Results related to building industry paint workers exposure to oil-based paints and coatings include increase in cancer rates of lungs, throat and larynge, leukemia, altered function of nervous system, kidney and hepatic disease, blood diseases or blood forming organs and on oil; and congenital defects (IARC, 1989).

Some toxicologically important constituents in automotive paint systems and chasis hardeners include ethyl acrylate, ethylbenzene, benzene, cumene, several acetates, isocyanates, glycols, ethers, epoxy resine and amines (Boutin et al., 2005) Polyurethane based paints can be easily found in automotive bodyshops thanks to their outstanding characteristics, such as durability, color stability, abrasion resistance, resistance to chemical products and to extreme weather conditions (Pronk et al., 2005). Thermal degradation of these paints during cutting, grinding and orbital sanding operations cause release of isocyanates whose exposure can cause bronchial asthma, bronchitis, rhinitis, conjuntivitis and dermatitis (Raulf-Heimsoth and Baur, 1998).

On the other hand, anticorrosion coatings are usually used to prevent rusting by isolating the surface from atmospheric oxygen (Jouen et al., 2004). This internal layer is very important since it is responsible for assuring coating adhesion to metal and acts as an agglutinant between the metallic surface and external layers. Historically, lead has been used in anticorrosion coatings, in particular in automotive manufacturing and protection. During the last decades, concerns about its long term toxic effects in human beings and natural systems have risen. Chemistry involved in processing some conversion coatings has been known for more than a 100 years. However, recent research in chemical processing, changes in components used for vehicle chasis structures, the ambient concern and the involved costs had forced the automotive industry to adopt new conversion coating technologies. To counteract ambient and human health impacts related to lead use, PPG companies have focused their coating developments by anode and cathode electrodeposition of lead free acrylics and epoxy (e-coat) for automobile manufacturing. This method avoids the long term use and exposure related to lead based products. In addition, in comparative studies made by Mercedes, the new e-coat epoxy formulas offer yield equivalent to the previous e-coats containing lead.

Other coatings used in the automotive industry have been replaced with metallic phosphate by zirconium oxide (ZrO_2), that besides reducing weight in the body design process, also reduces environmental impact, however, the difference in operation cost is not enough to boost an immediate replacement of processes involving phosphates in the automotive industry.

The new designs lead to the adoption of new pretreatment technologies. Electrophoretic deposition of a polymer film is a reference method involving different materials used in production of internal layer (Streitberger and Dossel, 2008). This process is produced by the application of the electric field and high polarity on metal electrode (Neirinck et al., 2013). This method has several advantages, such as its high efficiency, lack of drag and easy to control and optimize. Cathodic electrodeposition based on epoxy resins modified with partially blocked organic amine and isocyanates like reticulant and tin lead based metallic salt guarantees excellent corrosion resistance and adhesion to metal surface and different dispersions. However, this last formula implies use of toxic salts like catalysts, high curing temperatures (160–180°C) and high heating losses. Although solutions have been proposed to overcome these limitations using compatible products, profitability and volatility of these organic substances do not favor its implementation. As a consequence, several coating formulas have been adapted to water-based technologies (Van der Biest and Vandeperre, 1999; Unruh et al., 2011). Mainly based on acrylic/epoxy materials and, although they have a high homogeneity and enough corrosion resistance, (Gubbels et al., 2014; Cheraghi et al., 2015; Liao, et al., 2015; Toncelli et al., 2016) their content of volatile organic compounds is near 50 g/L; and curing time can last several hours.

Nanomaterials

Nanomaterials in different shapes (nanoparticles, nanofilms, nanoflakes, nanotubes, nanofibers and nanocompounds) have been used in the automotive industry for several purposes, such as improvements on mechanical, electrical, thermal, optical, of corrosion, self-cleaning, antistatic among other properties. Although these materials have enhanced properties for several automotive applications, they can be dangerous if they are not properly used. The toxicity of many nanomaterials is still not completely identified, and most of the studies had focused on pulmonary and hepatic toxicity of nanomaterials. The long-term toxicity and the assessment of chronic exposure must be studied in detail to uncover the underlying mechanisms and cells interaction according to its nanoparticle nature. The automotive industry uses several additives at nanoscale such as black carbon, CNT and oxide of different metals in refill matrixes, coating and compounds, as described above (with ZrO_2). Recent studies show that these nanomaterials can emerge in air, water and soil, and then, in humans and animals, which has caused a public debate on toxicological and ambiental effects of nanomaterials (Brayner et al., 2006). Risks can occur during manufacturing, transport, handling, use, residues elimination and recycling (Marchant et al., 2008). The paths through which pollution is caused are: inhalation in the pulmonary system, adsorption through skin and ingestion through the gastrointestinal system. In the last years, many toxicological studies have corroborated that particles with less than 100 nm in size are dispersed in the respiratory system faster than microscale particles and can move within the body to epithelial and endothelial cells, as well as lymphatic circulation to reach vital organs in the body, such as the nervous system, bone marrow, brain, lymph nodes and spleen in which the nanoparticle has the possibility to cause reactivity and toxicological effects.

The method of risks control is a chain of assessment, planification, implementation and corrective actions with the goal to minimize exposure of nanoparticles to employees allowing them to work in acceptable environments.

The main components of risks control are shown in Fig. 14.2, each block in the pyramid represent the effectivity to manipulate nanoparticles. The elimination is the first step and it must be implemented in the design stage in order to become more profitable. This can achieve substantial savings in the retrofit or acquisition of equipments and procedures to reduce the risk. Risks elimination is the highest level in the pyramid. This is Followed by substitution, with the purpose to replace the conditions that have a high risk by a lower risk. For example, replacing a material based on solvents with a material based on water, or to implement a process of clean technology that reduces use of organic solvents such as microwave, plasma or ultrasound. Replacement of nanomaterial can be difficult since it was probably introduced by its particular properties; however, some replacement might be possible. To replace nanoparticles in suspension by a dry dust version will reduce the aerosolization and will provide protection level for workers handling the material. Engineering controls protect workers when eliminating dangerous conditions and placing a barrier between the worker and danger. Well designed engineering controls can be very effective to protect employees and, in general, will be passive, this means that they will be independent of employee interactions. And such a design will also not interfere with their productivity and, ideally, will ease the operation. Administrative controls are often used in existing processes where risks are not under complete control, occurring when engineering control measurement are not feasible or not reducing exposure to an acceptable level. In the long term, administrative controls tend to be expensive to maintain and require an important effort from involved employees. Finally, personnel protection equipment is the less effective for controlling workers exposure to dangerous substances compared to other components in the hierarchy. It is the last line of defense after engineering controls and administrative controls that will support an employee against a contamination risk.

Finally, adequate information about capacity and explosive reaction, as well as nanomaterials ignition should be produced. The recommendation to counteract fire risk and explosion used when defining caution measurements are the following:

- Dust clouds will be difficult to visualize.
- Design of electrical equipment protected against dust.
- Reduction of potential ignition sources.

- Metalic and metal oxides nanoparticles can explode, thus inert environments or controlled environments in the storage and handling zones (attention to these zones is very important because in these kinds of environments the risk of oxygen reduction exists).
- Use of anti-static shoes to avoid static electricity and as a potential ignition source.

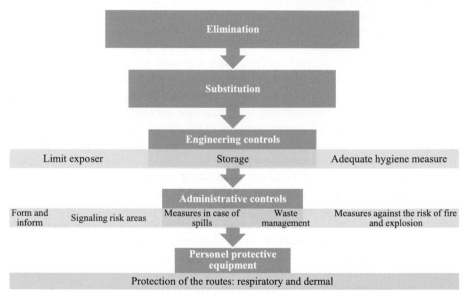

Figure 14.2   Graphical representation of effective controls hierarchy applies to nanomaterials.

The textile industry has been one of the most contaminant sectors since its beginning, however, environmental deterioration, people and animal health risks due to the severity of world's global contamination has brought attention to it in the last few years. Some strategies used in this industry to stop its harmful effect is avoiding unnecessary processes in order to achieve energy savings, implementing mechanical processes instead of chemical, giving preference to closed systems to reduce contaminant emissions to the environment and replacing the use of dangerous chemical substances by safe and low environmental impact options. All these actions prevent ambient pollution and avoid chemical accidents such as burns, explosions and fires.

In order to minimize risks and accidents it is necessary to identify the existing risks in the process, to seek, assess and compare alternatives that prevent and/or decrease them, to perform tests and based on results to implement and improve action strategies. These are the benefits obtained by implementation of risks and accidents prevention measurements:

Improvement for workers, consumers and living beings in general health and safety, as well to reduce the risk of major accidents.

Fulfill laws and standards required in the chain value: OEKO-TEX® Certifications (Oeko-tex.com, 2019) and REACh conformity (European Chemical Agency, 2019).

Improve approval of products from consumer markets, ensuring that textile products are manufactured with materials that lack dangerous chemical substances (for example, herbicide and pesticide free cotton, anitimonium free polyethylene tetraphthalate fibers, carcinogenic and mutagenic free colorants, etc.), besides differences with their competitors.

Reduce costs on contingency measurements for technical control, storage and disposal of materials due to their low risk nature, minimizing chemical danger by exposure and for potential accidents.

With regards to the textile and clothing industry, there are several preventive approaches taking into account that it consumes energy, water and chemical additives, causing abundant effluent discharges. Effluent composition in the textile industry usually is a mixture of pigments, colorants of highly toxic metallic compounds, solvents, optical brighteners, agents to increase tear resistance, flame retardants, heavy metals, antibacterials, dissolved solids, pesticides and biocides (Flick, 1990), these latter are used in vegetal and animal source natural fibers.

Many biological, chemical, physical or a mixture of these processes (chemical oxidation, coagulation, adsorption, etc.) can be used for the elimination of effluents pollutants, however, since it is not a simple task, prevention is the best tool for its solution. Some common strategies consist on reducing water usage; to eliminate or replace substances, for example, replace highly flammable liquids by other less flammable or carburant liquids, replace chemicals (of highly risk) carcinogenic, mutagenic, hypersensibilizing by less dangerous additives; to use diluted acids and bases instead of concentrated; to use products that are not dust, but rather in pulp or pellet form; replace materials of less volatility to reduce chemical risk which in turn will reduce installation and maintenance cost of the ventilation system, as well as vapors control (Starovoytova, 2019; Sharma, 2015).

NON-ECO-FRIENDLY SUBSTANCES

Non-eco-friendly substances used in the textile industry are classified as follows: non-biodegradable organic materials, dangerous substances and substances prone to accidents (Choudhury, 2017). The presence of dangerous substances and prone to accidents in the environment, can cause chemical emergencies due to their toxicity and reactivity. A chemical accident implies huge human and economical costs, as well as environmental damage. Some of the effects on people exposed to dangerous substances can be cancer, genetic disorders, spontaneous miscarriage, injury and death due to toxic chemical reagents or explosions of these. Some can occur instantaneously, but others can develop several years after exposure to such substances (Accidentes Químicos: Aspectos Relativos a la Salud: Parte II, 1998).

The risk of substances used in textile industry, can be caused by different factors: the presence of toxic metals in their chemical structure, being solvents harmful to health and explosive substances used in textile processes, or even the same colorants and additives by their toxic chemical nature. In majority, the substances used in this type of industry imply a health risk from different points of view. It is practically impossible to eliminate efluents pollutants before discharging them to the environment, a common goal is to find eco-friendly options to minimize use of such substances. Taking into account the above, a classification of toxic substances and their existing eco-friendly alternatives is proposed in order to decrease the environmental impact.

SUBSTANCES USED IN THE TEXTILE INDUSTRY AND THEIR ECO-FRIENDLY ALTERNATIVES

Toxic Metals

The toxic metals existing in the processes of textile industries come from diverse sources: from chemical fibers generation, from dyeing process (colorants and auxiliaries), fabrics finishing, plumbing, thread baths, of vegetal origin natural fibers. In the last ones, the plants during their growth adsorb the metals in the soil where they are sowed (Thornton, 1981).

After metals are discarded in the environment, they can pass into the bodies of animals and people by the trophic chain, because of being bioaccumalative. However, they can also enter by the water consumed, the air breathed or by skin contact. Within the organism, metals can form organo-metallic compounds soluble in lipids that toxically act within cells and internal organs, mainly the liver and kidney, affecting their functions (Choudhury, 2017 and Hazardous chemicals in textiles, 2013). Next, some heavy metals used in textile industries are listed and their associated health risks.

- **Antimony (Sb)**: Antimony trioxide may be used as a flame retardant in textiles. Antimony can cause damage to the skin, by simple contact like sitting on furniture fabrics that contains it (Cooper and Harrison, 2009).
- **Lead (Pb)**: Lead compounds are used as stabilizers, also lead acetate trihydrate is used as a mordant in textile printing and dyeing. Lead can damage the brain, nervous system and kidneys. In mild cases, it causes insomnia, anxiety, loss of appetite and gastrointestinal problems (Morrison and Murphy, 2006).
- **Cadmium (Cd)**: Is present in colorants, pigments and stabilizers, in particular camdium chloride is used as a mordant in textile printing and dyeing. Can cause disorders in the respiratory system, kidneys and the lungs. Cadmium and cadmium oxide are classified as carcinogenic, besides they are acute and chronic toxic elements in water (Nordberg, n.d.).
- **Chromium (Cr VI)**: Chromium trioxide, Sodium chromate, Sodium dichromate, Potassium dichromate and Chromium compounds are used in the textile industry as mordants (to fix dye colors in fabrics), as well as in pigments manufacturing. They can cause respiratory disorders, skin ulceration and cancer in respiratory airways by inhalation (Sueker, 2006).

- **Arsenic (As)**: Arsenic trioxide (As_2O_3) or white arsenic is used as a mordant. Sodium arsenite ($NaAsO_2$) is used as a drying agent in the textile industry, but is also used like a herbicide, which explains the presence of arsenic in vegetal fibers used in this industry. Can cause skin cancer, hyperpigmentation and black foot disease (Nordberg, n.d.).
- **Nickel (Ni)**: Nickel is used in alloys for buttons, zippers and rivets. Nickel sulfate is used in textile printing and dyeing. People allergic to nickel will have severe skin irritation when in contact with accessories with nickel (Choudhury, 2017).
- **Vanadium (V)**: Vanadium pentoxide (V_2O_5) and ammonium metavanadate (NH_4VO_3) is used in fabrics dyeing and printing. The inhalation of vanadium compounds can cause health problems even after a short exposure (for example, 1 hour), the initial symptoms are: excessive tearing, burning sensation in conjunctive, hemorrhagic or allergic rhinitis, sore throat, cough, bronchitis, cough, and chest pain. Severe exposure can cause pneumonia, with fatal results (Nordberg, n.d.).

As mentioned above, metals are present in the process of obtaining textiles, as part of diverse substances and functions, therefore in order to stop their use it is necessary to find alternative substances with the same functions, but less risk.

Dyes

Some colorants can cause discomfort and hypersensitivity to the respiratory system, causing a hypersensitive individual exposed to small amounts of colorant to exhibit allergic symptoms such as wheezing, tightness in the chest, lack of breath and even cause asthma. In 2008, the textile industry was identified as a manufacturing sector with increased carcinogenic risk since some of their used substances when metabolized in the intestine walls and liver produced potentially carcinogenic and mutagenic aromatic amines (Chequer et al., 2011).

It is worth noting that although many colorants are noxious by containing metals in their chemical structure, others are intrinsically risky such as azoic colorants, which were banned since 1986 by the Law of Protection to Environment in India due to their carcinogenic properties. To date (that is the year-2019), not only in India, but also the European Union lists 197 "substances of very high concern" (SVHC) and among them are the azoic colorants, which can not be used in products marketed in the EU unless specific use authorization is granted. Germany has a law stating that any clothing or any other article that gets in contact with the skin will have none of the 20 aromatic amines listed in such a document. In the same way, other countries had adopted similar restrictions. At this point, it is worth noting that if a nation other than from the above mentioned produces a textile product and wants to market it on OEKO-TEX® certified countries, in conformity to the REACh and or SVCH list, the product during its process needs certifications from the (European Chemical Agency (ECHA), 2019; Hong Kong Trade Development Council, 2008).

The toxicity of many substances used in the textile industry occurs when users come in contact wearing the textile fabrics, however, their toxic effect is detected in the earlier stages: during handling of workers in the manufacturing plants, and initially during substances production). By law, textile industries must not discharge the water they discard during their processes to the environment, but they must apply treatments in order to eliminate most of the pollutants. However, due to the complexity of colorants and chemical products isolation, as well as the lack of biodegradability of some of them, it is necessary to use several methods in conjunction, increasing their treatment cost. This causes an increasing interest in natural and organic colorants that are harmless to health and the environment, and that besides the textile industry, can also be used in industries such as tanning, plastic, paper, decoration, etc. The use of such colorants increases over time since they can be found in forest sources; vegetal, fungi or bacteria cells are non toxic or expensive chemicals, their process temperature is low (near 30°C), their synthesis process in general is performed at neutral pH, with high yield and purity, and they are biodegradable (Chouhan et al., 2013). The following are natural colorants used in the textile industry and are classified according to the kind of fibers they can dye. Such characteristics are important since threads and fabrics can be formed by mixtures of fibers, it is also useful if the same colorant dyes more than one type of fabric. It is worth mentioning that metallic mordants marked with an asterisk can be toxic metals, depending on the required tonality, and accordingly the proper mordant that is harmless to health must be preferred (Chouhan et al., 2013; Har and Kumar, 2014):

a) **Dyes for dyeing cotton**: Alizarin (yellow color) and Haematoxylin* (red)

b) **Dyes for dyeing silk**: Citronetin (yellow), Delphinidin (red) and Gamboge (reddish-yellow or brownish- orange)

c) **Dyes for dyeing wool**: Carminic acid (red), Chrysophanic acid* (yellow), Emodin* (yellow), Fisetin* (yellow), Galiosin (red), Kermesic acid (red), Maclurin* (yellow), Myricetin* (yellow) and Rubiadin (pale yellow)

d) **Dyes for dyeing wool and silk**: Berberine (yellow color), Brazilein (red), Ellagic acid (yellow), Embelin (yellow), Flemingin (orange-red), Hypericin (deep violet-red), Lawsone (reddish-brown), Orchil or Archil (purple or red-violet) and Rottlerin (orange red)

e) **Dye for dyeing wool and nylon**: Blueberry (red)

f) **Dyes for dyeing cotton and silk**: Bixin (red), Carthamin (red), Isorhamnetin* (yellow)

g) **Dyes for dyeing cotton and wool**: Coreopsin (yellow), Galloflavin and Isogalloflavin (yellow), Juglone (brownish-yellow), Purpurin (red), Munjistin (orange dye), Rhamnetin (yellow), Santalin* (brick red) and Vitexin (yellow)

h) **Dye for dyeing cotton and nylon**: Black tea (yellow)

i) **Dyes for dyeing cotton, wool and rayon**: Indigo or Indigotin (blue)

j) **Dyes for dyeing cotton, silk and wool**: Adinin (yellow color), Alkannin (red color), Canadian golden rod plant (olive-green), Catechin (brown), Chrysin* (yellow), Crocin and Crocetin (yellow), Curcumin (orange-yellow), Datiscetin (yellow), Erythroaphins (red), Euxanthic acid (yellow), Gardenin (yellow), Kaempferol* (yellow), Laccaic acid and Erythrolaccin (red), Morin* (yellow), Morindone (red), Patuletin (yellow), Pratol (yellow), Pseudopurpurin (red), Quercetagetin* (yellow), Rutin (yellow), Soranjidiol (dark reddish-brown), Tectoleafquinone* (yellow or red) and Ventilagin (purple-red)

As can be noted, most natural colorants have an affinity and are capable of dyeing natural origin fibers, but not synthetic fibers. Therefore, despite the existing low risk and toxicity of natural colorants, synthetic colorants are still used that cause toxic wastes together with excessive water consumption. During the cotton dyeing process, despite the addition of salts, alkali and auxiliaries it is recorded that such fibers can use only 75% of the colorant, making it necessary for many washes to completely remove the 25% of colorant that did not react (Ghaly et al., 2014). The pollutants from wastewater coming from the textile industry affect human health causing hemorrhages, skin irritation and ulceration, nausea, etc. In the environment such chemical products block sunlight, thus inhibiting photosynthesis and reoxygenation of aquifer system, besides increasing the biological demand of oxygen (Ghaly et al., 2014).

Reduction of Water Consumption by Textile Industry

In order to reduce pollution load on effluents, Huntsmann Company has developed new cotton dyes, which are p-chloroaniline (PCA) free dyes, a dangerous chemical used as an intermediary in the manufacturing of azoic colorants and pigments (REI and Levi Strauss & Co., brands restrict the use of PCA and other amines colorants). The dyes contain greater amount of reactive groups bonded to chromophore in comparison to the amount of reactive groups in common cotton dyes. With Avitera dyes the stain phase lasts only 4 hours instead of 7 hours needed for conventional dyes since they adhere better and easily to the fiber, therefore they require less water and energy to eliminate the non consumed dye (Bomgardner, 2018).

Cotton pretreatment.   With this technology a cationization of the raw cotton fiber is performed by the permanent bond of quaternary ammonium compound and cellulose fibers. With this treatment natural attraction between the dye and fiber is achieved resulting in 90% savings of water, 75% of energy, 90% of auxiliary chemical products and almost 50% of dye in comparison to processes that require salts in the stain bath (Bomgardner, 2018).

Pre-reducing synthetic indigo.   The dye used to make blue jeans is the soluble synthetic indigo, and it releases unreacted chemicals products. Indigo is different to most of dyes since it is insoluble in its non-reduced form. This lack of solubility results that from 70,000 metric tons of used indigo, around 400 metric tons are not added to the fabric during dyeing, and from these, 267 metric tons go to waste waters, in workers and in the air. Archroma is a company that found a solution to the insolubility of synthetic indigo by their conversion to pre-reduced solutions that are more soluble in water, that create stained fabrics without a detectable quantity of aniline, compared with the 2,000 ppm of remaining chemical substances remaining in the competitors' affluents (Bomgardner, 2018).

Natural bacterial indigo.   The natural indigo molecule coming from legumes is exactly the same to its synthetic version, with a single difference: synthetic indigo has tighter crystal formation that makes it more difficult to reduce, but not impossible. To reduce it bacterial fermentation vats or conventional reducing systems can be used. This allows obtaining a sustainable option to produce the popular textile blue denim worn worldwide (Hsu et al., 2018).

Dyes from engineered microbes. These are high yield renewable dyes, with stability to temperature and resistance to light; suitable for general use textiles, synthesized by genetically modified microbes. To achieve coloring of fabrics, microbes come in contact with a solution with the textile fabric, the microbes adhere to fibers and promote their growth. Then heat is applied in order to break the organism membranes allowing color to chemically adhere to the fiber with the help of metallic ions and salts in the microbes cytoplasm. With these dyes, only a single washing process is required, causing water and energy savings of 90 and 20% respectively, in comparison to standard processes (Bomgardner, 2018).

The **digital textile printing** is a more ecological alternative compared to conventional textile printing because it is not necessary to use water in order to remove non-impregnated dye, in particular when dyeing cotton. This method uses large scale printers with special versions of jet printing heads designed to work with textile inks (Bomgardner, 2018).

The substitution of water like a reaction medium during the dyeing process is the key for its sustainability, since large amounts are used, it is calculated that between 120 and 400 tons of water for each ton of dyeing fiber are needed (Clarke et al., 2018; Clark et al., 2016). For this reason, there is a lot of interest for alternatives of water usage. The following options besides replacing water usage, can represent a decrease in auxiliary costs as well as on generated contaminants.

Waterless dyeing using supercritical carbon dioxide drying. With this technology the dyeing process of natural and synthetic textiles is performed with CO_2, using non-polar colorants and waterless, avoiding pollution of the vital liquid. However, this alternative has the disadvantage of low solvation power, making it necessary to use cosolvents and additives, reducing the non contaminant advantage. Also, the high operational pressure levels required for using supercritical carbon dioxide, demand high energy, as well as rough systems that withstand such pressures, making it unviable to retrofit this solution to actual infrastructure of many companies (Clarke et al., 2018).

Dimethyl sulfoxide dyeing. The dyeing process using a reactive method with organic dimethyl sulfoxide (DMSO) solvent has several advantages compared to conventional dyeing: 1) avoids water usage, 2) requires up to 40% less colorant, 3) no need to add salts to reaction medium, 4) it is not necessary to pressurize the system, since it works at ambient pressure, 5) for cotton dyeing, no modifications to the fabric or colorants in order to use them (Chen et al., 2015).

The most studied **Ionic Liquids** (IL) are the systems containing phosphonium, imidazolinium or tricaprylmethyl ammonium cations, with different heteroatom functionality. A very promising IL is 1-(2-hydroxyethyl)-3-methylimidazolium dicyanamide, which has proven to be very versatile: 1) useful to dye polyester, cotton and wool, 2) no need of dispersants, swelling agents, carriers, mordants, antifoams and deaerators, and others, 3) the dyeing process is performed only with three components: pure dye, water, and the IL, 4) has good colorant yields, 5) due the complete exhaustion of the dyebath after dyeing, the residual dye baths can be used again, 6) no leaching of dye fabric when it is washed in water, avoiding aqueous pollutants, 7) the process is at a lower temperature and with less pressure than the conventional process, it also saves energy (Bianchini et al., 2015).

Potentially Explosive Substances

In the textile industry the explosion risks are related to flammability and explosivity of solvents used at high temperatures, with which explosions and fires can occur. However, an explosion can also occur with products that are non-flammable, but that are highly reactive. The following are examples of potentially explosive substances in the textile industry (Choudhury, 2017; Chemical Safety in the Workplace. Guidance Notes on Chemical Safety in Textile Finishing, 2003):

- Sodium hydrosulfite is a reducing agent which,when gets in contact with water, causes a highly exothermal reaction that can spontaneously ignite.
- Acetic acid is flammable at high concentrations.
- The hydrogen peroxide, sodium hypochlorite and sodium dichloroisocyanurate are oxidizing agents used to bleach fibers, however, if they are used without strict control, can result in violent and uncontrolled reactions, where an oxygen gas volume greater than the capacity of equipment that contains it, resulting in catastrophic failures.
- The solvent base for resins used in the lamination process for wet-coating quickly evaporates creating a flammable atmosphere.

As it can be noted, there are many potentially explosive chemical substances, therefore, it is necessary to obtain optional substances of a less hazardous nature.

Besides the use of colorants from natural and eco-friendly sources, the use of high temperatures for colorant synthesis changes during the long period in organic solvents using eco-friendly synthesis methods like microwave radiation in the presence of base or acid reagents and without organic solvents and sonochemistry techniques (Petkova et al., 2014) or ultrasonic radiation (Yachmenev et al., 1998). favoring shorter reaction times.

Volatile Organic Solvents

The Volatile Organic Solvents (VOCs) are organic chemical substances with a low boiling point, therefore their molecules can easily change from a liquid or solid phase to a gas phase, as a result of high vapor pressure in normal conditions at ambient temperature. Many VOCs are dangerous for living beings and/or cause damage to the environment, the symptoms in humans develop slowly as the concentration levels in general are low, hence their effects are seen over a long period. Among the most common diseases are respiratory ones, allergies of inmmunologial effects on new borns and children, irritation in the eyes, irritation of the nose and a sore throat, lack of coordination, nausea, damage to the liver, kidney and central nervous system. Some examples of VOCs are benzene, toluene, acetone, ethyl alcohol, per (or tetra) chloroethene (PERC), trichloroethene (TCE), acrylonitrile, acetaldehyde, decane, tetradecane and formaldehyde. The latter has a boiling point of $-19°C$, hence at ambient temperature changes to vapor phase and dissolves in the air unless it is kept in a closed environment (Bernstein et al., 2008).

The VOCs with their high flamability and/or toxic effects to health are not the only disadvantage in the use of solvents in the textile industry.

Chemicals and Auxiliaries

The hazard of chemical substances by their toxicity or potentially explosive nature has been explained. Some alternatives have been proposed, however, there are some substances of less danger useful in the textile industry, which are not listed. Table 14.1 lists some harmful textile chemicals and their eco-friendly substitutes (Table adapted from Choudhury, 2017; Shenai, 2001; Arputharaj, Raj Saxena, 2015; Carmo et al., 2017).

Table 14.1   Harmful textile chemicals and their eco-friendly substitutes

Existing chemicals	*Uses*	*Proposed substitutes*
Pentachlorophenol Formaldehyde	Size preservatives	Sodium silicofluoride
Nonyl phenyl ethylene oxide adducts (APEO)	Detergent Emulsifier	Fatty alcohol ethylene oxide adducts, alkyl polyglycosides
Silicones and amino-silicones + APEO emulsifier	Finishing; Softening, Water repelling	Eco-friendly anionic/cationic softeners, wax emulsions
C8 fluorocarbons	Water repellents	C6 fluorocarbons
Bleaching powder Calcium and Sodium hypochlorite	Cotton bleaching	Hydrogen peroxide, ozone at cold
Sodium silicate Phosphorus-based compounds	Hydrogen peroxide stabilizer	Nitrogen stabilizers
Dichlorobenzene Trichlorobenzene	Carriers in polyester dyeing	Butyl benzoate, benzoic acid
Kerosene	Emulsion thickener (Pigment printing)	Water-based thickeners
Formaldehyde	Crease resisting of cotton and its blend fabrics	Poly-carboxylic acid, non-form-aldehyde crosslinking agents
	Dye fixing for direct and reactive dyeings/prints	Non-formaldehyde based products
Sodium dichromate	Oxidation of vat and Sulfur dyes	Hydrogen peroxide, sodium perborate
Polyvinyl alcohol (PVA)	Yarn size	Potato starch or carboxy methylcellulose (CMC)
Carbon tetrachloride (CTC)	Stain removers	Enzymatic stain-removers Detergent stain-removers Detergent (non-ionic, ethoxylates) and water miscible solvent (glycol ethers) mixtures

Table 14.1   .... *(contd...)*

Table 14.1 Harmful textile chemicals and their eco-friendly substitutes (*contd...*)

Existing chemicals	Uses	Proposed substitutes
Synthetic non-biodegradable surfactants	Biodegrade or make the process sustainable	Sustainable and highly biodegradable surfactants from dextrins
Synthetic non-biodegradable surfactants+solvent	Coatings and degreasing	'Solvosurfactants' acting both solvent and surfactant, derived from glycerol (bio diesel)
Functional synthetic finish	Finishing	Bees wax, aloe vera and Vitamin A
Hydrochloric acid	Desizing	Amylases
Sodium hydroxide	Mercerization	Liquid ammonia
	Scouring of cotton	Pectinases
Sodium thiosulphate	Peroxide killer	Catalases
Acetic acid	Neutralization agent	Formic acid, Citric acid
Sodium sulphide	Reducing agents	Glucose, acetyl acetone, thiourea dioxide
Powder form of sulphur dyes	Dyeing	Pre-reduced dyes
Bromated diphenylethers	Flame retardant	Combination of inorganic salts and phosphonates
Chlorination	Shrink proofing	Plasma treatment
Toxic heavy metals	Color fastness	Chitosan
	Mordant	Tamarind seed coat

During the development of an eco-friendly approach in the textile industry, it can be concluded that harmful chemical substances actually used in their manufacturing processes can be replaced or modified only when less noxious alternative substances are available, which can perform the same functions or give the same or better results, keeping costs competitive and also being sustainable alternatives.

REFERENCES

Abidin, M.Z., Rusli, R., Khan, F., Shariff, A.M. Development of inherent safety benefits index to analyse the impact of inherent safety implementation. *Process Safety and Environmental Protection*, 2018, 117, 454–472.

Accidentes Químicos: Aspectos Relativos a la Salud: Parte II (1998). Guías prácticas: Capitulo 2. Generalidades: 2.3 Definición de "accidente químico". [online] Available at: http://cidbimena.desastres.hn/docum/ops/publicaciones/ops26s/ops26s.4.htm [Accessed 1 Feb. 2019]. ISBN: 92 75 32254 6.

Albrecht, M.A., Evans, C.W., Raston, C.L. Green chemistry and the health implications of nanoparticles. *Green Chemistry*, 2006, 8(5), 417–432.

Alvarado, C.C., Galindo, A.S., Berumen, C.P., López, L.L., Orta, C.Á., Garza, J.V., Donías, L.M. Modificación superficial de MWCNTs asistida por ultrasonido con ácido acético y ácido cítrico. *Afinidad*, 2014, 71(566).

Andrade Guel, M.L., López López, L.I., Sáenz Galindo, A. Nanotubos de carbono: Funcionalización y aplicaciones biológicas. *Revista mexicana de ciencias farmacéuticas*, 2012, 43(3), 9–18.

Arputharaj, A., Raja, A.S.M., Saxena, S. Developments in sustainable chemical processing of textiles. *In*: Muthu S., Gardetti M. (eds), Green Fashion. Environmental Footprints and Eco-design of Products and Processes. Springer: Singapore; 2015; pp. 217–252.

Asbestos Standard for General Industry U.S. Department of Labor Occupational Safety and Health Administration OSHA 3095 (1995) <https://www.osha.gov/Publications/osha3095.html>; January 31, 2019.

Ashford, N.A. Reducing physical hazards: Encouraging inherently safer production. *In*: Anastas, P.A. (ed.), Handbook of Green Chemistry, vol. 9. Weinheim: Germany; Wiley-VCH; 2013; pp. 485–500.

Asmatulu, R., Nguyen, P., Asmatulu, E. Nanotechnology safety in the automotive industry. *In*: Asmatulu, R. (ed.), Nanotechnology Safety. Elsevier; 2013; pp. 57–72.

Banach, M., Pulit-Prociak, J. Proecological method for the preparation of metal nano-particles. *Journal of Cleaner Production*, 2017, 141, 1030–1039.

Bernstein, J.A., Alexis, N., Bacchus, H., Bernstein, I.L., Fritz, P., Horner, E., Li N., Mason, S., Nel, A., Oullette, J., Reijula, K., Reponen, T., Seltzer, J., Smith, A., Tarlo, S.M. The health effects of nonindustrial indoor air pollution. *Journal Allergy & Clinical Immunology*. 2008, 121, 585–591.

Bianchini, R., Cevasco, G., Chiappe, C., Pomelli, C.S., Rodríguez-Douton, M.J. Ionic liquids can significantly improve textile dyeing: An innovative application assuring economic and environmental benefits. *ACS Sustainable Chemistry & Engineering*, 2015, 3, 2303–2308.

Bomgardner, M.M. These new textile dyeing methods could make fashion more sustainable. Large and small suppliers vow to help a resource-intensive, cost-sensitive industry change with the times. c&en [online], 2018, 96 (29). ISSN 0009-2347. Article archive. https://cen.acs.org/business/consumer-products/new-textile-dyeing-methods-make/96/i29 (accessed Feb, 01, 2019).

Boutin, M., Dufresne, A., Ostiguy, C., Lesage, J. Determination of airborne isocyanates generated during the thermal degradation of car paint in body repair shops. *Annals of Occupational Hygiene*, 2005, 50(4), 385-393.

Bouzembrak, Y., Marvin, H.J. Impact of drivers of change, including climatic factors, on the occurrence of chemical food safety hazards in fruits and vegetables: A Bayesian Network approach. *Food Control*, 2019, 97, 67–76.

Brayner, R., Ferrari-Iliou, R., Brivois, N., Djediat, S., Benedetti, M.F., Fiévet, F. Toxicological impact studies based on *Escherichia Coli* bacteria in ultrafine ZnO nanoparticles colloidal medium. *Nano Letter*, 200, 6(4), 866–870.

Cabello Alvarado, C.J., Saenz Galindo, A., Perez Berumen, C., Lopez Lopez, L., Avila Orta, C., Valdes Garza, J., Moran Donias, L.D. Modification surface of MWCNTs assisted by ultrasonic with acetic acid and citric acid. *Afinidad*, 2014, 71, 139–145.

Carmo, R.S.A., Arauz, L.J., Rosa, J.M., Baruque-Ramos, J., Araujo, M.C. Dyeability of polyester and polyamide fabrics employing citric acid. *Journal Fiber Science Technology*, 2017, 3, 31–44.

Catensson, S. Pharmaceutical waste. *In*: Kümmerer, K. (ed.), Pharmaceuticals in the Environment. Sources, Fate, Effects and Risks. 3rd Ed. Berlin, Heidelberg, New York; Springer; 2018; pp. 551–561.

Cave, S.R., Edwards, D.W. Chemical process route selection based on assessment of inherent environmental hazard. *Computers & Chemical Engineering*, 1997, 21, S965–S970.

ChemSafetyPRO. REACH SVHC List (2019). Available from: https://www.chemsafetypro. com/Topics/EU/REACH_SVHC_List_Substance_of_Very_High_Concern.html [accessed Feb, 01, 2019].

Chemical Safety in the Workplace. Guidance Notes on Chemical Safety in Textile Finishing Occupational Safety and Health Branch Labour Departament, 1st Ed., 2003, 5–26. [online] Available at: https://www.labour.gov.hk/eng/public/os/C/B127.pdf [Accessed 4 Feb. 2019].

Chen, L., Wang, B., Ruan, X., Chen, J., Yang, Y. Hydrolysis-free and fully recyclable reactive dyeing of cotton in green, non-nucleophilic solvents for a sustainable textile industry. *Journal Clean Production*, 2015, 107, 550–556.

Chequer, F.M.D., Dorta, D.J., de Oliveira, D,P. Azo dyes and their metabolites: Does the discharge of the azo dye into water bodies represent human and ecological risks? Advances in Treating Textile Effluent, Peter J. Hauser, IntechOpen, 2011. DOI: 10.5772/19872.

Cheraghi, S., Naghash, H.J., Younesi, M. Synthesis and characterization of two novel heterocyclic monomers bearing allylic function and their applications in the epoxy-acrylic-amine waterborne coatings. *Progress in Organic Coatings*, 2015, 88, 116–126.

Choudhury, A.K.R. Green Chemistry and Textile Industry. *Journal of Textile Engineering & Fashion Technology*, 2017, 2, 1–12. DOI: 10.15406/jteft.2017.02.00056

Chouhan, N. Kumar, A. Sharma, A. Ameta, R. Eco-friendly products. *In*: Amenta S.C., Amenta, R. (eds), Green Chemistry. Fundamentals and Applications. Mistwell Crescent Oakville, CRC Press, Taylor & Francis Group; 2013; pp. 61–64. ISBN: 978-1-4665-7826-5

Clark, J.H., Farmer, T.J., Herrero-Davila, L., Sherwood, J. Circular economy design considerations for research and process development in the chemical sciences. *Green Chemistry*, 2016, 18, 3914–3934.

Clarke, C.J., Tu, W.-C., Levers, O., Bröhl, A., Hallett, J.P. Green and sustainable solvents in chemical processes. *Chemistry Reviews*, 2018, 118(2), 747–800.

Cooper, R.G., Harrison, A.P. The exposure to and health effects of antimony. *Indian Journal of Occupational and Environmental Medicine*, 2009, 13(1), 3.

Dagwa, I.M. Development of Automobile Disk Brake Pad from Local Materials. Ph.D. (Manufacturing Engineering) Thesis, University of Benin, Benin City; 2005.

Daniell, W., Stebbins, A., Kalman, D., O'Donnell, J.F., Horstman, S.W. The contributions to solvent uptake by skin and inhalation exposure. *American Industrial Hygiene Association Journal,* 53(2), 1992, 124–129. DOI: 10.1080/15298669291359384

De Marco, B.A., Rechelo, B.S., Tótoli, E.G., Kogawa, A.C., Salgado, H.R.N. Evolution of green chemistry and its multidimensional impacts: A review. *Saudi Pharmaceutical Journal*, 2018, 27, 1–14.

Edwards, D.W., Lawrence, D. Assessing the inherent safetyof chemical process routes: is there a relation between plantcosts and inherent safety? *Process Saf. Environ. Prot. Trans.IChemE*, 1993, 71(B4), 252–258.

Efendy, H., Wan Mochamad, W-M., Yusuf, N.B.M. Development of natural fiber in non-metallic brake friction material. *Seminar Nasional Tahunan Teknik Mesin (SNTTM).*, 2010, 13–15, Ke-9 Palembang.

Eldin, A.B., Shalaby, A., Abdallah, M.S., Shaldam, M.A., Abdallah, M.A. Applying green analytical chemistry (GAC) for development of stability indicating HPLC method for determining clonazepam and its related substances in pharmaceutical formulations and calculating uncertainty. *Arabian Journal of Chemistry*, 2014, 12(7), 1212–1218. https://doi.org/10.1016/j.arabjc.2014.10.051

European Chemical Agency (ECHA). Comprensión de REACH. [online] Available at: https:// echa.europa.eu/es/regulations/reach/understanding-reach [Accessed 9 Feb. 2019]., 2019.

European Chemical Agency (ECHA). REACH SVHC List in Excel Table, Total Number: 197 Substances (Updated on 16 Jan 2019). [online] Available at: http://www.chemsafetypro. com/Topics/EU/REACH_SVHC_List_Excel_Table.xlsx [Accessed 1 Feb. 2019]., 2019.

Flick, E.W. Textile Finishing Chemicals: An Industrial Guide. New Jersey: Noyes Publications, 1990. ISBN: 0-8155-1234-1.

Gaggero, L., Ferretti, M. The self-sustained high temperature synthesis (SHS) technology as novel approach in the management of asbestos waste. *Journal Envirormental Manager*, 2018, 216, 246–256.

García-Serna, J., Pérez-Barrigón, L., Cocero, M.J. New trends for design towards sustainability in chemical engineering: Green engineering. *Chemical Engineering Journal*, 2007, 133(1–3), 7–30.

Ghaly, A.E., Ananthashankar, R., Alhattab, M., Ramakrishnan, V.V. Production, characterization and treatment of textile effluents: A critical review. *Journal of Chemical Engineering & Process Technology*, 2014, 5, 1–18.

Gilson, J.C. Asbestos. *In*: Parmeggiani L. (ed.), Enciclopedia de salud y seguridad en el trabajo. Madrid: Ministry of Work and Social Security, 1989.

Gómez, M.G., Castañeda, R., López, V.G., Vidal, M.M., Villanueva, V., Espinosa, M.E., Evaluation of the national health surveillance program of workers previously exposed to asbestos in Spain (2008). *Gaceta Sanitaria*, 2012, 26(1), 45–50.

Gubbels, E., Drijfhout, J.P., Posthuma-van Tent, C., Jasinska-Walc, L., Noordover, B.A., Koning, C.E. Bio-based semi-aromatic polyesters for coating applications. *Progress Orgorganic Coatings*, 2014, 77(1), 277–284.

Gudmand-Høyer, L., Bach, A., Nielsen, G.T., Morgen, P. Tribological properties of automotive disc brakes with solid lubricants. *Wear*, 1999, 232, 168–175.

Har, B.S., Kumar, A.B. Handbook of Natural Dyes and Pigments. New Delhi; Woodhead Publishing India Pvt. Ltd., 2014. ISBN: 978-93-80308-54-8

Hendershot, D.C. Inherently safer chemical process design. *Journal of Loss Prevention in the Process Industrie*, 1997, 10, 151–157.

Hooiveld, M., Haveman, W., Roskes, K., Bretveld, R., Burstyn, I., Roeleveld, N. Adverse reproductive outcomes among male painters with occupational exposure to organic solvents. *Occupation Environmental Medical*, 2006, 63(8), 53–544.

Hong Kong Trade Development Council. REACH Regulations. How they apply to Textile and Leather articles. Available from: http://info.hktdc.com/productsafety/200808/psl_ leather_080804.html (accessed Feb, 01, 2019).

Hsu, T.M., Welner, D.H., Russ, Z.N., Cervantes, B., Prathuri, R.L., Adams, P.D., Dueber, J.E. Employing a biochemical protecting group for a sustainable indigo dyeing strategy. *Natural Chemistry Biology*, 2018, 14, 256–261.

Idris, U.D., Aigbodion, V.S., Abubakar, I.J., Nwoye, C.I. Eco-friendly asbestos free brake-pad: Using banana peels. *Journal King Saud University*, 2013, 1–8.

Ikpambese, K.K., Gundu, D.T., Tuleun, L.T. Evaluation of palm kernel fibers (PKFs) for production of asbestos-free automotive brake pads. *Journal King Saud University Engineering Science*, 2016, 28(1), 110–118.

International Agency for Research on Cancer. Some organic solvents, resin monomers and related compounds, pigments and occupational exposures in paint manufacture and painting. Lyon, France: IARC, IARC Monogr. Eval. Carcinog. Risks Humans, 1989.

Issaabadi, Z., Nasrollahzadeh, M., Sajadi, S.M. Efficient catalytic hydration of cyanamides in aqueous medium and in the presence of Naringin sulfuric acid or green synthesized silver nanoparticles by using Gongronema latifolium leaf extract. *Journal of Colloid and Interface Science*, 2017, 503, 57–67.

Jayjock, M.A., Levin, L. Health hazards in a small automotive body repair shop. *Journal Occupation Environmental Hygiene*, 1984, 28(1), 19–29.

Jouen, S., Jean, M., Hannoyer, B. Atmospheric corrosion of nickel in various outdoor environments. *Corrosion Science*, 2004, 46, 499–514.

Kadziński, M., Cinelli, M., Ciomek, K., Coles, S.R., Nadagouda, M.N., Varma, R.S., Kirwan, K. Co-constructive development of a green chemistry-based model for the assessment of nanoparticles synthesis. *European Journal of Operational Research*, 2018, 264(2), 472–490.

Kakooei, H., Sameti, M., Kakooei, A.A. Asbestos exposure during routine brake lining manufacture. *Industrial Health*, 2007, 45, 787–792.

Kalliokoski, P. Solvent containing processes and work practices: Environmental observations. *Progress in Clinical and Biological Research*, 1986, 220, 21–30.

Karmaus, A.L., Osborn, R., Krishan, M. Scientific advances and challenges in safety evaluation of food packaging materials: Workshop proceedings. *Regulatory Toxicology and Pharmacology*, 2018, 98, 80–87.

Khalid, S.M.N. Food safety and quality management regulatory systems in Afghanistan: Policy gaps, governance and barriers to success. *Food Control*, 2016, 68, 192–199.

Khan, M.N., Khan, T.A., Khan, Z., Al-thabaiti, S.A. Green synthesis of biogenic silver nanomaterials using raphanus sativus extract, effects of stabilizers on the morphology, and their antimicrobial activities. *Bioprocess and Biosystems Engineering*, 2015, 38(12), 2397–2416.

Kidam, K., Hussin, N.E., Hassan, O., Ahmad, A., Johari, A., Hurme, M. Accident prevention approach throughout process design life cycle. *Process Safety and Environmental Protection*, 2014, 92(5), 412–422.

Kirchhoff, M.M. Promoting green engineering through green chemistry. *Environmental Science & Technology*, 2003, 37(23), 5349–5353.

Kletz, T.A. Inherently safer design—Its scope and future. *Process Safety and Environmental Protection*, 2003, 81(6), 401–405.

Koenig, S.G., Dillon, B. Driving toward greener chemistry in the pharmaceutical industry. *Current Opinion in Green and Sustainable Chemistry*, 2017, 7, 56–59.

Leder, C., Rastogi, T., Kümmerer, K. Putting benign by design into practice-novel concepts for green and sustainable pharmacy: Designing green drug derivatives by non-targeted synthesis and screening for biodegradability. *Sustainable Chemistry and Pharmacy*, 2015, 2, 31–36.

Li, C.J., Trost, B.M. Green chemistry for chemical synthesis. *Proceedings of the National Academy of Sciences*. 2008, 105(36), 13197–13202.

Liao, H., Zhang, B., Huang, L., Ma, D., Jiao, Z., Xie, Y., Cai, X. The utilization of carbon nitride to reinforce the mechanical and thermal properties of UV-curable waterborne polyurethane acrylate coatings. *Progress Organic Coatings*, 2015, 89, 35–41.

Limón, M.D., Aguilar, A., Aragón, M.P., Argemí, C., Cañedo, D., Carbonell, M. Límite de exposición profesional para agentes químicos en España 2013. INSHT. Instituto Nacional de Seguridad e Higiene en el Trabajo Ediciones. Madrid, 2013.

Ljungberg, L.Y., Edwards, K.L. Design, materials selection and marketing of successful products. *Material Design*, 2003, 24(7), 519–529.

Mageswari, A., Srinivasan, R., Subramanian, P., Ramesh, N., Gothandam, K.M. Nanomaterials: Classification, biological synthesis and characterization. *In*: Ranjan, S., Dasgupta, N., Lichtfouse, E., (eds), Nanoscience in Food and Agriculture, 3. Cham: Springer; 2016; pp. 31–71.

Mali, A.C., Deshmukh, D.G., Joshi, D.R., Lad, H.D., Patel, P.I., Medhane, V.J., Mathad, V.T. Facile approach for the synthesis of rivaroxaban using alternate synthon: Reaction, crystallization and isolation in single pot to achieve desired yield, quality and crystal form. *Sustainable Chemical Processes*, 2015, 3(1), 11.

Manley, J.B., Anastas, P.T., Cue, Jr, B.W. Frontiers in green chemistry: Meeting the grand challenges for sustainability in R&D and manufacturing. *Journal Clean Production*, 2008, 16(6), 743–750.

Marchant, G.E., Sylvester, D.J., Abbott, K.W. Risk management principles for nanotechnology. *Nanoethics.* 2008, 2(1), 43–60.

Marchiol, L., Mattiello, A., Pošćić, F., Giordano, C., Musetti, R. *In vivo* synthesis of nanomaterials in plants: Location of silver nanoparticles and plant metabolism. *Nanoscale Research Letters*, 2014, 9(1), 101.

Marioryad, H., Kakooei, H., Shahtaheri, S.J., Yunesian, M., Azam, K. Assessment of airborne asbestos exposure at an asbestos cement sheet and pipe factory in Iran. *Regulation Toxicology Pharmacology*, 2011, 60, 200–205.

Martin, B., Lehmann, H., Yang, H., Chen, L., Tian, X., Polenk, J. Schenkel, B. Continuous manufacturing as an enabling tool with green credentials in early-phase pharmaceutical chemistry. *Current Opinion in Green and Sustainable Chemistry*. 2018, 11, 1–104.

Matchaba-Hove, R. Allergies and Work Editorial. African Newsletter on Occupational Health and Safety, 1996.

Mohamed, H.M. Green, environment-friendly, analytical tools give insights in pharmaceuticals and cosmetics analysis. *TrAC Trends in Analytical Chemistry*, 2015, 66, 176–192.

Morrison, R.D., Murphy, B. (eds). Environmental Forensics. Contaminant Specific Guide. USA: Elsevier; 2006. ISBN 13: 978-0-12-507751-4

National Institute for Occupational Safety and Health (NIOSH). Method 7400: Fibers. *In*: NIOSH. Manual of Analytical Method, 3th Ed., Cincinnati: U.S. Department of Health, Education and Welfare; 1998.

Neira, M. Chrysotile Asbestos. World Health Organization, 2015.

Neirinck, B., Van der Biest, O., Vleugels, J.A. Current opinion on electrophoretic deposition in pulsed and alternating fields, *Journal Physics Chemistry B*, 2013, 117, 1516–1526.

Nordberg, G. Metales: propiedades químicas y toxicidad. *Instituto Nacional de Seguridad e Higiene en el Trabajo* (INSHT), 2012, Enciclopedia, OIT, 2.

Occupational Safety and Health Administration (OSHA). Asbestos in air OSHA method ID-160., 1997.

Occupational Safety and Health Branch. Labour Departament. Chemical Safety in the Workplace. Guidance Notes on Chemical Safety in Textile Finishing. 1st Ed., 5–26. Available from: https://www.labour.gov.hk/eng/public/os/C/B127.pdf (accessed Feb., 04, 2019), 2003.

Oeko-tex.com. OEKO-TEX® | STANDARD 100 by OEKO-TEX®. [online] Available at: https://www.oekotex.com/en/business/certifications_and_services/ots_100/ots_100_start.xhtml [Accessed 9 Feb. 2019]., 2019.

Olawoyin, R. Nanotechnology: The future of fire safety. *Safety Science*, 2018, 110, 214–221.

Osman, T.M. Environmental, health, and safety considerations for producing nanomaterials. *Jom*, 2008, 60(3), 14–17.

Palaniappan, C., Srinivasan, R., Tan, R. Selection of inherently safer process routes: A case study. *Chemical Engineering and Processing: Process Intensification*, 2004, 43(5), 641–647.

Panahi, D., Kakooei, H., Marioryad, H., Mehrdad, R., Golhosseini, M. Evaluation of exposure to the airborne asbestos in an asbestos cement sheet manufacyuring industry in Iran. *Environmental Monitoring Assessement*, 2011, 178, 449–454.

Petkova, P., Francesko, A., Fernandes, M.M., Mendoza, E., Perelshtein, I., Gedanken, A. Tzanov, T. Sonochemical coating of textiles with hybrid ZNO/chitosan antimicrobial nanoparticles. *ACS Applied Material Interfaces*, 2014, 6(2), 1164–1172.

Price, G.J., Nawaz, M., Yasin, T., Bibi, S. Sonochemical modification of carbon nanotubes for enhanced nanocomposite performance. *Ultrasonics Sonochemistry*, 2018, 40, 123–130.

Pronk, A., Tielemans, E., Skarping, G., Bobeldijk, I., Van Hemmen, J., Heederik, D., Preller, L. Inhalation exposure to isocyanates of car body repair shop workers and industrial spray painters. *Annales Occupation Hygiene*, 2005, 50(1), 1–14.

Pye, A.M. A review of asbestos substitute materials in industrial applications. *Journal of Hazardous Materials*, 1979, 3(2), 125–147.

Rajasekhar, C., Kanchi, S. Green nanomaterials for clean environment. *In*: Torres-Martinez, L.M., Kharissova, O.V., Kharisov, B.I. (eds), Handbook of Ecomaterials, Cham: Springer; 2018; pp. 1–18.

Ramanarayanan, R., Bhabhina, N.M., Dharsana, M.V., Nivedita, C.V., Sindhu, S. Green synthesis of zinc oxide nanoparticles using extract of *Averrhoa bilimbi* (L) and their photoelectrode applications. *Materials Today: Proceedings*, 2018, 5(8), 16472–16477.

Rao, R.U., Babji, G.A. Review paper on alternate materials for asbestos brake pads and its characterization. *Internacional Research. Journal Engineering Technology*, 2015, 2(2), 556–562.

Rastogi, T., Leder, C., Kümmerer, K. Designing green derivatives of β-blocker Metoprolol: A tiered approach for green and sustainable pharmacy and chemistry. *Chemosphere*, 2014, 111, 493–499.

Raulf-Heimsoth, M., Baur, X. Pathomechanisms and pathophysiology of isocyanate-induced diseases-summary of present knowledge. *American Journal Industrial Medicine* 1998, 34(2), 137–143.

Rongo, L.M.B. Can information dissemination workshops reduce allergy among small-scale industry workers in dar es salaam? *African Newsletter.*, 2004, 14(3), 52.

Rosato, D.V. Asbestos; Its Industrial Applications. Reinhold Publishing Corp., 1959.

Sadri, R., Hosseini, M., Kazi, S.N., Bagheri, S., Ahmed, S.M., Ahmadi, G., Ahmadi, G., Zubir N., Sayuti M., Dahari, M. Study of environmentally friendly and facile functionalization of graphene nanoplatelet and its application in convective heat transfer. *Energy Conversion and Management*, 2017, 150, 26–36.

Shameli, K., Bin Ahmad, M., Jazayeri, S.D., Sedaghat, S., Shabanzadeh, P., Jahangirian, H., Mahnazi, M., Abdollahi, Y. Synthesis and characterization of polyethylene glycol mediated silver nanoparticles by the green method. *International Journal of Molecular Sciences*, 2012, 13(6), 6639–6650.

Sharma, S.K. (ed.). Green Chemistry for Dyes Removal from Wastewater Research Trends and Applications. Salem, Massachusetts: Scrivener Publishing LLC; 2015. ISBN 978-1-118-72099-8.

Shenai, V.A. Non-ecofriendly textile chemicals and their probable substitutes—An overview. *Indian Journal Fibre Text*, 2001, 26, 50–54.

Simijonović, D., Vlachou, E.E., Petrović, Z.D., Hadjipavlou-Litina, D.J., Litinas, K.E., Stanković, N., Mihovic N., Mladenović, M.P. Dicoumarol derivatives: Green synthesis and molecular modelling studies of their anti-LOX activity. *Bioorganic Chemistry*, 2018, 80, 741–752.

Slušná, Ľ. Balog, M. Automobilový Priemysel Na Slovensku a Globálne Hodnotové Reťazce, Slovenská Inovačná a Energetická Agentúra. Bratislava, 2015.

Spiric, D., Jovanovic, D.R., Palibrk, V.P., Bijelovic, S., Djuragic, O., Reddy, P.G. Convergence on EU and USA food safety Regulation approach, regarding foodborne outbreaks. *Procedia Food Science*, 2015, 5, 266–269.

Starovoytova, D. Hazards and risks at rotary screen printing (part 6/6): Control of chemical hazards via cleaner production approaches. *Chemical and Process Engineering Research* [Online] 2019, 57, 17–45. Available from: https://iiste.org/Journals/index.php/CPER/article/view/41829/43078. ISSN 2224-7467 (Paper); ISSN 2225-0913 (Online).

Stave, C., Törner, M. Exploring the organisational preconditions for occupational accidents in food industry: A qualitative approach. *Safety Science*, 2007, 45(3), 355–371.

Stoycheva, S., Marchese, D., Paul, C., Padoan, S., Juhmani, A.S., Linkov, I. Multi-criteria decision analysis framework for sustainable manufacturing in automotive industry. *Journal Clean Production*, 2018, 187, 257–272.

Streitberger, H.-J., Dossel, K.-F. Automotive Paints and Coatings, 2nd Ed., Weinheim: WILEYVCH Verlag, 2008.

Sulivan, J.B. Krieger, G.R. Clinical Environmental Health and Toxic Exposure. USA: Lippincott, Williams & Wilkins, 2001.

Sueker, J.K. Chromium. *In*: Morrison, R., Murphy, B. (eds), Environmental Forensics. Contaminant Specific Guide. USA: Elsevier; 2006; pp. 81–95.

Sun, F., Yun, D.A.I., Yu, X. Air pollution, food production and food security: A review from the perspective of food system. *Journal of Integrative Agriculture*, 2017, 16(12), 2945–2962.

Swedish Chemicals Agency. Hazardous chemicals in textiles. *Report of a Government Assignment*. Publisher: Swedish Chemicals Agency: Stockholm, ISSN: 0284-1185., 2013.

Talawar, M.B., Sivabalan, R., Mukundan, T., Muthurajan, H., Sikder, A.K., Gandhe, B.R., Rao, A.S. Environmentally compatible next generation green energetic materials (GEMs). *Journal of Hazardous Materials*, 2009, 161(2–3), 589–607.

Thornton, I. Geochemical aspects of the distribution and forms of heavy metals in soils *In*: Lepp, N.W. (ed.), Effect of Heavy Metal Pollution on Plants. Pollution Monitoring Series, vol 2. London and New Jersey: Applied Science; 1981; pp 1–33. DOI: 10.1007/978-94-009-8099-0.

Tobiszewski, M., Namieśnik, J. Greener organic solvents in analytical chemistry. *Current Opinion in Green and Sustainable Chemistry*, 2017, 5, 1–4.

Toncelli, C., Schoonhoven, M.J., Broekhuis, A.A., Picchioni, F. Paal-Knorr kinetics in waterborne polyketone-based formulations as modulating cross-linking tool in electro-deposition coatings. *Material Design*, 2016, 108, 718–724.

Unruh, D.A., Pastine, S.J., Moreton, J.C., Fréchet, J.M. Thermally activated, single component epoxy systems. *Macromolecule*, 2011, 44, 6318–6325.

Van der Biest, O.O., Vandeperre, L.J. Electrophoretic deposition of materials. *Annual Review Material Science*, 1999, 29, 327–352.

Veerapandian, S., Nasar, A.S. Amine-and blocked isocyanate-terminated polyurethane dendrimers: Integrated synthesis, photophysical properties and application in a heat curable system. *RSC Advance*, 2015, 5(5), 3799–3806.

Villaseñor, M.J., Ríos, Á. Nanomaterials for water cleaning and desalination, energy production, disinfection, agriculture and green chemistry. *Environmental Chemistry Letters*, 2018, 16(1), 11–34.

Virta, R.L. Mineral Commodity Profiles, Asbestos. Reston, Virginia, USA: US Geological Survey, 2005.

Wang, Y., Shi, Z., Yin, J. Facile synthesis of soluble graphene via a green reduction of graphene oxide in tea solution and its biocomposites. *ACS Applied Materials & Interfaces*, 2011, 3(4), 1127–1133.

Warnasooriya, S., Gunasekera, M.Y. Assessing inherent environmental, health and safety hazards in chemical process route selection. *Process Safety and Environmental Protection*, 2017, 105, 224–236.

Warner, J.C., Cannon, A.S., Kevin, M. Dye environmental impact assessment review. *Green Chemistry*, 2004, 24, 775–799.

Wieczerzak, M., Namieśnik, J., Kudłak, B. Bioassays as one of the green chemistry tools for assessing environmental quality: A review. *Environment International*, 2016, 94, 341–361.

Yachmenev, V.G., Blanchard, E.J., Lambert, A.H. Use of ultrasonic energy in the enzymatic treatment of cotton fabric. *Industrial Engineering Chemistry Research*, 1998, 37, 3919–3923.

Yang, S., Peng, S., Xu, J., Luo, Y., Li, D. Methane and nitrous oxide emissions from paddy field as affected by watersaving irrigation. *Physics and Chemistry of the Earth* (Parts A/B/C), 2012, 53, 30–37.

Yilmaz, E. Use of hydrolytic enzymes as green and effective extraction agents for ultrasound assisted-enzyme based hydrolytic water phase microextraction of arsenic in food samples. *Talanta*, 2018, 189, 302–307.

15

Green Precipitation with Polysaccharide as a Tool for Enzyme Recovery

Débora A. Campos, Ezequiel R. Coscueta
and Maria Manuela Pintado*

Universidade Católica Portuguesa, CBQF–Centro de Biotecnologia e
Química Fina–Laboratório Associado, Escola Superior de Biotecnologia,
Rua Diogo Botelho 1327, 4169-005 Porto, Portugal.

INTRODUCTION

The two major types of macromolecules in biological systems are proteins and polysaccharides, which are responsible for cell structure, energy storage/production or enzymatic reactions. Moreover, these two macromolecules exert essential structural functions in the food systems, acting as stabilizers for emulsions and foams, and being the most important gelling and thickening agents (Tolstoguzov, 1991). However, such structural functions are hugely affected by the interactions between these important macromolecules and other components in food systems. The interactions between protein-polysaccharide was initially described by Bungenberg de Jong and co-workers in 1949, where they studied the complex formed by gelatine-acacia gum coacervation and published through the years several updates on this complex, drawing attention to protein/polysaccharide complexes and coacervates and the possibility of their application to foods (Khamanga and Walker, 2017). It was only after 1981 that studies were published describing the complex formation between proteins and polysaccharides, as well as, their interactions. Larionova, et al. in 1981 described the features of the formation of

*For Correspondence: mpintado@porto.ucp.pt

soluble polyelectrolytes complexes between carboxymethyl ethers of cellulose and dextran with basic proteins.

Until the year 2000 more than 40 scientific studies were published describing the interaction between different proteins and different polysaccharides, showing the increasing interest in water-soluble polymeric complexes. After this era, the scientific community-maintained an interest in the interaction between these macromolecules, but the field of study was reoriented towards the development of technologies to be applied in the food industry. One of the most cited studies reported by Ye in 2008 described the importance of the application of proteins and polysaccharide in the food industry and the key role that these complexes have in the structure and stability of processed food. In addition, the authors analyzed the development of novel food processes and products by giving a brief overview of recent research work on the complexation between milk proteins and polysaccharides and the application of the complexes in the dairy industry. On the other hand, Le Bourvellec and Renard in 2012 went one step further and studied the interactions between polyphenols and both proteins and polysaccharides, by giving a review of the methods used to study the interactions, as well as, the extension of each kind of interactions.

In recent years, the primary interest has been focused on the downstream process through the application of polysaccharides for complex formation and their application for protein isolation and purification. Thus, studies in scaling up have grown significantly by the demand for large amounts of enzymes to be used in biotechnological processes.

The enzyme industry has been growing in the past few years because of the increasing interests in the application of most environmentally friendly techniques. Enzymes are suitable for applying in several industries, food, pharmaceuticals, animal feed, and others with less impact on the market. In enzyme production, the scaling process represents the major problem since the extraction of enzymes comes from a large amount of biomass from the fermentation of homogenates and microbial suspension. On the other hand, other classes of enzymes can be extracted from natural sources, such as fruits and vegetables, since these sources naturally produce enzymes as natural secondary metabolites. Bioseparation steps for the recovery of the final product can account for up to 70% of overall production costs. Thus, many of the traditional processes are no longer applied to such industries, since they present several negative aspects, such as high costs, negative environmental impact, the presence of hard chemicals and a short life in the final products.

The recent trend of replacing linear economy by circular economy has opened the opportunity to explore the valuable compounds present in industrial waste. An important source of such waste is the agro-food industry that produces and processes tons of food daily, generating a high percentage of waste rich in bioactive compounds. By transforming this waste into a by-product prone to be valued new economic income can be generated, for example, functional ingredients can be obtained for various applications.

For establishing competitive biotechnological processes for enzyme extraction and purification, several research groups have devoted scientific research on the improvement of more "natural" techniques, such as bioseparation based on polysaccharides and natural polyelectrolytes, using their capacity to produce non-soluble complexes with different protein groups, the enzymes being the most important. Nowadays, biotechnology has been improving the application of bioseparation in scaling up for application in large amounts for industrial application.

This chapter reviews the enzyme's isolation and purification through precipitation using polysaccharides and natural polyelectrolytes for production of non-soluble complexes. Additionally, future trends in this field and the applications of such techniques to industries for bioseparation of industrial enzymes will also be covered.

POLYSACCHARIDES

Polysaccharides are carbohydrate molecules composed of long chains of monosaccharide units bound together by glycosidic linkages, which by hydrolysis release the constituent monosaccharides, such as glucose. Since polysaccharides are one of the significant macromolecules, they are present in all structures having different functions in nature. Furthermore, in addition to its natural functions and nutritional properties, polysaccharides have different physicochemical properties that make them very interesting from a technological standpoint.

Many polysaccharides are used to store energy in organisms, typically fold together and can contain many monosaccharides in a dense area. Moreover, as the side chains of the monosaccharides establish as many hydrogen bonds as possible between themselves, water cannot intrude the molecules, making them hydrophobic. They not only store the energy but allow for a change in the concentration gradient, which can influence the holding capacity of nutrients and water. All polysaccharide preparations are polydisperse, containing molecules with different degrees of polymerization, hence they can be organized by structure, by shape (linear or branched), by monomeric units (homoglycans, di-heteroglycans, tri-heteroglycans, tetra-heteroglycans, and penta-heteroglycans) and by charge (neutral, anionic and cationic). The charge has a significant effect on the ionic polysaccharides since these macromolecules adopt a fully extended shape (due to Coulombic repulsion) and thus impart high viscosity to their solutions. The viscosity and other properties of such molecules have a great interest for further applications in different areas.

Anionic polysaccharides are polymers of sugar acid (e.g., alginate); sugar units biosynthesized with anionic substituents such as sulfate groups (e.g., carra-geenan); or carboxyl groups that are substituted through a chemical reaction (e.g., carbo-xymethylcellulose). The presence of these anionic groups increases the polarity and water solubility, though the intrinsic charge weakens intermolecular associations between polymer chains due to repulsion (Nieto, 2016).

Alginates are linear, unbranched polymers, containing β-(1-4)-linked L-guluronic acid (G) units, and are therefore highly anionic and very polar polymers. These polymers are present in the cell wall of brown algae as calcium, magnesium and sodium salts of alginic acid.

Carrageenan is a collective term for sulfated polysaccharides extracted from certain species of red seaweed of the family, Rhodophyceae. Major commercial sources are *Eucheuma spinosum, Eucheuma cottonii, Gigartina* spp., *Chondrus crispus*, and *Hypnea* spp. In terms of grades, four commercial types of carrageenan extracts are available: kappa I, kappa II, iota and lambda types. Therefore, different seaweeds produce different carrageenans fractions, with one type being more predominant in any species. Carrageenans are made up of alternating galactopyranosyl dimer units linked by alternating β-(1-4) and α-(1-3) glycosidic bonds. The sugar units are sulfated either at C2, C3 or C6 of the galactose or C2 of the anhydrogalactose unit.

Carboxymethylcellulose or CMC is cellulose ether produced by reacting the alkali cellulose with sodium monochloroacetate, under controlled conditions. There are many degrees of substitution producing different properties of CMC. Hence, many grades of CMC are in the market to perform different functions. The structure of CMC involves carboxymethyl substitution of the native cellulose polymer at the C2, C3, or C6 positions of anhydroglucose units.

These described polysaccharides are very interesting for the industry given the different applications for which they have been studied. Likewise, due to the greater understanding in this field, other anionic polymers can be considered for technological applications. Until now some polysaccharides such as xanthan gum and pectin have been reported. Xanthan gum is a polysaccharide produced industrially by a microbial fermentation of bacterium *Xanthomonas campestris* involving a carbohydrate substrate and other growth-supporting nutrients. The molecular structure consists of a linear glucose chain linked by β-(1-4) glycosidic bonds like cellulose and possesses a trisaccharide side chain attached through O3 of alternate glucose units in the main chain (Fan and Chen, 2007).

Pectin is a soluble heterogeneous polysaccharide containing linear chains (Sonia and Sharma, 2014). It is mainly produced commercially from citrus peels and apple pomace. It has been agreed that pectin is not a regular chain but has a backbone made up of α(1-4)-linked D-galacturonic acid residues, and a more significant proportion of these galacturonic acid residues is methyl esterified naturally (Pérez et al., 2000).

On the other hand, there are only a few examples of cationic polysaccharides, such as chitosan, where the amino group confers a positive charge when the polysaccharide is dissolved in an acidic solution. Several other grades of powders such as low methoxy, amidated pectin and cationically modified guar that contain $-NH_2$ groups are also available in the market (Nieto, 2016).

Chitosan is commercially produced by the deacetylation of chitin, which is the structural element in the exoskeleton of crustaceans (such as crabs, shrimps, oysters and other mollusk shells) and cell walls of fungi. Nearly all chitin and chitosan produced commercially are chemically extracted from the sources mentioned above; chitin can also be produced from shell waste fermentation with microorganisms or with the aid of enzymes. Chitosan is poly-β-(1-4)-2-amino-deoxy-D-glucopyranose.

PROTEINS

Proteins are complex macromolecules that makeup almost 70% by weight of living cells and play roles in their fundamental structure and function. A large number of proteins are known (they have been isolated, purified and studied), of which the majority varies between 5000 to several million Daltons in weight, and are present in most processed and natural foods.

Proteins have an extraordinary diversity of functions and can be arbitrarily classified into two main categories: "structural proteins" and "proteins with biological activity". Structural proteins are present in all tissues of animals and some vegetables, internal organs, cell membranes and intracellular fluids. Its function depends mainly on its fibrous structure. Among them, can be mentioned collagen, elastin, keratin, etc. Proteins with biological activity have an active role in all biological processes, with enzymes being their most notable exponents.

Among proteins, enzymes are presented as the most interesting from the industrial standpoint, interest that increases year by year. Enzymes act as catalysts for the chemical reactions in biological systems. These offer much more competitive processes compared to chemical catalysts. Several enzyme-based processes have been developed to produce different valuable products since biocatalysis was first introduced nearly a century ago (Bruggink et al., 1998). Consequently, the use of enzymes has expanded to the manufacture of pharmaceutical intermediates and fine chemicals (Griengel et al., 2000; Hills, 2003).

The rise in the prices of traditional commodities such as oil has led industries to look for alternative sources of raw materials, such as biomass. Besides, an increase in social awareness concerning environmental problems has led to public pressure on green technologies, actively demanding the replacement of chemical processes by cleaner, safer and more ecological catalytic processes (Benkovic and Hammes-Schiffer, 2003). In this sense, enzymatic biocatalysis has rapidly been replaced by traditional chemical processes in many areas, and it is expected that this substitution may be accelerated through the development of new technologies in enzymatic engineering (Choi et al., 2015).

The application of enzymes to produce various types of chemical and biological substances have become a proven technology in the chemical, pharmaceutical, food, cosmetic, textile and paper industries. The chemical and pharmaceutical industries found that biocatalysis is faster, less expensive and a safer process, with a reduction in the number of reaction steps and the amount of produced waste (Huisman and Collier, 2013; Tomsho et al., 2011). In the food industry, biocatalysis has been used to produce raw materials and final products for a long time (Fernandes, 2010). However, most uses of biocatalysis have focused on hydrolytic reactions for scaling, solubility improvement and clarification. With the increasing demand for nutritional aspects, a great deal of attention has been paid to the functionality of foods beyond the primary function of nutrient supply. A recent trend in the food industry is to develop functional foods such as prebiotics, antioxidants, low-calorie sweeteners and rare sugars (Nagaoka and Akoh, 2008; Coscueta et al., 2016; Hajfathalian et al., 2017; Anal, 2017). Likewise, recently the cosmetic industry has faced a challenge due to the growing demands of consumers of natural and ecological cosmetics (Ansorge-Schumacher and Thum, 2013).

Consequently, the cosmetic industry promotes basic research and ecological processes using enzymes to develop more effective cosmetic products. In the textile industry, before the conversion into cloth and yarns, cotton undergoes several processes, including refining, bleaching, dyeing and polishing (Queiroga et al., 2007). These processes consume large amounts of energy, water and resources, discharging vast amounts of waste. For the development of cleaner processes, the use of enzymes is proliferating. In the pulp and paper industries, different enzymes are used to improve the quality of the pulp by eliminating pitch in wood, lignin and hemicellulose, which are typical impurities (Maijala et al., 2008). Also, the chemical process of pulp manufacture requires the addition of a large number of alkaline chemicals and chlorine (Fu et al., 2005), which has been replaced with enzymatic processes, avoiding the generation of elemental chlorine and significantly reducing the amount of waste that causes ozone depletion and acidification, as well as high energy consumption.

The growing demand for enzymes has led to the emergence of an industry around these macromolecules. Enzymes are not only obtained from natural sources but also are designed and produced in small biological factories. Enzyme engineering was due to the discovery of a genetic code, which opened the possibility of obtaining proteins on a large scale from various organisms in which they do not occur naturally. The proteins obtained in this way are known as recombinant proteins. These can be produced by microorganisms such as bacteria, fungi, viruses and yeasts, as well as in cultivable cell lines of insects, plants and mammals (Janasson et al., 2002; Palomares et al., 2004), protein expression systems "without cells" and animals and transgenic plants (Farrokhi et al., 2009). Overexpression of recombinant proteins offers the advantage of producing large amounts of proteins of interest with characteristics such as natural protein and with relative ease and quickly.

Enzymes are water-soluble molecules that are naturally mixed with many other compounds of a different nature, along with many other similar proteins (some of them with undesired catalytic activity). The catalytic activity of contaminating enzymes may decrease the regio or enantioselectivity as well as the specificity of the desired "biocatalyst". Thus, to be used, the enzymes of interest must be separated from their natural matrix. The separation strategies used for this purpose can involve long and tedious processes, or include a single chromatographic step (Barbosa et al., 2015). However, even in the best of cases, this can have a negative economic impact on the final cost of the biocatalyst. Even in the overexpression of recombinant proteins, despite the aforementioned advantages, the same complications of downstream processing occur. That is why in recent decades much work has been done with a focus on the search for simpler and more effective alternative purification operations depending on the final objective of the enzyme application (Campos et al., 2019; Barbosa et al., 2011; Rocha et al., 2012; Woitovich et al., 2016).

COMPLEX FORMATION BETWEEN POLYSACCHARIDES AND PROTEINS

The complex formation between polysaccharides and proteins is a phenomenon that allows reaching the energy equilibrium, leading to a decrease in the total electrostatic free energy of the mixture (Tolstoguzov, 1991). The electrostatic interactions that occur between macromolecules with net opposite charges lead to the formation of complexes, which aggregate up to reach sizes and superficial properties that lead to phase separation (Schmitt and Turgeon, 2011).

Interactions between proteins and polysaccharides may result in three different thermodynamic consequences: co-solubility, incompatibility or complexing. Generally, the decrease in free energy results from a positive enthalpic contribution occurring from the electrostatic interactions between the biopolymers, and the negative entropic contribution arises from the release of water molecules due to the compaction of biopolymers. Therefore, the complex formation is driven mainly by electrostatic interactions and short-range interactions such as hydrophobic, van der Waals forces or hydrogen bonds might be secondarily involved in the formation of the protein-polysaccharide complexes (Schmitt et al., 2009). Therefore polyelectrolytes interact strongly with proteins due to the substantial electrical charge. Initially, the studies mainly looked at the interactions with synthetic polyelectrolytes, but later natural polyelectrolytes (polysaccharides) were incorporated into these studies, increasing the possibilities of processing and the interest in the formation of complexes between natural polysaccharides and proteins.

The attraction between the natural polysaccharides and proteins can be affected by physicochemical parameters, such as pH value, ionic strength, protein/polysaccharide ratio, polysaccharide linear charge density, protein surface charge density, and stiffness of the polysaccharide chain (Schmitt et al., 2009; Schmitt and Turgeon, 2011; Ru et al., 2012). The interaction between protein/polysaccharide occurs when the pH > pI (pI – isoelectric point) and initiates at the first critical pH (pH$_c$). Usually, the complex formation occurs within a pH range, therefore, in the second critical pH (pH$_{\varphi1}$) there is an abrupt increase in turbidity showing the continued aggregation of soluble complexes into insoluble protein/polysaccharide complexes due to charge neutralization (Cooper et al., 2005). A complex formed by the electric charges can transfer between the soluble and non-soluble state by the "simple" shift of pH, as shown in Fig. 15.1. With an acid-base titration curve, it is possible to see the turbidity formation and identify the critical pH of the complex formation. Another way to study these interactions is through dynamic light scattering, Campos et al., in 2016 studied the pH shift that occurs in complex systems between bromelain (*Ananas comosus*, pineapple enzyme) and carrageenan (a natural anionic polysaccharide), and described the hydrodynamic radius (Hd), as well as, the intensity percentage of the complexes formed. A Hd maximum and intensity maximum was achieved at pH between 4.6 and 4.8, which was indicative of complex formation; this range of pH was identified as the first critical pH (pH$_c$). Two other critical pH values were identified. The second critical point was at pH ≤ 3.75 (pH$_{\varphi1}$) when the association between the macromolecules started. The third critical point pH ≥ 6.7 (pH$_{\varphi2}$) was the point where insoluble complexes dissociate into soluble complexes or interaction between protein molecules and polysaccharides chains do not occur. Between these two critical pH points the range of non-soluble complex formation (Campos et al. in 2016). The same behavior was reported by Duran et al., in 2018 who studied the interaction between carrageenan and quinoa proteins (*Chenopodium quinoa* Willd). They reported that the main protein of the quinoa seed isolate was soluble at alkaline pH due to its negative net charge and had stability over a wide range of pH. In addition, they reported a precipitation process by secondary aggregation. This complexation occurs through hydrogen bonds and hydrophobic interactions, but also through covalent disulfide bridges that are known to be formed by quinoa proteins.

Several studies also showed the evaluation of enzymes complexed with polysaccharides by fluorescence spectroscopy, thus analyzing the effect of complexation on the structure of the enzyme. These studies should be added to this kind of experimental research, to better understand the implication of the folding of the enzymes on their activities. Campos et al., 2016 evaluated the complex between carrageenan and bromelain by fluorescence spectroscopy and the results showed that carrageenan did not negatively affect the microenvironment of tryptophan residues present in bromelain since the fluorescence spectrum did not change when compared with control bromelain(Campos et al., 2016; 2019).

In summary, and considering the specific case of enzymes, it can be said that the enzyme-polysaccharide complex formation induces a more ordered structure, so the thermodynamic stability of the complex is higher than that of the polysaccharide or enzyme alone. As a result, an elevation of the protein melting temperature on complexation with polysaccharide is observed, it is influenced by the hydrophobicity and electric charge density of polysaccharide and protein. As polysaccharides are excluded from the protein domain, increasing the hydration of the macromolecule, leads to thermal stability being observed. Moreover, this stabilization preserves the enzymatic activity through time especially when the enzyme is stored in solution for long periods.

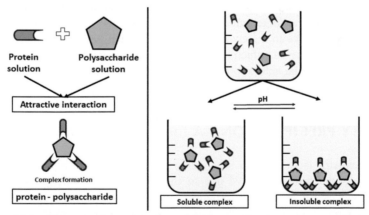

Figure 15.1   Diagram of protein-polysaccharide complex formation, transfer between soluble complex to insoluble complex (*Adapted from* Campos et al , 2016).

APPLICATION OF BIOSEPARATION IN THE ISOLATION OF ENZYMES

A process of enzyme bioseparation refers to the recovery, isolation and purification of these proteins from a complex matrix, from animal, vegetable or microbial sources.

In the last years, the bioseparation problem has become relevant due to the growing need to have large quantities of enzymes in different production processes, since it has been described as a viable alternative for replacing the traditional process in the industrial environment. The purification process of enzymes always involves four principal steps: i) removal of cell and cell fragments from the environment in which they are found (clarification); ii) concentration or purification of low resolution (primary recovery); iii) high-resolution purification (allows the separation of the interesting molecule from others with similar features); iv) packaging of the final product.

An ideal bioseparation process should be mainly simple, fast and of low cost since it will represent about 70% of the total product cost (Liam and Mota, 2003). Moreover, it should combine the high extraction yield and selectivity, but also, must ensure reproducibility at the industrial scale up. Besides the selectivity in bioseparation would depend mainly on the final application of the product. This is what will determine the purity, the required concentrations and even the permitted levels for different types of impurities. In this sense, for example, the enzymes to be applied in the medical and pharmaceutical areas present more restrictive directives, so the selectivity in the process must be more significant.

Enzyme bioseparation can be divided into two main groups: i) high yield and high purity grade, such as affinity chromatography; or ii) high yield and low purity grade, such as precipitation, aqueous two-phase systems, ultrafiltration, among others. The latter is the easiest to scale up with a lower cost of application, and mainly refers to precipitation, because of the natural ingredients applied, as polysaccharides, and the simple equipment required for the process. Another aspect to consider in bioseparation processes, mainly on a macro scale, is the ecological one. When handling a large amount of biomass, it is necessary to apply methods that are not aggressive to the environment, that is, they generate the least possible amount of waste material, which is, also biodegradable or non-toxic. In ideal cases, these methods must be able to produce by-products instead of waste, i.e., material that can be valorized when implemented in other processes or as a product in itself.

AFFINITY PRECIPITATION—A tool for green downstream

Green chemistry is the key to sustainable development. Green technologies are emerging as the best alternative due to their reduction in energy consumption, allowing the use of alternative resources and guaranteeing high quality and safe products. A process can be classified as a green process if integrated technologies allow the reduction of negative impacts on the environment, by reduction of water and energy requirements, as well as, reduction of processing time throughout the productions steps, which includes the downstream stage. As an example, natural enzyme isolation can be very difficult because usually it is present in a very low concentration and has to be separated from a complex biomass, requiring several stages in a process for enzyme separation and isolation, and is one of the crucial steps of precipitation. It is important to highlight that precipitation is a part of green chemistry, that is, using natural compounds as precipitants in sustainable conditions, we can refer to this methodology as "green precipitation".

In recent years, the search for solutions for the scaling of protein isolation and purification methods for biotechnological processes has increased significantly as a result of the growing demand for enzymes. In traditional methods for enzyme separation in large volumes, approximately 90% use a high percentage of inorganic salts or organic solvents, with ammonium sulfate being the most common precipitant. This salt is used in the range of 40–70% of the saturation level to separate 50% of the total enzyme content. The problem of the application of this salt is that it has a high cost and a high negative impact in the process, as well as on the environment.

Affinity precipitation was first described in 1970 by Mosbach, and in its work was based on the observation of an insoluble complex formed by lactate dehydrogenase (LDH) with a polyelectrolyte, which was added in such a way as to precipitate, this being a direct consequence of the affinity interaction between a multivalent enzyme and a bifunctional ligand.

The precipitation with natural polyelectrolytes (green precipitation), considering the desirable conditions for the enzyme bioseparation, is a method that has been favorably positioned in recent years. It allows the recovery of a macromolecule of a complex mixture using certain natural polymeric ligands capable of interacting specifically with the target molecule and reversibly precipitate with certain stimuli. This bifunctionality of the ligand is the basis of the purification strategy (Hilbrig and Freitag, 2003), since in a first step, the ligand in its soluble form interactions with the protein of interest, producing the formation of the complex and therefore becomes insoluble and separates by decantation. Then, by means of an appropriate stimulus (change of some environmental conditions, i.e., pH, temperature, ionic strength and the presence of certain co-solutes) the dissociation of the complex is induced, allowing the precipitation of the free ligand (for recycling) and the recovery of the protein in soluble form.

Green precipitation presents several advantages such as easy scaling, a possibility of continuous processing, simplicity in the necessary equipment and low necessary concentrations of ligand (polysaccharide). On the other hand, its main disadvantage is the low sensitivity in some instances, since co-precipitation can occur with other molecules, which can be considered as impurities, which require coupling with other methodologies to obtain a purer product. In addition, green precipitation can be combined with other novel and green methods such as aqueous two-phase systems, taking advantage of the combined properties of both methodologies, resulting in an increase in selectivity and also allowing, in the same step, to extract and separate more than one compound of interest (Teotia and Gupta, 2004; Leong et al., 2017; Rocha et al., 2016).

Conclusion

During the past few years the production and commercialization of industrial enzymes have been growing, due to the enormous progress in the upstream process. The projections show continuous growth in the enzyme production, more prominently in those of natural origin. This is mainly related to the development of new strategies to overcome the problems related to the downstream operations of the production process (extraction, separation and purification). The recent innovation in bioseparation, more precisely in green precipitation, has been driven both by the imperative need to transform food industrial waste into by-products and in turn add value, as well as by the need to reduce operational costs by increasing the general yield of the processes. Thus, it is necessary to implement these new integrated techniques to further optimize yields. Although, a greater interdisciplinary contribution is still needed to reach an efficient scale up that allows transferring these technologies in an adequate way to the industry. The recovery of high-value natural ingredients, mainly enzymes, from by-products through a successful application of green precipitation operations will result in a better and more efficient economy. This application will lead to savings that will benefit manufacturers and consumers, and also the environment. In a perfect system, where everything is framed under sustainable practices bioprocessing should set an example to other industrial sectors.

REFERENCES

Anal, A.K., Food Processing By-products and Their Utilization. John Wiley & Sons, 2017.

Ansorge-Schumacher, M.B., Thum, O. Immobilised lipases in the cosmetics industry. *Chemical Society Reviews*, 2013, 42(15), 6475–6490.

Barbosa, J.M.P., Souza, R.L., Fricks, A.T., Zanin, G.M., Soares, C.M.F., Lima, Á.S. Purification of lipase produced by a new source of *Bacillus* in submerged fermentation using an aqueous two-phase system. *Journal of Chromatography*, 2011, 879(32), 3853–3858.

Barbosa, O., Ortiz, C., Berenguer-Murcia, Á., Torres, R., Rodrigues, R.C., Fernandez-Lafuente, R. Strategies for the one-step immobilization–purification of enzymes as industrial biocatalysts. *Biotechnology Advances*, 2015, 33(5), 435–456.

Benkovic, S.J., Hammes-Schiffer, S. A perspective on enzyme catalysis. *Science*, 2003, 301(5637), 1196–1202.

Bungenberg de Jong, H.G. Crystallisation-coacervation-flocculation. *Colloid Science*, 1949, 2, 232–258.

Bruggink, A., Roos, E.C., de Vroom, E. Penicillin acylase in the industrial production of β-lactam antibiotics. *Organic Process Research & Development*, 1998, 2(2), 128–133.

Campos, D.A., WoitovichValetti, N., Oliveira, A., Pastrana-Castro, L.M., Teixeira, J.A., Pintado, M.M., Picó, G. Platform design for extraction and isolation of Bromelain: Complex formation and precipitation with carrageenan. *Process Biochemistry*, 2016, 54, 156–161.

Campos, D.A., Coscueta, E.R., Valetti, N.W., Pastrana-Castro, L.M., Teixeira, J.A., Picó, G.A., Pintado, M.M. Optimization of bromelain isolation from pineapple byproducts by polysaccharide complex formation. *Food Hydrocolloids*, 2019, 87, 792–804.

Choi, J.-M., Han, S.-S., Kim, H.-S. Industrial applications of enzyme biocatalysis: Current status and future aspects. *Biotechnology Advances*, 2015, 33(7), 1443–1454.

Cooper, C., Dubin, P., Kayitmazer, A., Turksen, S., Polyelectrolyte–protein complexes. *Current Opinion in Colloid & Interface Science*, 2005, 10(1–2), 52–78.

Coscueta, E.R., Amorim, M.M., Voss, G.B., Nerli, B.B., Picó, G.A., Pintado, M.E. Bioactive properties of peptides obtained from Argentinian defatted soy flour protein by Corolase PP hydrolysis. *Food Chemistry*, 2016, 198, 36–44.

Duran, N.M., Galante, M., Spelzini, D., Boeris, V. The effect of carrageenan on the acid-induced aggregation and gelation conditions of quinoa proteins. *Food Research International*, 2018, 107, 683–690.

Fan, K., Chen, S. Bioprocessing for value-added products from renewable resources. *Shang, TY Elsevier Science*, 2007, 684.

Farrokhi, N., Hrmova, M., Burton, R.A., Fincher, G.B. Heterologous and cell-free protein expression systems. *In*: Gustafson, J., Langridge, P., Somers, D. (eds), Plant Genomics. Methods in Molecular Biology™ (Methods and Protocols), vol 513. LLC: Humana Press, a part of Springer Science+Business Media; 2009; pp 175–198.

Fernandes, P. Enzymes in food processing: A condensed overview on strategies for better biocatalysts. *Enzyme Research*, 2010, 2010, Article ID 862537, 19 pages. http://dx.doi.org/10.4061/2010/862537

Fu, G.Z., Chan, A., Minns, D. Preliminary Assessment of the Environmental Benefits of Enzyme Bleaching for Pulp and Paper Making (7 pp). *The International Journal of Life Cycle Assessment*, 2005, 10(2), 136–142.

Griengl, H., Schwab, H., Fechter, M. The synthesis of chiral cyanohydrins by oxynitrilases. *Trends in Biotechnology*, 2000, 18(6), 252–256.

Hajfathalian, M., Ghelichi, S., García-Moreno, P.J., Moltke Sørensen, A.-D., Jacobsen, C. Peptides: Production, bioactivity, functionality, and applications. *Critical Reviews in Food Science and Nutrition*, 2017, 1–33.

Hilbrig, F., Freitag, R. Protein purification by affinity precipitation. *Journal of Chromatography B*, 2003, 790(1), 79–90.

Hills, M., Fan Cultures. Routledge, 2003.

Huisman, G.W., Collier, S.J. On the development of new biocatalytic processes for practical pharmaceutical synthesis. *Current Opinion in Chemical Biology*, 2013, 17(2), 284–292.

Jonasson, P., Liljeqvist, S., Nygren, P.A., Ståhl, S. Genetic design for facilitated production and recovery of recombinant proteins in *Escherichia coli*. *Biotechnology and Applied Biochemistry*, 2002, 35(2), 91–105.

Khamanga, S.M., Walker, R.B. Science and practice of microencapsulation technology. *In*: Rajabi-Siahboomi, A.R. (ed.), Multiparticulate Drug Delivery: Formulation, Processing and Manufacturing. Switerzland; Springer; 2017; pp 119–154.

Larionova, N., Unksova, L.Y., Mironov, V., Sakharov, I.Y., Kazanskaya, N., Berezin, I. Study of the complex-formation between soluble carboxymethyl ethers of polysaccharides and proteins. *VYSOKOMOLEKULYARNYE SOEDINENIYA SERIYA A*, 1981, 23(8), 1823–1829.

Larsson, P.-O., Mosbach, K. Affinity precipitation of enzymes. *FEBS Letters*, 1979, 98(2), 333–338.

Le Bourvellec, C., Renard, C. Interactions between polyphenols and macromolecules: Quantification methods and mechanisms. *Critical Reviews in Food Science and Nutrition*, 2012, 52(3), 213–248.

Leong, Y.K., Lan, J.C.-W., Loh, H.-S., Ling, T.C., Ooi, C.W., Show, P.L. Cloud-point extraction of green-polymers from Cupriavidus necator lysate using thermoseparating-based aqueous two-phase extraction. *Journal of Bioscience and Bioengineering*, 2017, 123(3), 370–375.

Lima, N., Mota, M. Biotecnologia dos ambientes aéreos e confinados. *Biotecnologia: Fundamentos e Aplicações*, 2003, 321–338.

Maijala, P., Kleen, M., Westin, C., Poppius-Levlin, K., Herranen, K., Lehto, J., Reponen, P., Mäentausta, O., Mettälä, A., Hatakka, A. Biomechanical pulping of softwood with enzymes and white-rot fungus *Physisporinus rivulosus*. *Enzyme and Microbial Technology*, 2008, 43(2), 169–177.

Mosbach, K. Matrix-bound enzymes. *Acta Chemica Scandinavica*, 1970, 24(6), 2084–2092.

Nagaoka, H., Akoh, H. Decomposition of EPS on the membrane surface and its influence on the fouling mechanism in MBRs. *Desalination*, 2008, 231(1–3), 150–155.

Nieto, M.B. Edible film and packaging using gum polysaccharides. *In*: Cerqueira, M.A.P.R., Pereira, R.N.C., Ramos, O.L.da-S., Teixeira, J.A.C., Vicente, A.A. (eds.), Edible Food Packaging, Materials and Processing Technologies. Boca Raton: CRC Press; 2016; pp. 31–101.

Palomares, L., Estrada-Mondaca, S., Ramírez, O. Production of recombinant proteins: Challenges and solutions. *Methods in Molecular Biology*, 2004, 267, 15–52.

Pérez, S., Mazeau, K., du Penhoat, C.H. The three-dimensional structures of the pectic polysaccharides. *Plant Physiology and Biochemistry*, 2000, 38(1–2), 37–55.

Queiroga, A.C., Pintado, M.M., Malcata, F.X. Novel microbial-mediated modifications of wool. *Enzyme and Microbial Technology*, 2007, 40(6), 1491–1495.

Rocha, M.V., Romanini, D., Nerli, B.B., Tubio, G. Pancreatic serine protease extraction by affinity partition using a free triazine dye. *International Journal of Biological Macromolecules*, 2012, 50(2), 303–309.

Rocha, M.V., Di Giacomo, M., Beltramino, S., Loh, W., Romanini, D., Nerli, B.B. A sustainable affinity partitioning process to recover papain from Carica papaya latex using alginate as macro-ligand. *Separation and Purification Technology*, 2016, 168, 168–176.

Ru, Q., Wang, Y., Lee, J., Ding, Y., Huang, Q. Turbidity and rheological properties of bovine serum albumin/pectin coacervates: Effect of salt concentration and initial protein/polysaccharide ratio. *Carbohydrate Polymers*, 2012, 88(3), 838–846.

Schmitt, C., Aberkane, L., Sanchez, C. Protein-polysaccharide complexes and coacervates. *Handbook of Hydrocolloids*, 2009, 420–476.

Schmitt, C., Turgeon, S.L. Protein/polysaccharide complexes and coacervates in food systems. *Advances in Colloid and Interface Science*, 2011, 167(1–2), 63–70.

Sonia, T.A., Sharma, C.P. Oral Delivery of Insulin. Elsevier, 2014.

Teotia, S., Gupta, M. Purification of phospholipase D by two-phase affinity extraction. *Journal of Chromatography A*, 2004, 1025(2), 297–301.

Tolstoguzov, V. Functional properties of food proteins and role of protein-polysaccharide interaction. *Food Hydrocolloids*, 1991, 4(6), 429–468.

Tomsho, J.W., Pal, A., Hall, D.G., Benkovic, S.J. Ring structure and aromatic substituent effects on the pKa of the benzoxaborole pharmacophore. *ACS Medicinal Chemistry Letters*, 2011, 3(1), 48–52.

Woitovich Valetti, N., Brassesco, M.E., Picó, G.A. Polyelectrolytes–protein complexes: A viable platform in the downstream processes of industrial enzymes at scaling up level. *Journal of Chemical Technology & Biotechnology*, 2016, 91(12), 2921–2928.

Ye, A. Complexation between milk proteins and polysaccharides via electrostatic interaction: Principles and applications–A review. *International Journal of Food Science & Technology*, 2008, 43(3), 406–415.

16

Conclusions

Aidé Sáenz-Galindo*

School of Chemistry. Autonomous University of Coahuila. Blvd-Venustiano
Carranza s/n Colonia República, Saltillo, Coahuila, México ZC 25290.

Green chemistry is a philosophy based on 12 principles, that can be adapted to any
chemical transformation process. It has been three decades since Green chemistry
was introduced in the area of chemistry and related fields, with this philosophy,
any chemical transformation process can be improved and/or innovated, while
respecting the environment. At the end of the 90's, when the 12 precepts of green
chemistry were announced, there was hardly any acceptance, as it was used only by
a handful of experts in the field of chemistry. However, at present, it is noteworthy
that publications, such as scientific articles, books, chapters books, patents and
different means of disseminating results about chemical transformation processes,
and approaches are designed with the concept of being friendly to the environment.
This philosophy is of vital importance, because of the great deterioration of the
planet during the last century, where a high percentage of contamination in soil,
air and water was observed. Unfortunately, chemical and industrial processing are
some of those that contribute to these types of contamination, as their applications
in areas are very necessary for human beings, who cannot live without the
development of medicine, technology, transportation, to mention some areas where
chemistry has an impact. However, thanks to the philosophy of green chemistry,
each one of the processes involving chemistry can be adapted as much as possible
to green precepts, in order to reduce contamination as much as possible.

The 12 precepts of green chemistry are well presented which include
prevention of contamination, the design of the process, trying to incorporate each
of the components involved in the process to this philosophy and having a good
atomic economy, thus minimizing contamination through the whole development

*For Correspondence: aidesaenz@uadec.edu.mx

of products. It is also considered that all synthetic methodology should be environmentally friendly, for this reason it is important to involve raw material as inoculants as well, the chemical products should be designed to be the least polluting in its whole, another aspect to consider is to trying to avoid the use of unnecessary solvents, which involves another stage of purification in the process, scientists should try to use green solvents such as water, alcohols, and others that are classified as green solvents.

The use of emerging energies such as ultrasound and/or microwaves that have been proven to reduce process times, as well as energy requirements or to try to bring the process to room temperature with normal atmospheric pressures. Another important aspect to consider is the use of renewables products as raw materials, as well as trying to avoid as much as possible the use in the process of unnecessary blocking components, because this can cause another stage in the process. On the other hand, use of catalysts is an excellent way to accelerate the processes, which help to reduce processing time, and some of them could be recovered to be used in other steps or processes.

The precepts of Green chemistry also involve degradation of products once they have fulfilled their function to cover any necessity, that is to say that they should be disposed off in the most environmentally friendly manner, as well as the process must be monitored in real time, during the process and at its end and finally it is important to emphasize that the process must be monitored to minimize the risk of chemical accidents including emissions and fires. As it can be seen , Green chemistry is a well-designed work philosophy, designed to encompass each one of the variables that can affect or influence a chemical transformation process, from its design to its degradation. The most remarkable thing is that this philosophy can be applied in all areas related to chemistry.

Index